PARASITES AND PATHOGENS

PARASITES AND PATHOGENS

EFFECTS ON HOST HORMONES AND BEHAVIOR

EDITED BY

NANCY E. BECKAGE
University of California, Riverside

CHAPMAN & HALL

ITP International Thomson Publishing

New York • Albany • Bonn • Boston • Cincinnati • Detroit • London • Madrid • Melbourne
Mexico City • Pacific Grove • Paris • San Francisco • Singapore • Tokyo • Toronto • Washington

Cover design: Trudi Gershenov
Cover photos: elephant: Ben Hart, University of California-Davis; caterpillar: Marianne Alleyne and Nancy Beckage, University of California-Riverside; rat: J.F. Mueller, provided by S.N. Thompson, University of California-Riverside

Printed in the United States of America

For more information, contact:

Chapman & Hall
115 Fifth Avenue
New York, NY 10003

Thomas Nelson Australia
102 Dodds Street
South Melbourne, 3205
Victoria, Australia

International Thomson Editores
Campos Eliseos 385, Piso 7
Col. Polanco
11560 Mexico D. F.
Mexico

International Thomson Publishing Asia
221 Henderson Road #05-10
Henderson Building
Singapore 0315

Chapman & Hall
2-6 Boundary Row
London SE1 8HN
England

Chapman & Hall GmbH
Postfach 100 263
D-69442 Weinheim
Germany

International Thomson Publishing - Japan
Hirakawacho-cho Kyowa Building, 3F
1-2-1 Hirakawacho-cho
Chiyoda-ku, 102 Tokyo
Japan

1 2 3 4 5 6 7 8 9 10 XXX 01 00 99 98 97
Library of Congress Cataloging-in-Publication Data

Parasites and pathogens : effects on host hormones and behavior /
edited by Nancy E. Beckage.
p. cm.
Includes bibliographical references and index.
ISBN 0-412-07401-X (hb : alk. paper)
1. Host-parasite relationships. I. Beckage, N.E. (Nancy E.)
QL757.P276 1997
591.52'49--dc20 96-41749
CIP

British Library Cataloguing in Pubication Data available

To order this or any other Chapman & Hall book, please contact **International Thomson Publishing, 7625 Empire Drive, Florence, KY 41042.** Phone: (606) 525-6600 or 1-800-842-3636.
Fax: (606) 525-7778. e-mail: order@chaphall.com.

For a complete listing of Chapman & Hall's titles, send your request to **Chapman & Hall, Dept. BC, 115 Fifth Avenue, New York, NY 10003.**

Table of Contents

Part 2: Parasitism and Reproduction

Part 3: Parasites, Pathogens, and Host Behavior

To Ross, John, and Ian

Miriam Rothschild:
Her Legacy to the Emerging Field of Host-Parasite Relationships

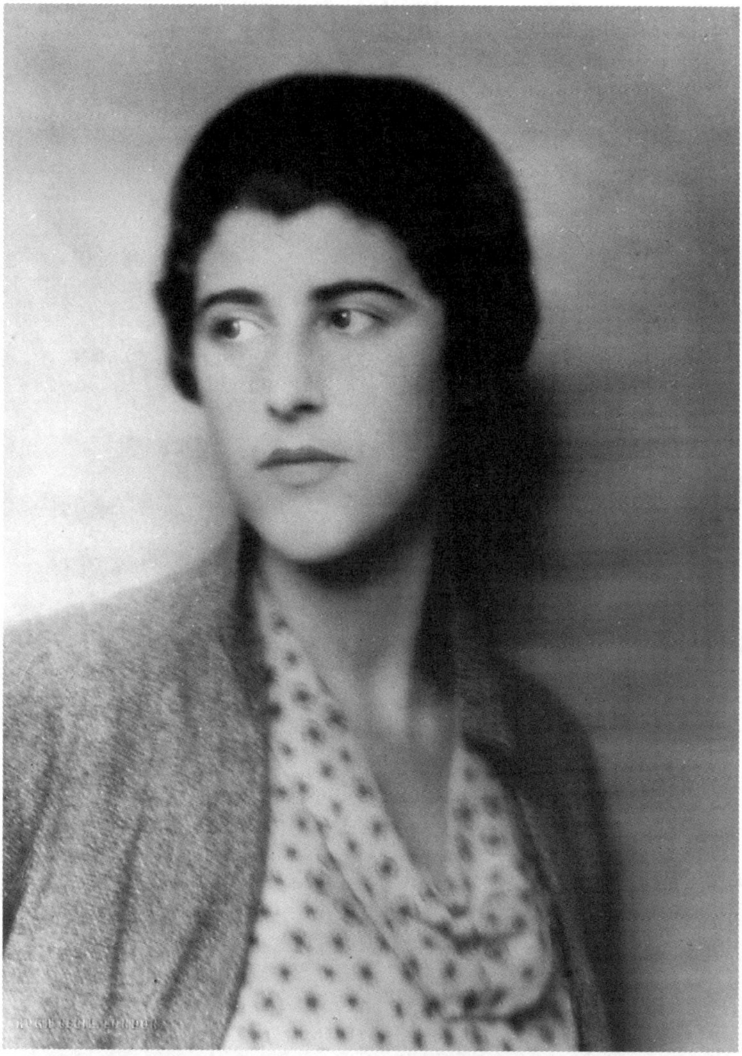

The authors dedicate this volume to the Honorable Miriam Rothschild, whose pioneering observations inspired many of their own studies delineating how parasites communicate with hosts. This photograph shows the Honorable Miriam Rothschild at the time she conducted her elegant studies of the intricate physiological relationships between rabbits and fleas in the 1960s.

Dedication

Since its inception, the field of host-parasite relationships has been dominated by biologists who have a demonstrably intense "feeling for the organism." The Honorable Miriam Rothschild was one of the first experimentalists to provide documentation of hormonal host-parasite interactions. Her elegant studies showed that the reproductive synchronization between female rabbit hosts and rabbit fleas have a hormonal basis in that the fleas respond to their host's hormones, facilitating their multiplication to occur soon after that of their mammalian hosts.

She deciphered how the reproductive cycle of a female rabbit host was synchronized with that of the fleas feeding upon it, showing that the hormones of the rabbit doe regulate the reproductive cycles of her attendant female fleas. Ovarian maturation in the flea temporally tracks that of the rabbit. After the doe's parturition, both male and female fleas transfer to the newborn young. Pairing is stimulated by a kairomone and some unknown blood factor produced by the nestlings, and the fleas copulate in the nest and oviposit; the spent fleas then return to the doe. After hatching in the nest, the immature fleas feed upon the blood present in the adult fleas' feces; following pupation and development to the adult stage, the freshly ecdysed fleas then infest the young rabbits.

The physiological mechanism explaining how mammalian steroids regulate the cycle of ovarian maturation in these arthropods still remains an enigma to challenge future specialists in the field of molecular endocrinology. While the female flea's cycles of sexual receptivity, ovarian maturation, and regression are clearly linked to the host's circulating levels of estrogen, corticosteroids, and progesterone, we are still unclear as to whether the flea's hormone receptors and ovarian tissues are directly stimulated by the vertebrate's molecules or, alternatively, are stimulated indirectly via more subtle mechanisms, such as conversion of mammalian hormones to insect-active molecules.

The Honorable Miriam Rothschild also was one of the first to document that the presence of larval trematodes causes gigantism in snails, which are rendered partially or completely castrated by parasitism. The mechanisms causing parasitic castration have only recently been identified, as described in this volume.

References

Holloway, M. (1996). Profile: Miriam Rothschild, a natural history of fleas and butterflies. *Scientific American* 274:36–38.

Rothschild, M. (1965). Fleas. *Scientific American* 213:44–53.

Rothschild, M., and Ford, B. (1964). Maturation and egg-laying of the rabbit flea (*Spilopsyllus cuniculi* Dale) induced by the external application of hydrocortisone. *Nature* 203:210–211.

Rothschild, M., and Ford, B. (1964). Breeding of the rabbit flea (*Spilopsyllus cuniculi* (Dale)) controlled by the reproductive hormones of the host. *Nature* 201:103–104.

Rothschild, M., and Ford, B. (1966). Hormones of the vertebrate host controlling ovarian regression and copulation of the rabbit flea. *Nature* 211:261–266.

Rothschild, M., and Ford, B. (1969). Does a pheromone-like factor from the nestling rabbit stimulate impregnation and maturation in the rabbit flea? *Nature* 221:1169–1170.

Rothschild, M., Ford, B., and Hughes, M. (1970). Maturation of the male rabbit flea (*Spilopsyllus cuniculi*) and the oriental rat flea (*Xenopsylla cheopis*): Some effects of mammalian hormones on development and impregnation. *Trans. Zool. Soc. Lond.* 32:105–188.

Rothschild, M., Schlein, Y., and Ito, S. (1986). *A colour atlas of insect tissues via the flea.* London: Wolfe Science Books.

Nancy E. Beckage

Foreword

When Nancy Beckage and I first met in Lynn Riddiford's laboratory at the University of Washington in the mid 1970s, the fields of parasitology, behavior, and endocrinology were thriving and far-flung—disciplines in no serious danger of intersecting. There were rumors that they might have some common ground: *Behavioural Aspects of Parasite Transmission* (Canning and Wright, 1972) had just emerged, with exciting news not only of the way parasites themselves behave, but also of Machiavellian worms that caused intermediate hosts to shift fundamental responses to light and disturbance, becoming in the process more vulnerable to predation by the next host (Holmes and Bethel, 1972). Meanwhile, biologists such as Miriam Rothschild (see Dedication), G. B. Solomon (1969), and Lynn Riddiford herself (1975) had suggested that the endocrinological ramifications of parasitism might be subtle and pervasive. In general, however, parasites were viewed as aberrant organisms, perhaps good for a few just-so stories prior to turning our attention once again to real animals.

In the decade that followed, Pauline Lawrence (1986a,b), Davy Jones (Jones et al., 1986), Nancy Beckage (Beckage, 1985; Beckage and Templeton, 1986), and others, including many in this volume, left no doubt that the host-parasite combination in insect systems was physiologically distinct from its unparasitized counterpart in ways that went beyond gross pathology. At the same time, Roy Anderson and Robert May (1978a,b; 1979a,b), Dan Brooks (1979a,b), Peter Price (1980), and others opened the world of parasitology to ecologists and evolutionary biologists. Our understanding of host-parasite interactions expanded considerably, but the study of mechanism and the study of behavior, ecology and evolution occupied separate, albeit nominally parallel, universes.

In this book those universes do not exactly collide, but they do move closer to a tantalizing unity. The potential benefit is great, for knowledge about mechanisms underlying behaviors can help evolutionary biologists understand the history of those behaviors just as developmental patterns can illuminate morphological evolution. The more we understand mechanism, the more confidently we can place the behavioral aspects of parasitism in an evolutionary—that is, predictive—context. From such a context, we can in turn test increasingly sophisticated hypotheses about the ways that parasites and hosts interact at every level.

This evolutionary framework must be the ultimate goal of much biological research, unless we wish to describe multitudes of individual cases for decades to come. A comprehensive, evolutionary approach is especially important in the behavioral study of host-parasite interactions. The behavioral changes that occur as a result of parasitism can have profound effects on ecological interactions such as competition, predator-prey interactions, foraging decisions, and mate choice. They often are reflected in altered habitat selection, activity levels, thermal preference, and responses to a variety of stimuli. These changed behaviors not only shift the distribution and abundance of hosts, but they also affect parasite transmission and survival, which in turn can mediate future levels of virulence. From the exploration of the intricacies of coevolution to the development of biological control agents, the union of parasitology, mechanism, behavior, and evolution will serve us well.

The mixture of these fields is taking place in this book, and the result is hardly a set amalgam. The combinations are delightful, stimulating, and perhaps a little volatile. This is worth some cheering. In the spacious world of biology and of the intellect, fences rarely make good neighbors.

References

Anderson, R. M., and May, R. M. (1978a). Regulation and stability of host parasite population interactions I. Regulatory processes. *J. Anim. Ecol.* **47**:219–247.

Anderson, R. M., and May, R. M. (1979a). Population biology of infectious diseases, Part I. *Nature* **280**:361–367.

Beckage, N. E. 1985. Endocrine interactions between endoparasitic insects and their hosts. *Ann. Rev. Entomol.* **20**:371–413.

Beckage, N. E., and Templeton, T. J. (1986). Physiological effects of parasitism by *Apanteles congregatus* in terminal stage tobacco hornworm larvae. *J. Insect Physiol.* **32**:299–314.

May, R. M., and Anderson, R. M. (1978b). Regulation and stability of host parasite population interactions II. Destabilizing processes. *J. of Anim. Ecol.* **47**:249–267.

May, R. M., and Anderson, R. M. (1979b). Population biology of infectious diseases, Part II. *Nature* **280**:455–461.

Brooks, D. R. (1979a). Testing the context and extent of host-parasite coevolution. *System. Zool.* **28**:299–307.

Brooks, D. R. (1979b). Testing hypotheses of evolutionary relationships among parasitic helminths: The digeneans of crocodilians. *Amer. Zool.* **19**:1225–1238.

Canning, E. U. and Wright, C. A. (1972). *Behavioural aspects of parasite transmission.* London: Academic Press.

Holmes, J. C., and Bethel, W. M. (1972). Modification of intermediate host behaviour by parasites. In *Behavioural aspects of parasite transmission* (E. U. Canning and C. A. Wright, eds.). London: Academic Press.

Lawrence, P. A. (1986a). Host-parasite hormonal interactions: An overview. *J. Insect Physiol.* **32**:295–298.

Lawrence, P. A. (1986b). The role of 20-hydroxyecdysone in the moulting of *Biosteres longicaudatus,* a parasite of the Caribbean fruit fly, *Anastrepha suspensa.* J. Insect Physiol. **32**:329–337.

Price, P. W. (1980). *Evolutionary biology of parasites.* Princeton, NJ: Princeton University Press.

Riddiford, L. M. (1975). Host hormones and insect parasites. In *Invertebrate immunity* (K. Marmarosh and R. E. Shope, eds.). New York: Academic Press.

Janice Moore
Department of Biology
Colorado State University
Fort Collins, CO 80523

Preface

The field of host-parasite relationships represents a challenging new direction in parasitology, which recently has attracted the attention of endocrinologists, physiologists, immunologists, ethologists, neurobiologists, and ecologists. Current levels of inquiry, while they often begin at the population and organismal levels, often extend downward to the focus of suborganismal and molecular biological sciences, to explore mechanisms of parasite-mediated regulation of host gene expression, as well as direct parasite-induced manipulation of host physiology. Simultaneously, the perspectives of evolutionary biologists and epidemiologists are greatly enhancing our appreciation of the adaptive significance of the physiological host-parasite relationship for the parasites involved and their respective hosts (see Foreward).

A major goal of this volume is to provide a survey of how parasites and pathogens influence host endocrinology, development, reproduction, and behavior. Additionally, we assess how the physiological host environment influences the success of many parasites. Clearly, many parasite-derived molecules (e.g., hormones, peptides, enzymes, cytokines) influence the host's physiology and development, and vice versa.

The virulence of parasites and pathogens for their hosts ranges widely. The insect parasitoids that usually, if not invaribly, kill their host clearly rank among the most highly virulent organisms. Interestingly, in many of these systems it is not a simple matter of the parasite physically consuming the host, thereby causing its demise; instead, a delicate interplay of hormones seems responsible for induction of host arrest, whereupon the parasitoid completes its own development and metamorphosis at the expense of the host.

Aside from showing variation in their mechanisms of physiological virulence, invaders have evolved mechanisms to either manipulate the hormonal milieu of their host or respond to it. Parasitic induction of derangements in development and reproduction in host organisms is common. Even ectoparasites such as ticks, mites, and fleas exhibit these types of hormonal interactions with their host. Parasite mimicry of host hormones is one of the most intriguing characteristics seen in many host-parasite relationships. Yet another important well-documented phenomenon is the tight developmental synchronization of parasites with hosts, often mediated by the parasites' response to their host's hormones. Simultaneously, the parasites may secrete hormonally active molecules, facilitating physiological "cross-talk" between partners.

Parasitic castration is such a common phenomenon that its evolutionary significance seems globally important. Indeed, limiting one's host population presumably might have a negative fitness effect when parasites are host-limited, which may be compensated for by a net positive effect of partial or complete host sterility. Presumably, host castration allows a redirection of nutrients to fuel the production of parasites instead of progeny. Interestingly, not only the adult host form is affected. Even in immature male insects testicular atrophy occurs, so the gonads fail to proliferate normally. In these instances, an intriguing question arises: Why does the parasitoid castrate the host if indeed it already plans to kill it?

As several chapters in this volume document in exquisite detail, invertebrates such as insects and snails, as well as many vertebrates, serve as useful experimental models for studying parasitism-induced developmental and behavioral anomalies. In snails, the immunological responses of the host to schistosomes may indeed be linked to endocrine factors and reproductive disturbances. Many vertebrate parasites (including the schistosomes just mentioned) obviously spend part of their life cycle in invertebrate hosts and hence interact with two taxonomically distant classes of hosts, showing very different interactions with each taxon. In vertebrates, it is well recognized that males are more susceptible to a variety of parasites than females because of a complex interplay of the immune and endocrine systems involving corticosteroids and testosterone, as well as estrogen and progesterone.

Daunting questions continue to challenge us. For example, precisely how do parasites and pathogens effect changes in neuronal regulation of behavior? The direct invasion of the mammalian nervous system by viral agents likely explains why some viruses drastically affect their host's behavior to evoke an enhancement of the rate of disease transmission. How parasitic macro-invaders, which are seemingly unable to gain entry to the nervous system itself, invoke changes in their host's behavior is less clear. In invertebrates, the mechanisms whereby parasites and pathogens affect host feeding, circadian rhythms, photo- and geotropisms, and the whole suite of behaviors associated with reproduction, including calling, courtship, mating, egg maturation, and oviposition, have yet to be dissected at a mechanistic level. A complex cocktail of hormones as well as other neuromodulators such as serotonin, produced either by the parasites or by the host, may likely be involved.

We only now are initiating the critical experimental studies to answer these questions, and many invertebrates have played key roles as model systems in which to unravel these mechanisms. Clearly, many parasitism-induced changes in the behavior of intermediate hosts enhance the rate of parasite transmission to the final definitive host. Indeed, many evolutionary benefits may accrue in situations in which the host's behavior is substantively modified, perhaps thus explaining the frequency at which these phenomena occur in animal systems.

Ideas presented by speakers in the January 1995 American Society of Zoology symposium "Parasitic Effects on Host Hormones and Behavior" generated the initial framework for this book. The National Science Foundation Program in Integrative Biology and Neuroscience and the American Society of Zoology (now the Society for Integrative and Comparative Biology) provided financial support for the symposium and are gratefully acknowledged.

Our intended audience is a large one, encompassing undergraduate and graduate students, as well as professional researchers, in the fields of parasitology, entomology, endocrinology, developmental biology, evolutionary ecology, neurobiology, and ethology. We anticipate that many scientists who currently know very little about parasites will enjoy this book.

Aside from the authors themselves, several esteemed colleagues provided helpful peer reviews. They included Peter Bryant, Diana Cox-Foster, Ken Davey, Dale Gelman, Anthony James, Marcia Loeb, Andres Pope Moller, Janice Moore, Lynn Riddiford, Jerry Theis, Nelson Thompson, James Truman, John Wiens, and Marlene Zuk. I thank them heartily for ensuring the production of a high-quality volume. Lastly, I owe a great debt to our editor, Henry Flesh, at Chapman and Hall, for his ever-patient and expert assistance during preparation of this volume.

Nancy Beckage
Department of Entomology
University of California-Riverside
Riverside, CA 92521

List of Contributors

Shelley A. Adamo, Department of Psychology, Dalhousie University, Halifax, Nova Scotia, Canada B3H 4j1

Coen M. Adema, Department of Biology, 167 Castetter Hall, University of New Mexico, Albuquerque, NM 87131

Nancy E. Beckage, Department of Entomology, 5419 Boyce Hall, University of California, Riverside, CA 92521

Marijke de Jong-Brink, Faculty of Biology, Department of Experimental Zoology, Vrje Universiteit De Boelalaan 1087, 1081 HV, Amsterdam, The Netherlands.

John J. Brown, Department of Entomology, Washington State University, Pullman, WA 99164.

Ian Fairweather, School of Biology and Biochemistry, The Queen's University of Belfast, Belfast BT7 1NN, Northern Ireland.

Miodrag Grbic', Department of Entomology, University of Wisconsin, Madison, WI 53706.

Benjamin L. Hart, Department of Anatomy, Physiology and Cell Biology, School of Veterinary Medicine/Animal Behavior Program, University of California, Davis, CA 95616.

Carolyn G. Hatalski, Department of Anatomy and Neurobiology, University of California, Irvine, CA 92697-4290.

Lynn A. Hertel, Department of Biology, 167 Castetter Hall, University of New Mexico, Albuquerque, NM 87131.

Diana K. Hews, Department of Zoology, Arizona State University, Tempe, AZ 85287

Nigella Hillgarth, Department of Zoology and Burke Museum NJ-15, University of Washington, Seattle, WA 98195

Robert M. Hoek, Faculty of Biology, Department of Experimental Zoology, Vrje Universiteit De Boelelaan 1087, 1081 HV, Amsterdam, The Netherlands

Hilary Hurd, Centre for Applied Entomology and Parasitology, Department of Biology, University of Keele, Keele ST5 5BG, United Kingdom

Armand Kuris, Department of Biological Sciences and Marine Science Institute, University of California, Santa Barbara, CA 93106

Kevin D. Lafferty, Institute of Marine Science, University of California, Santa Barbara, CA 93106

Wessel Lageweg, Faculty of Biology, Department of Experimental Zoology, Vrje Universiteit De Boelelaan 1087, 1081 HV, Amsterdam, The Netherlands

W. Ian Lipkin, Department of Anatomy and Neurobiology, University of California, Irvine, CA 92697-4290

Eric S. Loker, Department of Biology, 167 Castetter Hall, University of New Mexico, Albuquerque, NM 87131

Michael C. Moore, Department of Zoology, Arizona State University, Tempe, AZ 85287

C. Kirk Phares, Department of Biochemistry and Molecular Biology, University of Nebraska Medical Center, 600 South 42nd Street, Omaha, NE 68198-4525

Michal Polak, Department of Biology, Syracuse University, Lyman Hall, 108 College Place, Syracuse, NY 13244-1270

Darcy A. Reed, Department of Entomology, Washington State University, Pullman, WA 99164

August B. Smit, Faculty of Biology, Department of Experimental Zoology, Vrje Universiteit De Boelelaan 1087, 1081 HV, Amsterdam, The Netherlands

Michael R. Strand, Department of Entomology, University of Wisconsin, Madison, WI 53706

Tracey Webb, Centre for Applied Entomology and Parasitology, Department of Biology, University of Keele, Keele ST5 5BG, United Kingdom

John C. Wingfield, Department of Zoology and Burke Museum NJ-15, University of Washington, Seattle, WA 98195

PARASITES AND PATHOGENS

PART 1

Host-Parasite Hormonal Interactions

1

New Insights: How Parasites and Pathogens Alter the Endocrine Physiology and Development of Insect Hosts

Nancy E. Beckage

I. Introduction

Many parasites and pathogens alter the growth, behavior, and development of insect hosts in a dramatic fashion. The invader often initially remains cryptic, until its presence later becomes obvious as the host suffers major growth, pigmentation, or morphological anomalies, shows alterations in sexual differentiation or behavior; or develops reproductive disturbances such as castration or loss of fecundity (see Chapters 8, 9, and 12; also Moore, 1993, 1995; Adamo et al., 1997).

Parasitoids are parasitic insects that invariably kill their host (Godfray, 1994). The mass of the parasite relative to the host may remain quite small, or, alternatively, the host itself may be ultimately transformed into little more than a "parasite in host disguise." With polyembryonic or solitary endoparasitoids of lepidopteran hosts, or nematode parasites of biting midges (Fig. 1-1; Mullens and Velten, 1994) and mosquitoes, all of the host's tissues, with the exception of its cuticle, may be consumed owing to the increase in mass of its developing parasite(s). In contrast, with gregarious endoparasitoids, several hundred of them may develop following a single oviposition, and the host's tissues such as the fat body are depleted markedly during the final stages of parasitism though they are not directly consumed by the developing parasitoids (Fig. 1-2). Nevertheless, the host still often shows marked developmental derangements, which we now know are symptomatic of the induction of major endocrine disturbances, reflecting redirection of the normal developmental program. These symptoms are particularly likely to appear during the final stages of endoparasitism preparatory to emergence of the parasites from the host. For example, tobacco hornworm larvae parasitized by *Cotesia congregata* (Fig. 1-2) are developmentally arrested in the larval stage. Although the weight of the larvae may exceed that required for metamorphosis in this species, the host fails to metamorphose (Alleyne and Beckage, 1997).

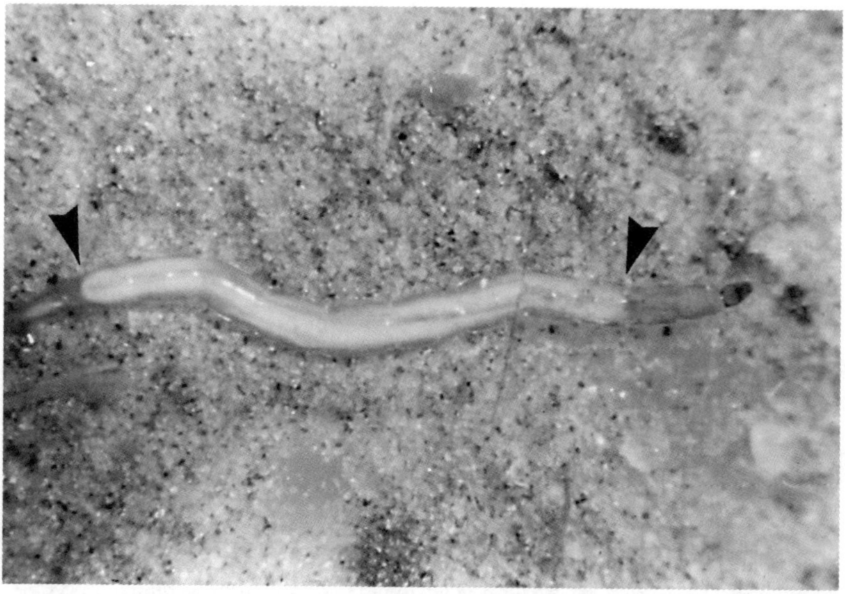

Figure 1-1. Appearance of a larva of the biting midge *Culicoides variipennis* infected by a nematode (*Heliodomermis magnapapula)* (Mullens and Velten, 1994). Note that the adult nematode (arrowhead) occupies almost the entire body cavity of the host fly larva (a "worm within a worm") and is about to emerge from the host. Reprinted with permission from Mullens and Rutz (1982). Photo courtesy of Brad Mullens (University of California–Riverside).

Many species of parasitoids are simultaneously host "conformers" and "regulators," with the parasitoid showing a response to its host's hormones, while its presence (or that of the accompanying polydnavirus and venom components injected along with eggs during oviposition) appears capable of eliciting hormonal changes in the host's endocrine status (Lawrence, 1986a). For example, the parasitoid *Diachasmimorpha (= Biosteres) longicaudatus* induces developmental arrest in its Caribbean fruitfly host, while molting of the parasitoid is simultaneously stimulated by the prepupal peak in host ecdysteroid titers, which triggers the first larval ecdysis of the wasp (Lawrence, 1986b). Indeed, the absence of any host–parasitoid endocrine relationship appears to be the exception rather than the rule.

In the past decade, many models of endocrine communication have been developed using insect–insect interactions to gain insights into the physiological dynamics of host–parasite relationships (Beckage et al., 1997). This chapter provides an overview of the mechanisms used by parasites and pathogens that result in the alteration or redirection of normal patterns of growth and development of insect hosts, with major emphasis on studies published during the past

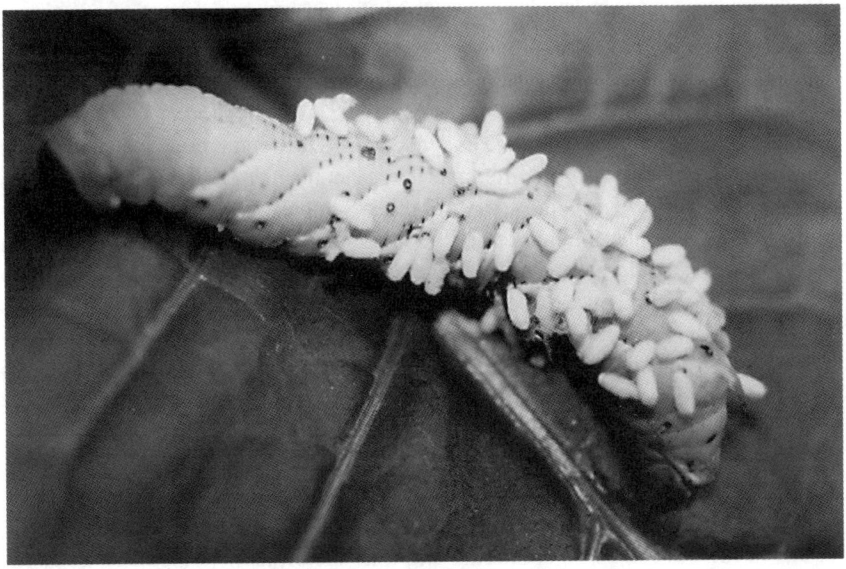

Figure 1-2. Fifth instar larva of the tobacco hornworm, *Manduca sexta*, with emergent braconid wasp parasitoids of the species *Cotesia congregata*. The wasps ecdyse from the second to the third larval instar as they egress from the host, spin cocoons, and undergo metamorphosis to the adult stages while the host is developmentally arrested. The host ceases feeding about 12–24 hours before the wasps emerge, shows a reduced level of spontaneous locomotion, and has an extremely low metabolic rate (Beckage and Templeton, 1986; Alleyne et al., 1997; Adamo et al., 1997). Photo taken by Marianne Alleyne (University of California–Riverside).

four years (see Beckage et al., 1993; Beckage, 1993a,b for summaries of earlier studies).

Two systems—the endocrine/neuroendocrine and nervous systems—frequently serve as critical control points in invoking such changes. A wealth of information exists documenting how parasitism and pathogenic infection influence the endocrine regulation of development in insect hosts. Yet we know considerably less about how these agents alter the neural regulation of development and behavior, either directly, by invasion of the nervous system itself, or indirectly, via parasite secretion of neuromodulators or other factors (Beckage et al., 1997).

Additionally, based on findings now well established in vertebrate systems (Kavaliers and Colwell, 1992; Ransohoff and Benveniste, 1996), we anticipate that the neural, endocrine, and immune systems will eventually be found to be functionally linked, but have as yet little evidence that such a three-way pathway of regulation exists in insect systems. The nervous and endocrine signaling mechanisms regulating molting and metamorphosis are clearly integrated, and imposed upon those interactions is a possible interface of hormones with the immune system. The molting fluid enzyme FAD-glucose dehydrogenase is developmentally regulated by ecdysteroids (Cox-Foster et al., 1990), but also is activated in an immune encapsulation response as shown by studies of Cox-Foster and Stehr (1994) showing that the enzyme is produced in hemocyte capsules that form on the surface of abiotic implants. Eicosanoids also play newly discovered important roles in insect immunity (Stanley-Samuelson and Pedibhotla, 1996) and may be subject to endocrine modulation. Conceivably, some immune-response genes could be hormone activated or otherwise regulated by endocrine factors or endogenous yet-to-be-discovered cytokines (Beck and Habicht, 1996). There are many molecules involved in the insect immune response whose regulation has yet to be studied (Gillespie et al., 1997). In vertebrates, many cytokines interface with components of both the endocrine and nervous systems (Ransohoff and Benveniste, 1996).

The findings of Marijke de Jong-Brink and her colleagues provide evidence that a neuro-immuno-endocrine regulatory pathway does in fact operate in the snail during schistosome infection. The snail's nervous system (Szmidt-Adjide et al., 1996), as well as its endocrine and immune systems (Amen and de Jong-Brink, 1992; chapters 3 and 4), experience pathological changes during infection. A similar link also likely exists in other invertebrates. For example, octopamine is a mediator of some cell-mediated insect immune responses such as nodulation (Dunphy and Downer, 1994), while it also has a stimulatory effect on mobilization of lipids and carbohydrates, the latter effect being documented in the house cricket, *Acheta domesticus* (Fields and Woodring, 1991). Octopamine also inhibits juvenile hormone (JH) production in another cricket, *Gryllus bimaculatus* (Woodring and Hoffmann, 1994), while, in contrast, it stimulates production of JH in the honeybee (Kaatz et al., 1994; Rachinsky, 1994). Hence, molecules such as JH or invertebrate cytokines (Beck and Habicht, 1996) may serve as potential links coordinating the response of the endocrine, immune, and nervous systems, as well as mechanisms of interaction.

Molecules that are well established as modulators in self/nonself recognition processes during embryonic development may also play a role in host–parasite/parasitoid recognition processes. For example, an insect counterpart of the mammalian NF-KB mediating transcriptional control of immune response genes has been identified, and the *dif* gene (*dorsal-related* immunity factor) is activated after wasp parasitization in some strains of *Drosophila* (Carton and Nappi, 1996). Hence, developmental modulators may serve a dual role by altering immune system function as well.

Given the large number of behavioral changes observed in insect hosts, we conclude that the nervous system is particularly likely to be affected by endoparasitism or pathogenic infection, despite our current lack of physiological information about the pathological mechanisms themselves. The host behavioral manipulation exhibited by the braconid *Cotesia glomerata* is not seen following parasitization by the closely related congeners *C. rubecula* (Brodeur and Vet, 1994) or *C. congregata,* indicative of interspecies differences in the capacity to interfere with post-emergence behavior of the host. Specifically, following egression of the *C. glomerata* parasitoids, the host pierid caterpillar spins a web over the wasp cocoons and aggressively protects the parasitoids when disturbed (Brodeur and Vet, 1994). Hence, the host can be considered to be transformed into a "parasitoid defender" morph. In *Manduca sexta*, the host remains as a developmentally arrested (Fig. 1-2) larva on the underside of a leaf, without resuming feeding or exhibiting spontaneous locomotion, while its emergent nonfeeding third instar parasitoids complete cocoon spinning, metamorphosis, and adult eclosion. The host eventually dies several weeks later, presumably of starvation or desiccation.

Parasitoids and parasites induce many other behavioral changes in insect hosts (Moore, 1993, 1995; chapter 12) that beg the question of hormonal involvement. For instance, the presence of tachinids may modify the circadian rhythmicity of the male's courtship song in a manner hypothesized to reduce the risk of attracting parasitoids (Zuk et al., 1993, 1995). The courtship behavior itself is known to be hormonally regulated, suggesting a hormonal pathway may be involved. Specifically, in regions where tachinids are present, the singing of the host population is restricted to the scotophase, whereas in areas where flies are absent, calling occurs during dusk as well, when the parasitoid could exploit visual as well as acoustic cues to locate hosts. The cricket's mating, aggression (male), and egg-laying (female) behavior patterns also show alterations during parasitism (Adamo et al., 1995). This is not surprising given that the parasitoids grow to a very large size relative to the host, and several may attack a single host; *en masse* they may occupy a large volume of the hemocele before they emerge and undergo puparium formation (Fig. 1-3). Amazingly, the male's calling behavior seems not to be silenced completely until shortly before the parasitoids emerge from the host (Cade, 1984).

Many virus-infected insects including many species of Lepidoptera and Orthoptera alter their behavior to show a negative geotaxis following virus replication (Fig. 1-4; Tanada and Kaya, 1993). This elevation-seeking behavior, combined with the "wilting" or "melting" of the insect due to expression of cathepsin-like and other virally encoded gene products (Beckage, 1996; Volkman, 1996), results in the mature virions being showered effectively from the dead carcass to uninfected insects feeding in the foliage below. Many fungal pathogens similarly induce host flies (Fig. 1-5), aphids, ants, grasshoppers, and planthoppers to assume a characteristic wing-elevated posture when spores are due to be released (Fig. 1-5; Hajek and St. Leger, 1994). As a result, the fungal spores, which are released in the early morning, are widely distributed on the uninfected insects that have sought shelter the previous evening in the vegetation below (Six and Mullens,

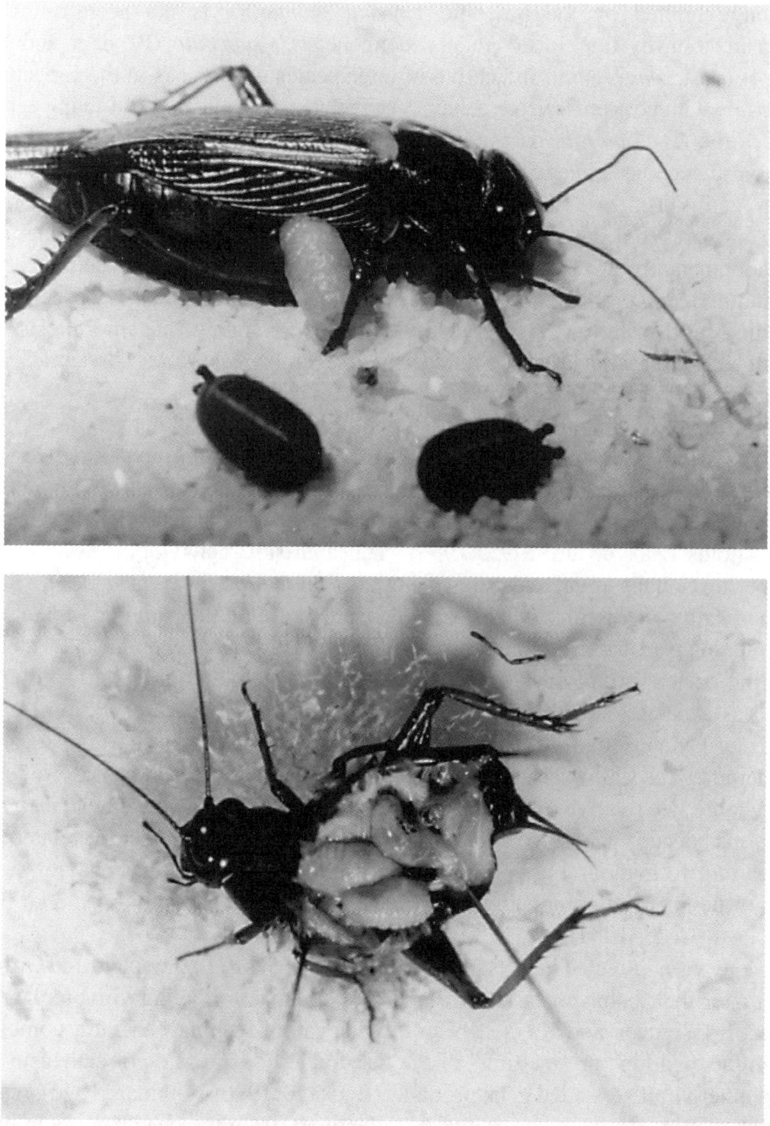

Figure 1-3. Photo showing a field cricket with emergent tachnid parasitoids (*Ormia ochracea*). The parasitoids occupy most of the body cavity of the host as shown (below) prior to their emergence and pupariation in the sand adjacent to the cricket (above). Mating, aggressive, and egg-laying behaviors are altered during parasitism (Adamo et al., 1995). Photo courtesy of Marie Read and Ron Hoy (Cornell University).

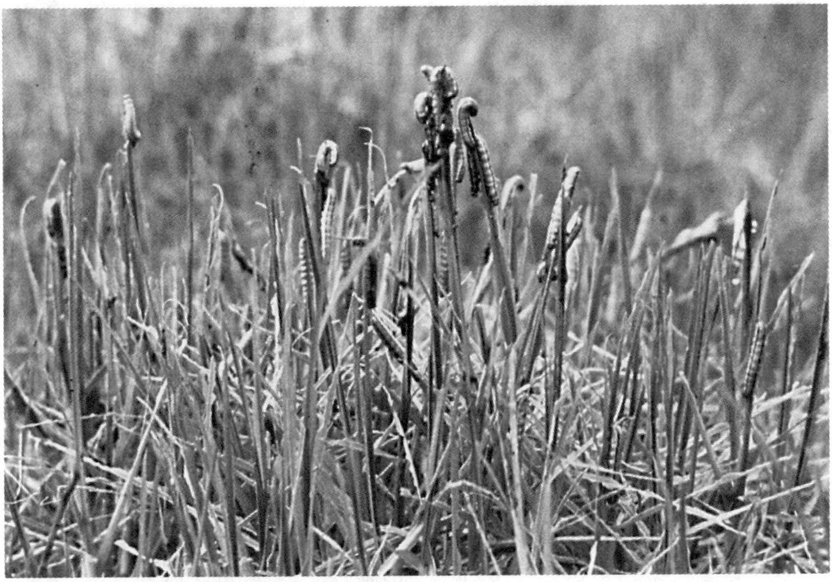

Figure 1-4. Late stages of baculovirus infection stimulates elevation-seeking behavior in the armyworm, *Pseudaletia unipuncata*. The larvae have migrated to the tops of grass blades, where the larvae attach and "melt" to shower virions on uninfected caterpillars resident in the vegetation below (Tanada and Kaya, 1993; reprinted with permission of Academic Press). Photo courtesy of Harry Kaya (University of California–Davis).

1996). Behavioral fevers are also elicited by these and other fungal pathogens, although the mechanism whereby changes in host thermal preferences is induced remains unknown (Watson et al., 1993; Horton and Moore, 1993).

Recently the fields of parasitology, bacteriology, and virology have witnessed a resurgence of interest in mechanisms of pathogenesis (Smith, 1995) and virulence (Ewald, 1994, 1995). Since the emphasis of this book is focused on how invaders influence host physiology and behavior, these topics will be emphasized, in the context of generating a description of mechanisms of virulence and pathogenesis evolved by parasites and pathogens to affect insect growth and development.

II. Pathological Effects on Host Growth and Development

The rate of development of infected insects often is slowed due to inhibitory effects on food consumption and growth. The feeding rate is frequently reduced during an infection, with most parasites and pathogens (parasitoids, fungi, bacteria, and viruses) causing a decline in the rate of food consumption and a lower growth rate. Thus, the host's development time frequently is extended, since the infected individual requires a longer period to reach the minimum critical size required to initiate the next larval ecdysis or a metamorphic molt.

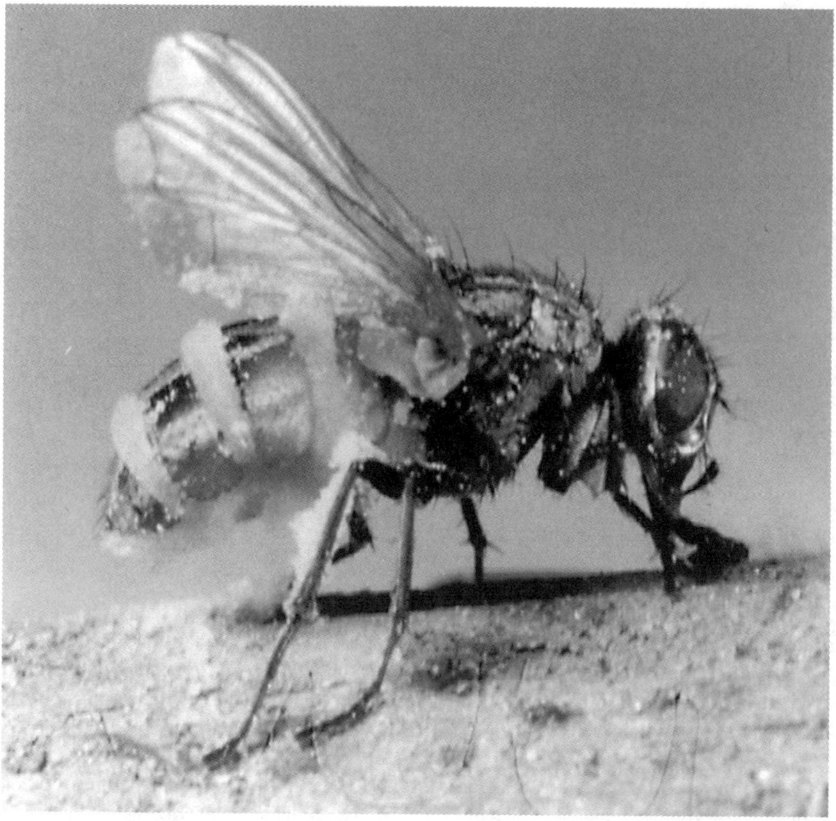

Figure 1-5. Typical appearance of the wing-elevated posture and extended proboscis of a housefly (*Musca domestica*) infected with mature spores of the fungus *Entomophthora muscae.* Note the presence of numerous fungal spores on the legs and abdomen. The fungi are disseminated up to several centimeters away from the insect host following the release of the conidial shower. Reprinted with permission from Mullens et al. (1987). Photo courtesy of Brad Mullens (University of California–Riverside).

Sublethal infection of many insects with nuclear or cytoplasmic polyhedrosis viruses leads to an extended development time in both the larval and pupal stages, with the degree of developmental delay often being proportional to the dose of virus administered (Goulson and Cory, 1995; Sait et al., 1994a,b; Rothman and Myers, 1996; Volkman, 1997). Longevity and fecundity are often decreased by infection. Effects vary according to the species of virus involved and the host. For example, vertical transmission of the *Mamestra brassicae* NPV occurs without signficant effects on pupal weight, sex ratio, fecundity, or egg viability, despite inducing significant delaying effects on developmental time, which may play an important role in the ecology and population dynamics of virus transmission

(Sait et al., 1994a,b; Goulson and Cory, 1995). Other authors report that sublethal virus infection has significant effects on adult size, fecundity, and egg viability (Goulson and Cory, 1995; Rothman and Myers, 1996; Sait et al., 1994a,b) indicative of interspecies variation in these effects.

Thus, pathogenesis-induced anorexia is not limited to vertebrate hosts (e.g., see Arneberg et al., 1996). Whether serotonin, octopamine, acetylcholine, proctolin, or other neurotransmitters and neuromodulators invoke feeding inhibition in insect hosts remains to be determined. A notable exception to the trend of decreased growth is the induction of supernumerary instars in insects parasitized by gregarious or polyembryonic species of parasitoid wasps (Beckage and Riddiford, 1978). In the latter cases, failure to attain the minimum head capsule width needed for metamorphosis during the penultimate stage prior to the last larval molt may result in supernumerary molting to an additional instar. This situation allows the host to ingest the minimum amount of nutrients needed for the parasitoids to emerge successfully. A period of indispensable host nutrition is required for successful development of the wasps, which, if not attained, results in the death of both the host larva and its parasitoids (Beckage and Riddiford, 1983). Indeed, the growth of the host appears to be monitored carefully and modulated by the parasitoids. If the host is not of sufficient size to permit the wasps to emerge, then molting to a supernumerary instar is induced (Beckage and Riddiford, 1978), thus permitting the parasitoids to attain their own minimum critical size required to complete their own larval development and pupation. Interestingly, at the very beginning of the host–parasitoid relationship, the adult female wasp is sometimes able to discern the developmental stage of the caterpillar at the time of parasitization, using cues detected following arrival on a caterpillar-infested leaf, and exhibits a preference for caterpillars of a particular instar/size range to best satisfy the needs of her progeny (Mattiacci and Dicke, 1995). The host's initial as well as its final size appears to be monitored by both the parent parasitoid and its progeny.

A growth-blocking peptide (GBP) is produced in armyworm larvae parasitized by the braconid parasitoid *C. kariyai,* which is not unique to parasitized larvae, since it also occurs in nonparasitized larvae of the same species, but it is produced in particularly high concentration shortly after the host is parasitized (Hayakawa and Yasuhara, 1993). Its presence suppresses JH esterase activity (Hayakawa, 1995) thereby prolonging the larval stage of the host. The action of the growth-blocking peptide is not restricted to the host armyworm species, since it acts on other noctuid species and *Bombyx mori.* Whether homologous-related (or even identical) peptides are produced in other species of parasitized insects remains to be determined. Presumably the peptide acts on the nervous or endocrine systems or may directly act on JH esterase activity, thereby delaying metamorphosis and pupation. Production of this host gene product appears adaptive for the parasitoid involved, since pupation of the host would interfere with the parasitoid's normal emergence process (Hayakawa, 1995). The GBP appears to elevate dopamine levels (Noguchi et al., 1995), which might depress feeding activity, thereby acting via the nervous system. How it acts to inhibit JH esterase activity is not known.

In vertebrates, host anorexia, or certainly the lack of adequate critical nutrients, results in immunosuppression, thus rendering the host more susceptible to infection. Hence, induction of host anorexia is highly adaptive for the invader. In invertebrates, the adaptive significance of the host's anorexic state has not been directly demonstrated, though presumably a depression of host immunological cellular and humoral defenses likely occurs when the host is severely nutritionally deprived. Nutritional stress, as occurs when insects are reared under crowded conditions in either the laboratory or the natural environment, increases susceptibility to parasites and pathogens, hence explaining why outbreaks of viral and bacterial pathogens are particularly likely to occur when insect populations undergo a rapid expansion (Tanada and Kaya, 1993).

The synchronization of cessation of host food consumption with wasp emergence is highly adaptive for the parasitoids because it allows the wasps to exit from their host without risking the consumption of emerging wasps or cocoons by an actively feeding host (Adamo et al., 1997). Often, owing to their failure to feed at the normal rate, parasitized insects are developmentally delayed relative to similarly aged nonparasitized cohort. Aside from their failure to feed at the normal rate, however, these insects often display developmental anomalies indicative that their JH and ecdysteroid titers are altered by parasitism, since their developmental program is either redirected or halted completely, resulting in developmental arrest of the host (see Section III below). The same is true for many pathogen-infected insects.

Interestingly, many gregarious endoparasitoids alter the growth and development of their host in a parasitoid "dose-dependent" fashion, further demonstrating that parasitoids modulate the growth of their host in a highly sophisticated manner. In tobacco hornworm larvae parasitized by *C. congregata*, the metabolic rate of the host is altered during parasitism (Alleyne et al., 1997). Moreover, the final weight of fifth instar hosts, initially parasitized as newly ecdysed fourth instar larvae, varies from 3 to 8 g, and the final dry weight of the host varies in direct proportion to the number of parasitoids developing within it (Alleyne and Beckage, 1997). Thus, hosts containing more parasitoids are induced to grow larger and feed for a slightly longer period prior to wasp emergence, compared to larvae with fewer parasitoids. In addition, parasitism enhances metabolic efficiency and the efficiency conversion of digested nutrients to body mass, although the rate of food consumption is reduced (Thompson, 1993; Alleyne and Beckage, 1997). Thus, while parasitism has an overall inhibitory effect on feeding, heavily parasitized individuals show less of an inhibition compared to hosts with fewer parasitoids. How this effect is mediated physiologically remains to be elucidated, but it may involve a combination of endocrine, neuroendocrine, and nervous system components. In contrast, in vertebrates, the depression of appetite is usually positively correlated with parasite intensity, with more heavily infected individuals showing a greater degree of feeding inhibition (Arneberg et al., 1996).

III. Hormonal Disuption during Infection

Several endocrine mechanisms operate to cause derangements of the normal developmental program during parasitism or pathogenic infection. These include alterations in the rates of hormone biosynthesis, catabolism, or inactivation. Alternatively, rates of hormone interaction with host cells or receptors could be affected, or there could be mechanisms involving a combination of these strategies. Some parasites exploit the host's sensitivity to hormones, growth factors, or other agents including neuromodulators, and produce hormonally active molecules that influence the host's system(s) directly. Many examples of these phenomena are described elsewhere in this volume.

Many parasitoids kill their host by causing its developmental arrest. Why influence host development to such a degree that host metamorphosis is delayed or completely prevented? One obvious reason is that synthesis of a tanned pupal cuticle might prevent emergence of parasitoids normally adapted to emerging from the host's larval stage. Moreover, the pupa is frequently enclosed in a cocoon or encased within a tanned pupal cuticle housed in a subterranean pupal cell. Thus, the cessation of host food consumption followed by arrestment of host development in the larval stage ensures a normal parasitoid emergence process, while onset of host metamorphosis would interfere with the completion of the parasitoid's life cycle.

The mechanisms by which parasitoids cause developmental arrest are in essence the primary factors contributing toward their immense value as biological control agents (Godfray, 1994). The feasibility of biological control depends upon the fact that the parasitoid kills the insect pest prior to its reproductive stage, thus curtailing the pest population (Van Dreische and Bellows, 1995). Certainly, many recent advances in our knowledge of how insect hormones modulate insect development on both the physiological and molecular levels (Cherbas and Cherbas, 1996; Gilbert et al., 1996; Lonard et al., 1996; Riddiford, 1996) will allow us to pinpoint more precisely how parasites and pathogens alter the development program of their host.

The biology of the braconid wasp *C. congregata* and its interactions with host larvae of the tobacco hornworm, *M. sexta*, illustrate many relevant points. The host's JH titer is elevated and the level of hemolymph JH esterase activity is extremely low, thereby preventing host metamorphosis (Beckage and Riddiford, 1982). If the host's cuticle is too thick when the parasitoids attempt to emerge, then their emergence is prevented, and the host develops characteristic cuticular "spots" where the parasitoids begin to erode the cuticle but fail to emerge, so host cuticular characteristics are critically important for "successful" parasitism

(Alleyne and Beckage, 1997). Similarly, the wasp *C. kariyai* emerges through the larval integument but not the pupal cuticle of its host, *Pseudaletia separata*, and parasitoids forced to oviposit in pupal hosts are developmentally 'stranded' within the host (Shimizu et al., 1995).

Parastism of the gypsy moth by *Glyptapanteles liparidis* serves as another excellent example of how parasitism by a braconid wasp halts the host's development prior to wandering. Recent studies by Schopf et al. (1996) have shown that the titer of JH III in hemolymph of unparasitized fifth instar larvae remains below 1 pmol/ml but attains a mean value of 89 pmol/ml in parasitized larvae; hemolymph titers of JH II are also increased during parasitism. Simultaneously, levels of JH esterase activity in the hemolymph are dramatically reduced in concert with detection of high JH titers, so that the JH titers and JH esterase levels are inversely related. In this system, whether the high levels of JH III detected in the hemolymph of the host originate from the corpora allata of the host, or reflect JH biosynthesis and release by the hymenopteran parasitoids (which normally would be expected to produce JH III rather than I or II) to cause an elevation of their host's JH titer, remains to be resolved. Parasitism of *Pseudoplusia includens* by *Microplitis demolitor* has similar effects on JH and JH esterase levels as shown by Balgopal et al. (1996).

Secretion of JH III by the parasitoid *Diachasmimorpha (=Biosteres) longicaudatus* was implicated as causing an elevation of the hemolymph JH titer in host larvae and pharate pupae by Lawrence et al. (1990). Secretion of ecdysteroids or metabolites of the steroid molting hormones has also been documented using parasitoids maintained *in vitro* to verify hormone production (Grossniklaus-Buergin et al., 1989; Brown et al., 1993; Lawrence and Lanzrein, 1993). Thus, wasp parasitoids are believed to manipulate their host's endocrine environment by secreting hormones and metabolites into the hemolymph circulating around them. Whether tachinid flies behave similarly remains to be determined, since we have considerably less information about the physiology of those species compared to parasitoid wasps. They do respond to ecdysteroid pulses produced during metamorphosis of their host, however, and molt in response to ecdysteroid peaks appearing within the host environment (Baronio and Sehnal, 1980).

Yet another strategy is exemplified by wasps of the genus *Chelonus* that induce the onset of precocious metamorphosis of their host during the penultimate instar, then emerge from the prepupal stage. A marked increase in hemolymph JH III esterase activity occurs to trigger host metamorphosis in noctuids including *Trichoplusia ni* and *Spodoptera exigua* and cause pupal cell formation an instar earlier than normal (Jones, 1985), while in the wasp itself, JH III appears to be catabolized almost exclusively to the diol metabolite (Lanzrein and Hammock, 1995). A two-way developmental interaction is evident, since parasitization induces a precocious onset of host metamorphosis and developmental arrest in the prepupal stage, while the wasps adapt their development so that the first instar parasitoids only molt to the second instar after their host has molted to the precocious last instar (Grossniklaus-Buergin et al., 1994). Thus, the parasitoid both modifies and responds to endocrine signals in its host's hemocoel.

Theoretically, an abnormal elevation of JH titers may also be caused by a reduction in the activity of JH epoxy hydrolase present in the tissues. This enzyme, which converts the juvenile hormones (JH I, II, and III) to their respective JH diols, was originally thought to play only a minor role in JH catabolism *in vivo*, but recently has been recognized as significantly contributing to the rate of JH metabolism in tissues other than hemolymph including the midgut, fat body, and integument (Kallapur et al., 1996). How this enzyme is affected by parasitism in those species that exhibit larval developmental arrest has yet to be examined. Nor have effects of parasitism on JH binding proteins present in lepidopteran hemolymph (Park et al., 1993) or tissues (Charles et al., 1996) been analyzed yet. Thus, many unexplored avenues exist for influencing the JH titer of the insect host. The endocrine physiology of the wasps themselves remains virtually unknown, perhaps due to their small size.

In some cases, the polydnavirus and/or venom injected by the parasitoid are sufficient, in the absence of the wasp parasitoid itself, to induce drastic developmental derangements in the host (Jones, 1985); however, the precise effect seen following artificial injection of these factors into nonparasitized larvae may differ from those symptoms induced by natural parasitism. For example, tobacco hornworms injected with very low dosages of purified *C. congregata* polydnavirus (Fig. 1-6a,b) develop a characteristic pale pink coloration and progress to the wandering or prepupal stage; in contrast, their development normally is halted prior to wandering when they receive parasitoid eggs as well as virus during oviposition (Dushay and Beckage, 1993; Beckage and Templeton, 1986; Beckage et al., 1994; Alleyne and Beckage, 1997). Hence, virus-injected nonparasitized animals progress farther developmentally than do their naturally parasitized counterparts. Also, polydnavirus and venom of *Chelonus inanitus* injected together into host eggs induce *Spodoptera littoralis* to undergo developmental arrest in the prepupal stage but fail to induce the onset of precocious metamorphosis prior to arrest (Søller and Lanzrein, 1996). Thus, some but not all of the effects of natural parasitism are reproducible with the virus, at least in many species examined thus far. Søller and Lanzrein (1996) demonstrated that both polydnavirus and venom of the wasp are needed to induce developmental arrest of prepupae of its noctuid host, *S. littoralis*, and a similar dual requirement was previously reported to be required to induce arrest in the armyworm *P. separata* parasitized by *C. kariyai* (Tanaka, 1987; Tanaka et al., 1986). The precise role of the venom remains unclear.

Ecdysteroid deficiency, or a reduction in levels relative to nonparasitized controls of the same chronological age, also was detected in nymphal and adult aphids parasitized by *Aphidius ervi* (Pennacchio et al., 1995). Hormonal changes, as well as onset of detection of parasitism-specific proteins, are temporally associated with parasitic castration of the adult host aphid. Whether hormonal factors are linked to castration of the host varies according to the species of parasitoid involved as well as the sex of the host (see other chapters in this volume; also Reed and Beckage, 1997).

In the hornworm, the host becomes developmentally arrested while the host's prothoracic glands remain intact (Beckage and Buron, 1993) and capable of producing ecdysteroid (Gelman et al., 1996; Kelly et al., 1996). In contrast, degeneration of the host's prothoracic glands occurs in *Heliothis virescens* larvae parasitized by *Campoletis sonorensis,* thus accounting for host developmental arrest (Dover and Vinson, 1990; Dover et al., 1995). A unique aspect is that the host's prothoracic glands only degenerate if the host is parasitized in the fifth instar; parasitization in the fourth instar does not induce gland degeneration. Moreover, the effect is induced by the injection of the polydnavirus purified from the female wasp's reproductive tract, and the effect is inducible in *Helicoverpa zea* and *Spodoptera frugiperda*, indicating the effect is not uniquely inducible in the natural host of *C. sonorensis,* namely *H. virescens* (Dover et al., 1995). The latter authors hypothesized that endocrine changes associated with the molt from the penultimate to the last stadium alter the response of the glands to the virus or viral gene products, thereby explaining the stadium-specific degenerative effect. If the virus is injected during or following apolysis and molting to the last instar, the programmed gland degeneration occurs. Moreover, gland atrophy appears solely virally mediated, and does not require the action of venom or other factors.

The next decade will certainly generate new insights as we identify new molecules and mechanisms of developmental disruption in normal as well as parasitized insects. For example, Neckameyer (1996) showed that in *Drosophila* reduced levels of dopamine result in akinesia, developmental retardation, and decreased fertility. We have no clue as to how parasitism affects dopamine levels in parasitized insects, except for one report by Noguchi et al. (1995), in which hemolymph levels of dopamine were shown to be elevated in armyworm larvae parasitized by *C. kariyai.* This finding was interpreted as indicating that the mechanism of developmental delay and arrest seen in parasitized larvae is likely to be mediated by the dopamine elevation induced by the polydnavirus and venom injected by the wasp parasitoid into the host. In this host species, a growth-blocking peptide (GBP) has been characterized that interferes with the host's normal growth and development and the injection of purified GBP was found similarly to elevate hemolymph dopamine levels in a manner mimicking that induced by natural parasitism (Noguchi et al., 1995).

Octopamine and serotonin influence corpora allata (CA) activity in honey bee larvae, and both of these agents cause a dose-dependent stimulation of JH release (as well as release of methyl farnesoate) under *in vitro* conditions (Kaatz et al., 1994; Rachinsky, 1994). Hence, though this phenomenon has yet to be demonstrated in other species, yet another avenue that parasitism may influence JH titers may be via octopamine, serotonin, and other CA modulating molecules.

Nervous control of JH biosynthesis has been demonstrated in *Locusta migratoria* (Horseman et al., 1994).

Recent evidence from unparasitized lepidopteran insects indicates that the prothoracic glands may produce 3-dehydroecdysone, rather than 20-hydroxyecdysone, which is then converted to the active form of the molting hormone (20-HE) elsewhere at the target site of activity (Gilbert et al., 1996). Moreover, recent evidence from the silkworm, *Bombyx mori,* suggests that ecdysone, or its metabolites other than 20-hydroxyecdysone, has a critical role in the molting process (Tanaka et al., 1994). Thus, the classic dogma in insect physiology, which holds that the prothoracic glands secrete ecdysone, which is then converted to 20-HE by the 20-monooxygenase localized in target tissues, recently has been challenged in several species. The hormone(s) are still produced by the prothoracic glands, however.

Pathogen-induced host developmental arrest also occurs, particularly in lepidopteran insect hosts infected with baculoviruses that fail to pupate once the larval stage is infected. The hormonal induction of metamorphosis is prevented because the baculovirus has a gene encoding the enzyme ecdysteroid UDP-glucosyltransferase (EGT), which efficiently catalyzes the transfer of galactose to ecdysone, causing inactivation of ecdysone or other ecdysteroids produced in the host (O'Reilly et al., 1992; O'Reilly, 1995). Low dosages of the virus are more effective at preventing the pupal molt compared to earlier larval molts. This finding suggests that the prewandering ecdysteriod peak and peak associated with pupation, rather than the larval–larval molting peaks, are specifically inactivated by the EGT viral gene product (Volkman, 1997). Overexpression of the gene encoding the molt-triggering prothoracicotropic hormone is effective in causing molting only when the gene is expressed in an *egt*-minus virus, since the ecdysteroid induced to be synthesized by the recombinant baculovirus is rapidly inactivated due to the action of the conjugating enzymes in the native virus (O'Reilly et al., 1995). Thus, viral infection may interrupt the normal larval developmental program owing to insect-active hormonal signals (or catabolizing enzymes) encoded by the virus itself.

A variation on this theme is hypothesized to occur in *Heliothis virescens* larvae parasitized by *Cardiochiles nigriceps.* The teratocytes of the parasitoid inactivate 20-hydroxyecdysone produced in the host, and the production of polar ecdysteroid metabolites prevents molting in response to any ecdysteroid produced; the host is thereby developmentally arrested (Pennacchio et al., 1994). Teratocytes have also been implicated as secreting inhibitors of JH esterase (Zhang et al., 1992; Dong et al., 1996) as well as playing a nutritive or biosynthetic role.

In addition, some viruses replicate in response to ecdysteroid signals produced in the host, particularly during metamorphosis and adult development (Stairs, 1965, 1970). Thus, a viral response to host hormones is implicated as being responsible for stimulation of replication, mediated by a hormone response element present in the virus or another mechanism. Similarly, the polydnaviruses begin replication during adult development of the female wasp carrying them in her ovaries (Webb and Summer, 1992; Buron and Beckage, 1992; Beckage, 1996) leading to the hypothesis that replication may be hormone-activated by ecdysteroids produced during adult development. The latter hypothesis was verified by Webb and Summers (1992) who found that replication could be stimulated by *in vitro* incubation of wasp ovaries in ecydsteroid-rich culture media, whereas no replication occurred in hormone-deficient media.

Accumulation of neuropeptides also occurs during parasitism of the tobacco hornworm by *C. congregata*, especially in the cerebral neuroendocrine system, midgut, and hindgut tissues of the host during its terminal stage when the host is developmentally arrested (Gelman and Beckage, 1995; Zitnan et al., 1995 a,b; Kelly et al., 1997). The peptide prothoracicotropic hormone (PTTH) is elevated in the cerebral neurosecretory cells of the brain of the host larva during its last instar prior to wasp emergence, and the levels of accumulation continue to be enhanced post-emergence for 7–10 days before death ensues (Zitnan et al., 1995b). Brain ecdysiotropic activity is also significant, indicating the PTTH content of the brain is significant despite the lack of molting by the host during the post-emergence period of developmental arrest (Kelly et al., 1997). Meanwhile, immunoreactive FMRF-amide-like peptides accumulate in the host's midgut endocrine cells, a phenomenon which may or may not be linked to the lack of feeding seen in the host (Zitnan et al., 1995a). The latter peptide is also detectable in cells localized at the midgut–hindgut juncture at the base of the anal vesicle in the wasp (Zitnan et al., 1995a), suggesting the peptides may be released into the hemolymph at this site.

IV. Metabolic Interactions: Possible Influence of Hormones

The metabolic effects of parastism still remain unexplained in most insect systems. Possibly, yet-to-be identified hormonal factors contribute to causing the observed enhancement of metabolic efficiency observed in many host insects attacked by wasp parasitoids (Thompson, 1993; Alleyne and Beckage, 1997; Alleyne et al., 1997). Alternatively, the observed alterations in JH and ecdysteroid titers detected in host insects may elicit metabolic redirection to serve the parasitoid's needs at the expense of proliferation of the host's tissues. The role of the parasitoid-associated viruses (Fig. 1-6a,b) in causing metabolic changes in the host's fat body and other tissues also remains to be clarified. Several proteins appear down-regulated during parasitism, including the hemolymph storage protein arylphorin (Beckage and Kanost, 1993). Shelby and Webb (1994, 1997) demonstrated that arylphorin synthesis is inhibited at the translational, rather than transcriptional, level in host *Heliothis virescens* larvae parasitized by *Campoletis sonorensis,* and that the polydnavirus carried by the wasp is the agent responsible.

During development of braconid parasitoids, teratocytes grow synchronously with the wasps and collectively constitute a large biomass in the hemolymph of

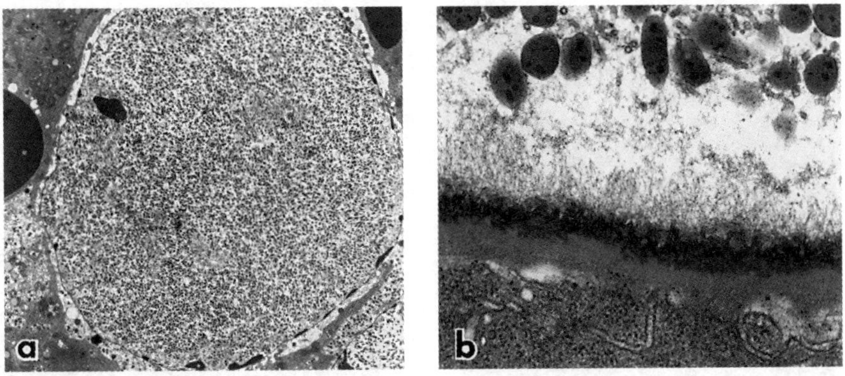

Figure 1-6. The nucleus of an ovarian calyx cell in *Cotesia congregata* showing abundant polydnavirus virions (*a*) that have completed replication prior to rupture of the cell and release of the virions into the calyx lumen. The nucleus occupies the bulk of the volume of the cell. In panel (*b*) the chorion and outer surface of an egg from the interior of the calyx is shown (below), together with the polydnavirus virions present in the lumen of the reproductive tract (above). The virions have tails, which are seen in cross-sections and longitudinal sections in (*b*). Reprinted from Buron and Beckage (1992) with permission from Academic Press. Photo courtesy of Isaure de Buron (University of South Carolina–Spartanberg).

the host. These large cells are notable for their large lipid inclusions and glycogen deposits, and have been variously reported as serving nutritive, immunological, and endocrine functions in the host–parasitoid relationship (Dahlman and Vinson, 1993). In *M. sexta*, an average of 160 cells develops from the serosal membrane of each parasitoid, and each cell is covered with abundant microvilli, suggestive of an absorptive role (Fig. 1-7a) (Buron and Beckage, 1997). Each cell undergoes characteristic developmental changes according to the stage of parasitism of the host. Each grows to a diameter of 150 to > 200 microns before large secretory blebs appear on the surface of the cell coincident with emergence of the wasps (Fig. 1-7b). Since the mean number of parasitoids deposited per oviposition is about 150, the total number of cells in the host ranges up to several hundred thousand. The presence of numerous lipid inclusions and glycogen granules in these cells, together with the parasitoids' biomass (Fig. 1-8) [which represents up to 20–25% of the dry weight of the host-parasitoid complex (Alleyne and Beckage, 1997)], suggests their combined presence may compete substantively with host tissues for nutrients ingested by the host. Their rapid growth during parasitism also suggests the normal pathways of lipid synthesis in the fat body and transport in the hemolymph (Blacklock and Ryan, 1994) may be altered in the host, possibly due to hormonal factors and redirection of lipid, glycogen, and protein storage from the host's tissues to favor those of the parasitoids and their accompanying teratocytes. Simultaneously, the cells may be secreting regulatory

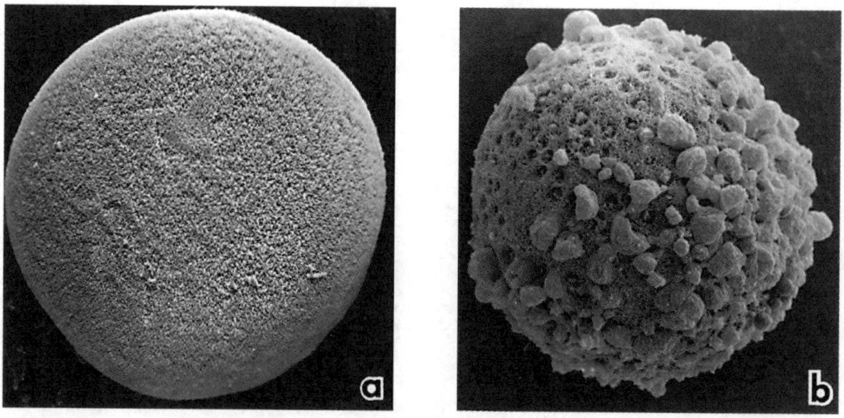

Figure 1-7. Scanning electron micrographs of teratocytes of *Cotesia congregata* recovered from the hemolymph of fifth instar host larvae of the tobacco hornworm (from Buron and Beckage, 1996). The cells approximate 150–200 microns in diameter; a few days prior to emergence of the parasitoids, abundant microvilli but no irregularities are present on the cell surface (*a*), and a few days later during emergence of the wasps the cells develop large secretory blebs, or outpocketings, on the surface (*b*). Each parasitoid gives rise to about 160 cells, which dissociate from the serosal membrane; several hundred thousand cells may be present in the hemolymph of each host. Photo courtesy of Isaure de Buron (University of South Carolina–Spartanberg).

proteins/peptides and other factors to facilitate growth and development of the parasitoids.

V. Induction of Intersex Formation

Another example of an intriguing phenomenon that has yet to be explained on a mechanistic physiological level is the induction of intersex formation in many arthropods including mayflies, Hymenoptera (ants, bees, and wasps), flies, and crabs. Intersexes occur in many other parasitized animals as well (Wülker, 1975). The taxon that acts as the parasitic species varies widely and may be a nematode (infecting mayflies, flies, and ants, for example), another insect (i.e., Strepsiptera in bees), or a rhizocephalan barnacle (as occurs in crabs). Fungal (Watson and Petersen, 1993; Møller, 1993) and other pathogens are also known to induce the formation of intersexes in their host(s).

Mermithid nematodes induce formation of intersexes in mayfly hosts. Vance and Peckarsky (1996) demonstrated that parasitized *Baetis bicaudatus* nymphs are significantly smaller than similarly aged unparasitized nymphs and emerge as adults up to 4 weeks later than unparasitized individuals. While the nematode *Gasteromermis* sp. does not interfere with emergence of the adult mayfly, it does induce formation of intersexes (with respect to the morphological characters

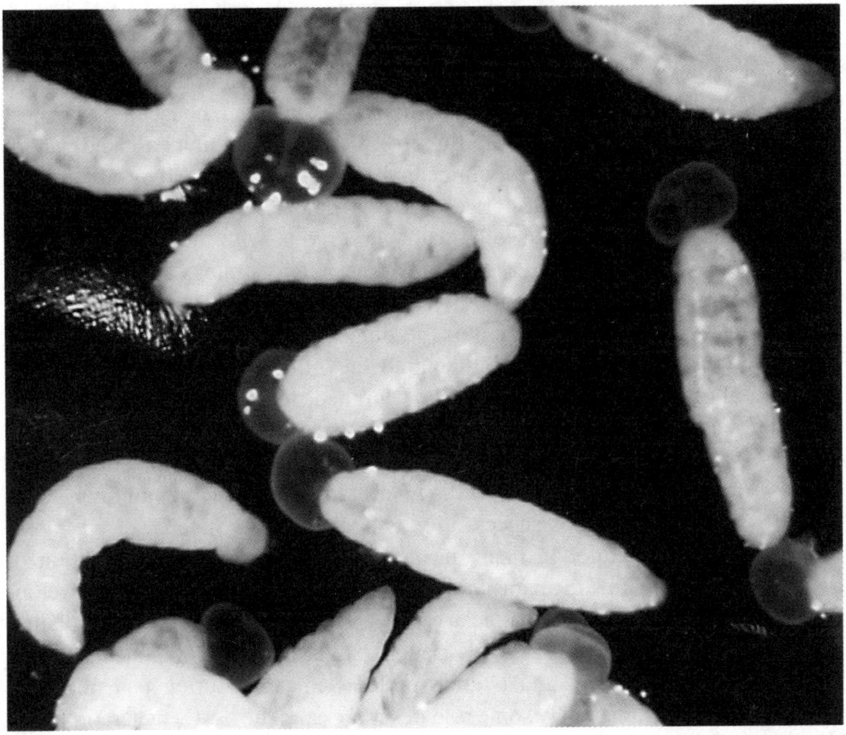

Figure 1-8. Appearance of mature second instar larvae of *Cotesia congregata* dissected from the hemocoel of a fifth-instar tobacco hornworm preparatory to emergence from the host. Note large anal vesicle at the posterior end of each larva; larvae are ca. 3 mm long before they undergo their second larval ecdysis and emerge from the host to complete cocoon spinning, as seen in Fig. 1-2. Total mass of the parasites and their accompanying teratocytes is substantial. Photo taken by Frances Tan (University of California–Riverside).

associated with the eyes and genitalia), which chromosome analysis shows represent feminized males, rather than masculinized females (Vance, 1996). The secondary sexual behaviors demonstrated by the intersexes are thought to be highly adaptive for the parasites involved, since the feminized male host shows directional flight upstream similar to nonparasitized females (whereas normal males do not) and returns to the water, mimicking the oviposition behavior of nonparasitized individuals, where the parasites complete their life cycle (Vance and Peckarsky, 1996). The intersex adults contain no eggs, ovaries, or testes. Their field studies demonstrated that nematode infections kill or castrate a significant proportion of the mayfly population studied and that infection rates appear constant from year to year, thus underscoring the potential importance of parasitic interactions to stream ecosystems.

VI. Other Developmental Interactions

Yet other interactions may involve direct physiological manipulation of the host by the invading parasite or pathogen. In fungus-infected houseflies, the host assumes a characteristic elevated posture (Fig. 1-5) and exhibits a gated pattern of mortality; the timing of death of infected flies appears governed by a biological clock within the fungus (Krasnoff et al., 1995). Thus, the temporal pattern of host mortality is hypothesized to be pathogen-driven, with the majority of flies dying just a few hours before the onset of darkness. This strategy appears adaptive for the pathogen since a high humidity environment as occurs during the scotophase is necessary for germination of the fungal spores.

In a few cases of parasitism by tachinid flies, the parasitoid fails to kill its host and allows the host to continue development to adulthood and reproduce after the parasitoid successfully emerges (DeVries, 1984; English-Loeb et al., 1990). Possibly in these instances, the normal endocrine manipulation of the host's development by the parasitoid is interfered with to such a minimal extent that the parasitoid develops successfully as does its host. Such cases are rarely reported in the literature. In many instances of "successful" encapsulation of wasp parasitoids, neither the parasitoid nor the host survives even though development of the parasitoid may be halted at a very early stage due to egg encapsulation (Dushay and Beckage, 1993). This pattern suggests that components of the cocktail of virus, venom, and ovarian proteins injected during parasitization, along with the eggs, may play some role in suppressing the host's metamorphosis.

Interestingly, parasitoids alone may also cause their host's developmental arrest. Transplanted parasitoids, washed free of host hemolymph and teratocytes, which are taken from the hemocoel of a parasitized animal, and implanted into nonparasitized larvae, may cause arrest of the surrogate "host" in the absence of virus, venom, and teratocytes (Adamo et al., 1997; Lavine and Beckage, 1995). This complex scenario suggests that multiple factors may contribute to the induction of host arrest.

VII. Future Glimpses

Viruses may also be used as tools for genetic manipulation and/or transformation of pest arthropods. Witness, for example, the recent progress achieved in stable transformation of cells of the malaria-carrying mosquito *Anopheles gambiae* using pantropic retroviral vectors genetically modified to broaden their host range to include insect cells (Matsubara et al., 1996). The mutualistic or symbiotic polydnaviruses carried by parasitoids (Lavine and Beckage, 1995; Strand and Pech, 1995) also provide an avenue for possible genetic manipulation of the insect carrying the virus (i.e., the parasitoid) or for transferring gene(s) to the insect host. The evolution of relationships such as that the PDV exhibits with wasp 'vectors' and lepidopteran hosts (Whitfield, 1994) and other types of mutualistic interactions (Douglas, 1994; Sapp, 1994; Connor, 1995), as well as the selective pressures maintaining such associations among partners, have given rise to much recent speculation about their modes of evolution and inheritance.

Recent advances in our understanding of developmental changes occurring in the parasites themselves, as exemplified by the developmental alterations occurring in malaria parasites as they move through vector mosquitoes (Boulanger et al., 1995), will also provide avenues for future manipulation of these organisms. Insights may also be gained from studies of the molecular genetics of development of *Drosophila*. For example, parasites and pathogens frequently alter host circadian rhythms, geotaxes, and other behaviors such as mating as well as temperature, humidity, and substrate preferences (Moore, 1993, 1995) but the mechanisms involved have yet to be identified, including those regulating thermoregulatory preferences and behavioral fevers (Horton and Moore, 1993). Recent progress in identifying genes important in determination of thermosensation and hygrosensation preferences in *Drosophila* (Sayeed and Benzer, 1996) may facilitate identification of mechanisms operating in parasitized insects to modulate those behaviors. In the latter study, ablation experiments showed that thermoreceptors are housed on the third antennal segment, whereas hygroreceptors are located on the distal antennal arista. Mutants defective in showing normal preferences may yield clues as to the nature of the genes involved in regulating these behaviors, as well as to probable modifications in their expression during parasitism and pathogenic infection. Molecular studies of genes modulating mating (Hall, 1994) and circadian rhythms in *Drosophila* (Hall, 1995; Meinertzhagen and Pyza, 1996) likely also will prove useful in determining how parasites and pathogens affect insect hosts.

Future studies directed at identification of molecules important in parasite growth and differentiation will also aid us in developing strategies to manipulate parasite development in insect hosts or vectors. For example, retinoic acid, which is an essential developmental signaling molecule in *Drosophila* (Oro et al., 1992) and other invertebrates as well as vertebrates, appears to be important in development and differentiation of *Brugia malayi*, a mosquito-borne nematode that causes human lymphatic filariasis (Wolff and Scott, 1995). *In vitro,* and presumably *in vivo*, this parasite takes up retinoic acid from its external milieu in a dose-dependent fashion.

Continued identification of parasite-secreted molecules as well as clarification of their effects on host cells will generate critical information about how parasites directly manipulate their host environment. Hormonally active parasite-secreted molecules have been better characterized in vertebrate systems as compared to invertebrates and insects (chapter 5; also Foster and Lee, 1996). Nevertheless, we already have evidence to show that JHs and ecdysteroids are secreted by insect parasitoids (Lawrence and Lanzrein, 1993) and future studies likely will indicate that neuropeptides and metabolic regulators, as well as immunologically active cytokines, are secreted by endoparasitoids as well. A caveat is that peptide hormones such as prothoracicotropic hormones may vary in the hormone-receptor binding domains, even amongst different species of Lepidoptera (e.g., *Manduca sexta* versus *Bombyx mori*; Rybczynski et al., 1996), hence interspecies variability in peptide hormone sequence structure appears likely. Molecules secreted by wasp parasitoids may or may not be host-active, depending on the host species harboring the parasitoid. The juvenile hormone produced by Hymenoptera (JH III) differs from the lepidopteran hormones JH I and II (Lanzrein and Hammock, 1995), and ecdysteroidlike molting hormones may also show interspecies variability in their structure and activity.

Other future strategies likely will involve combining parasitoid- or polydnavirus-encoded genes with those expressed by pathogens, to increase the level of pathogenicity expressed by a virus or other infectious agent. For example, the venom injected by the ectoparasitoid *Euplectrus comstockii* induces developmental paralysis of lepidopteran hosts and modifies the insect's susceptibility to NPV (Coudron et al., 1995). Parasitoid development and emergence from insect hosts may be accelerated (McCutchen et al., 1996) or delayed (or prevented) following treatment of the host with native or recombinant baculoviruses, and parasitized insects frequently are found to be more susceptible to NPVs compared to their nonparasitized cohorts (Fuxa, 1993). The precise effect likely depends upon the type of developmental interaction the parasitoid normally exhibits with its host, the nature of the baculovirus recombinant the insect host is treated with (i.e., whether a developmental regulator or a paralyzing venom is expressed by the virus), and when ingestion of occluded virions occurs relative to parasitization of the host (Brooks, 1993).

Polydnavirus-encoded gene products may be expressed in recombinant baculoviruses, thus enhancing virus virulence by causing immediate developmental arrest, knocking out the host's immune system or otherwise altering the host's physiology (Lavine and Beckage, 1995; Washburn et al., 1996) to enhance the efficacy of the baculovirus. Such enhancement has already been developed using other insect-selective toxins, that knock the feeding insect off the plant following onset of viral infection and replication (Hoover et al., 1995; Bonning and Hammock, 1996).

Yet another potential strategy may be to express antimicrobial compounds such as the cecropins, originally isolated as antibacterial compounds in lepidopteran species, in insect vectors to halt the development of malarial parasites and other human pathogens (Gwadz et al., 1989). For example, the cecropinlike synthetic peptide *Shiva*-3 reduces ookinete production and offers the prospect of using *Shiva*-like peptides to engineer malaria-resistant vectors (Rodriguez et al., 1995). Cercopins are also active in minute amounts (50 micromolar concentration) in causing attenuation of the motility and killing of microfilariae of the nematode of *Brugia pahangi in vivo* in female mosquitoes (Chalk et al., 1995). Hence, other human pathogens may likewise be affected.

Certainly, increasing the "cross-talk" among these overlapping disciplines, as well developing techniques to enhance "cross-species gene transfer," can only serve to improve our chances for development of novel biopesticides. Future research likely will demonstrate that parasites can be genetically modified to halt their development in insect hosts, and the hosts themselves hopefully can be genetically modified to reduce parasite success and increase vector refractoriness (Carlson et al., 1995; Pfeifer and Grigliatti, 1996). Other future strategies likely will include genetic engineering of beneficial arthropods themselves as transformation of beneficial species—phytoseiid predatory mites and wasp parasitoids—already has been achieved successfully (Hoy, 1994; Presnail and Hoy, 1992, 1994, 1996). Intracellular symbiotic organisms such as *Wolbachia* have also been implicated as agents capable of serving as mechanisms of interspecies gene transfer and genetic modification (Boyle et al., 1993; Braig et al., 1994; Hoy, 1994; Collins and Paskewitz, 1995). Retrotransposons may be exploitable as transformation tools (Friesen, 1993).

So too might the "genetic symbionts" of parasitoids (i.e., PDVs) be exploited as genetic tools for expression of foreign genes to increase their level of success within a given host. Genetic alteration of polydnaviruses may serve to facilitate the successful development of parasitoid species in a wider range of agricultural pest species, or increase their speed-of-kill, akin to virus recombination strategies already being developed for enhancement of baculovirus virulence. Since the viruses are integrated in the wasps' chromosomal DNA (Lavine and Beckage, 1995; Strand and Pech, 1995), this strategy obviously will involve genetic modification of the wasp itself. Baculovirus–polydnavirus hybrids may be constructed to better augment the virulence of baculoviruses, which sometimes take days or even weeks to kill the host insect.

Knowledge of the physiological mechanisms exploited by parasites and pathogens to regulate host development undoubtedly will bear additional fruit with respect to yielding information valuable in formulation of novel biorational pesticides, as well as generating fascinating glimpses into mechanisms of "developmental pathogenesis," which are of interest from evolutionary, as well as parasitological, perspectives. The strategies evolved by these organisms clearly include many different pathways of endocrine manipulation, as well as complex mechanisms of immunosuppression and evasion, which are also required to achieve successful development in insect hosts.

Host–parasite coevolution and the mechanisms whereby each partner "tracks" the other in a coevolutionary sense are attracting increasing attention from immunologists, endocrinologists, epidemiologists, and ethologists, as well as evolutionary biologists (Barnard and Behnke, 1990; Ewald, 1994; Clayton and Moore, 1996). Many of the interactions described here provide insights into how two taxonomically related, but nevertheless physiologically independent, insect partners coordinate their development.

Acknowledgments

Research support for out tobacco hornworm studies was provided by the National Science Foundation, the NRI Competitive Grants Entomology/Nematology Program, and the University of California-Riverside. I thank Frances Tan for assistance with the manuscript, and Ron Hoy, Marie Read, Brad Mullens, Harry Kaya, Isaure de Buron, Marianne Alleyne, and Frances Tan for providing photos. Harry Kaya and Shelley Adamo provided helpful editorial comments on the text. I dedicate this chapter to my son Ross Rhoades in recognition of the patience and understanding he graciously displayed during the preparation of this volume, which are gratefully acknowledged by his mom.

References

Adamo, S. A., Linn, C. E., and Beckage, N. E. (1997). Correlation between changes in host behaviour and octopamine levels in the tobacco hornworm, *Manduca sexta*, parasitized by the gregarious braconid parasitoid wasp *Cotesia congregata. J. Exp. Biol.* **200:**117–127.

Adamo, S. A., Robert, D., and Hoy, R. R. (1995). Effects of a tachinid parasitoid, *Ormia ochracea*, on the behaviour and reproduction of its male and female field cricket hosts (*Gryllus* spp.). *J. Insect Physiol.* **41:**269–277.

Alleyne, M., and Beckage, N. E. (1997). Parasitism-induced effects on host growth and metabolic efficiency in tobacco hornworm larvae parasitized by *Cotesia congregata. J. Insect Physiol.* **43**(4):407–424.

Alleyne, M., Chappell, M. A., Gelman, D. B., and Beckage, N. E. (1997). Effects of parasitism by the braconid wasp *Cotesia congregata* on metabolic rate in host larvae of the tobacco hornworm, *Manduca sexta. J. Insect Physiol.* **43**(2):143–154.

Amen, R. I., and de Jong-Brink, M. (1992). *Trichobilharzia ocellata* infections in its snail host *Lynmaea stagnalis:* An *in vitro* study showing direct and indirect effects on the snail internal defence system, via the host central nervous system. *Parasitology* **105:**409–416.

Arneberg, P., Folstad, L., and Karter, A. J. (1996). Gastrointestinal nematodes depress food intake in naturally infected reindeer. *Parasitology* **112:**213–219.

Balgopal, M. M., Dover, B. A., Goodman, W. G. and Strand, M. R. (1996). Parasitism by *Microplitis demolitor* induced alterations in the juvenile hormone titers and juvenile hormone esterase activity of its host, *Pseudoplusia includens. J. Insect Physiol.* **42:**337–345.

Barnard, C. J., and Behnke, J. M. (1990). *Parasitism and host behaviour*. London: Taylor & Francis.

Baronio, P., and Sehnal, F. (1980). Dependence of the parasitoid *Gonia cinerascens* on the hormones of its lepidopterous hosts. *J. Insect Physiol.* **26:**619–626.

Beck, G., and Habicht, G. (1996). Cytokines in invertebrates. In *New directions in invertebrate immunology* (K. Soderhall, S. Iwanaga, and G. R. Vasta, eds.), pp. 131–154, Fair Haven, N. J.: SOS Publications.

Beckage, N. E. (1993a). Endocrine and neuroendocrine host-parasite relationships. *Receptor* **3:**233–245.

———. (1993b). Games parasites play: The dynamic roles of proteins and peptides in the relationship between parasite and host. In *Parasites and pathogens of insects* (N. E. Beckage, S. N. Thompson, and B. A. Federici, eds.), vol. 1, pp. 25–57. New York: Academic Press.

———. (1996). Interactions of viruses with invertebrate cells. In *New directions in invertebrate immunology* (K. Soderhall, S. Iwanaga, and G. R. Vasta, eds.) pp. 375–399. Fair Haven, N.J.: SOS Publications.

Beckage, N. E., and Buron, I. de. (1993). Lack of prothoracic gland degeneration in developmentally arrested host larvae of *Manduca sexta* parasitized by the braconid wasp *Cotesia congregata. J. Inv. Pathol.* **61:**103–106.

Beckage, N. E., and Kanost, M. R. (1993). Effects of parasitism by the braconid wasp *Cotesia congregata* on host hemolymph proteins of the tobacco hornworm, *Manduca sexta. Insect Biochem. Molec. Biol.* **23:**643–653.

Beckage, N. E., Tan, F. F., Schleifer, K. W., Lane, R. D., and Cherubin, L. L. (1994). Characterization and biological effects of *Cotesia congregata* polydnavirus on host larvae of the tobacco hornworm, *Manduca sexta. Arch. Insect Biochem. Physiol.* **26:**165–195.

Beckage, N. E., and Riddiford, L. M. (1978). Developmental interactions between the tobacco hornworm *Manduca sexta* and its braconid parasite *Apanteles congregatus. Entomol. Exp. Appl.* **23:**139–151.

———. (1982). Effects of parasitism by *Apanteles congregatus* on the endocrine physiology of the tobacco hornworm *Manduca sexta. Gen. Comp. Endocrinol.* **47:**308–322.

———. (1983). Growth and development of the endoparasitic wasp *Apanteles congregatus:* Dependence on host nutritional status and parasite load. *Physiol. Entomol.* **8:**231–241.

Beckage, N. E., Tan, F. F., Adamo, S. A., Reed, D. A., Drezen, J. M., Linn, C. E., and Buron, I. de. (1997). The tobacco hornworm: a premier model for analyzing interactions of hosts, parasitoids, and polydnaviruses. *J. Insect Physiol.* (submitted).

Beckage, N. E., and Templeton, T. J. (1986). Physiological effects of parasitism by *Apanteles congregatus* in terminal-stage tobacco hornworm larvae. *J. Insect Physiol.* **32:**299–314.

Beckage, N. E., Thompson, S. N., and Federici, B. A. eds. (1993). *Parasites and pathogens of insects.* Vol. 1, *Parasites.* Vol. 2, *Parasites.* New York: Academic Press.

Blacklock, B. J., and Ryan, R. O. (1994). Hemolymph lipid transport. *Insect Biochem. Molec. Biol.* **9:**855–973.

Bonning, B. C., and Hammock, B. D. (1996). Development of recombinant baculoviruses for insect control. *Ann. Rev. Entomol.* **41:**191–210.

Boulanger, N., Charoenvit, Y., Krettli, A., and Betschart, B. (1995). Developmental changes in the circumsporozoite proteins of *Plasmodium berghei* and *P. gallinaceum* in their mosquito vectors. *Parasitol. Res.* **81:**58–65.

Boyle, L., O'Neill, S. L., Robertson, H. M., and Karr, T. L. (1993). Interspecific and intraspecific horizontal transfer of *Wolbachia* in *Drosophila. Science* **260:**1796–1799.

Braig, H. R., Guzman, H., Tesh, R. B., and O'Neill, S. L. (1994). Replacement of the *Wolbachia* symbiont of *Drosophila simulans* with a mosquito counterpart. *Nature* **367:**453–455.

Brodeur, J., and Vet, L. E. M. (1994). Usurpation of host behaviour by a parasitic wasp. *Anim. Behav.* **48:**187–192.

Brooks, W. M. (1993). Host-parasitoid-pathogen interactions. In *Parasites and pathogens of insects* (N. E. Beckage, S. N. Thompson, and B. A. Federici, eds.), vol. 2, pp. 231–272. New York: Academic Press.

Brown, J. J., Kiuichi, M., Kainoh, Y., and Takeda, S. (1993). *In vitro* release of ecdysteroids by an endoparasitoid, *Ascogaster reticulatus* Watanabe. *J. Insect Physiol.* **39:**229–234.

Buron, I. de, and Beckage, N. E. (1992). Characterization of a polydnavirus (PDV) and virus-like filamentous particle (VLFP) in the braconid wasp *Cotesia congregata* (Hymenoptera: Braconidae). *J. Invert. Pathol.* **59:**315–327.

———. (1997). Developmental studies of teratocytes of the braconid parasitoid wasp *Cotesia congregata* in host larvae of the tobacco hornworm, *Manduca sexta. J. Insect Physiol.* (submitted).

Cade, W. H., (1984). Effects of fly parasitoids on nightly duration of calling in field crickets. *Can. J. Zool.* **62:**226–228.

Carlson, J., Olson, K., Higg, S., and Beaty, B. (1995). Molecular genetic manipulation of mosquito vectors. *Ann. Rev. Entomol.* **40:**359–388.

Carton, Y., and Nappi, A. J. (1996). *Drosophila* cellular immunity against parasitoids. *Parasitol. Today* (in press).

Chalk, R., Townson, H., and Ham, P. J. (1995). *Brugia pahangi:* The effects of cecropins on microfilariae *in vitro* and in *Aedes aegypti. Exp. Parasitol.* **80:**401–406.

Charles, J. P., Wojtasek, H., Lentz, A. J., Thomas, B. A., Palli, S. R., Parker, A. G., Dorman, G., Hammock, B. D., Prestwich, G. D., and Riddiford, L. M. (1996). Purification and reassessment of ligand binding by the recombinant, putative juvenile hormone receptor of the tobacco hornworm, *Manduca sexta. Arch. Insect Biochem. Physiol.* **31:**371–393.

Cherbas, P., and Cherbas, L. (1996). Molecular aspects of ecdysteroid hormone action. In *Metamorphosis: Postembryonic reprogramming of gene expression in amphibian and insect cells.* L. I. Gilbert, J. R. Tata, and B. G. Atkinson, eds., pp. 175–221. San Diego: Academic Press.

Clayton, D. H., and Moore, J., eds. (1997). *Host-parasite evolution: General principles and avian models.* Oxford: Oxford University Press.

Collins, F. H., and Paskewitz, S. M. (1995). Malaria: Current and future prospects for control. *Ann. Rev. Entomol.* **40:**195–219.

Connor, R. C. (1995). The benefits of mutualism: A conceptual framework. *Biol. Rev.* **70:**427–457.

Coudron, T. A., Rice, W. C., Ellersieck, M. R., and Pinnell, R. E. (1995). Mediated pathogenicity of the baculovirus HzSNPV by the venom from *Euplectrus comstockii* (Hymenoptera: Eulophidae). *Biol. Cont.* **5:**92–98.

Cox-Foster, D. L., Schonbaum, D. P., Murtha, M. T., and Cavener, D. R. (1990). Developmental expression of the glucose dehydrogenase gene in *Drosophila melanogaster. Genetics* **124:**873–880.

Cox-Foster, D. L., and Stehr, J. E. (1994). Induction and localization of FAD-glucose dehydrogenase (GLD) during encapsulation of abiotic implants in *Manduca sexta* larvae. *J. Insect Physiol.* **40:**235–249.

Dahlman, D. L., and Vinson, S. B. (1993). Teratocytes. Developmental and biochemical characteristics. In *Parasites and pathogens of insects* (N. E. Beckage, S. N. Thompson, and B. A. Federici, eds.), vol. 1, pp. 145–165. New York: Academic Press.

DeVries, P. J. (1984). Butterflies and tachinids: Does the parasite always kill its host? *J. Nat. Hist.* **18:**323–326.

Dong, K., Zhang, D., and Dahlman, D. L. (1996). Down-regulation of juvenile hormone esterase and arylphorin production in *Heliothis virescens* larvae parasitized by *Microplitis croceipes. Arch. Insect Biochem. Physiol.* **32:**237–248.

Douglas, A. E. (1994). *Symbiotic interactions.* Oxford: Oxford University Press.

Dover, B. A., Tanaka, T., and Vinson, S. B. (1995). Stadium-specific degeneration of host prothoracic glands by *Campoletis sonorensis* calyx fluid and its association with host ecdysteroid titers. *J. Insect Physiol.* **41:**947–955.

Dover, B. A., and Vinson, S. G. (1990). Stage-specific effects of *Campoletis sonorensis* parasitism on *Heliothis virescens* development and prothoracic glands. *Physiol Entomol.* **15:**405–414.

Dunphy, G. B., and Downer, R. G. W. (1994). Octopamine, a modulator of the haemocytic nodulation sreponse of non-immune *Galleria mellonella* larvae. *J. Insect Physiol.* **40:**267–272.

Dushay, M. D., and Beckage, N. E. (1993). Dose-dependent separation of *Cotesia congregata*-associated polydnavirus effects on *Manduca sexta* larval development and immunity. *J. Insect Physiol.* **39:**1029–1040.

English-Loeb, G. M., Karban, R., and Brody, A. K. (1990). Arctiid larvae survive attack by a tachinid parasitoid and produce viable offspring. *Ecol. Entomol.* **15:**361–362.

Ewald, P. (1994). *Evolution of infectious disease.* Oxford: Oxford University Press.

———. (1995). The evolution of virulence: A unifying link between parasitology and ecology. *J. Parasitol.* **81:**659–669.

Fields, P. E., and Woodring, J. P. (1991). Octopamine mobilization of lipids and carbohydrates in the house cricket, *Acheta domesticus. J. Insect Physiol.* **37:**193–199.

Foster, N., and Lee, D. L. (1996). A vasoactive intestinal polypeptide-like protein excreted/secreted by *Nippostrongylus brasiliensis* and its effect on contraction of uninfected rat intestine. *Parasitology* **112:**97–104.

Friesen, P. D. (1993). Invertebrate transposable elements in the baculovirus genome: Characterization and significance. In *Parasites and pathogens of insects* (N. E. Beckage, S. N. Thompson, and B. A. Federici, eds.), vol. 2, pp. 147–178. New York: Academic Press.

Fuxa, J. R. (1993). Insect resistance to viruses. In *Parasites and pathogens of insects* (N. E. Beckage, S. N. Thompson, and B. A. Federici, eds.), vol. 2, pp. 197–209. New York: Academic Press.

Gelman, D. B., and Beckage, N. E. (1995). Low molecular weight ecdysiotropins in proctodaea of fifth instars of the tobacco hornworm, *Manduca sexta* (Lepidoptera: Sphingidae) and hosts parasitized by the braconid wasp *Cotesia congregate* (Hymenoptera: Braconidae). *Eur. J. Entomol.* **92:**123–129.

Gelman, D. B., Reed, D. A., and Beckage, N. E. (1997). Manipulation of fifth instar host *(Manduca sexta)* ecdysteroid levels by the parasitoid *Cotesia congregata. J. Insect Physiol.* (submitted).

Gilbert, L. I., Rybczynshi, R., and Tobe, S. S. (1996). Endocrine cascade in insect metamorphosis. In *Metamorphosis: Postembryonic reprogramming of gene expression in amphibian and insect cells* (L. I. Gilbert, J. R. Tata, and B. G. Atkinson, eds.), pp. 59–107. San Diego: Academic Press.

Gillespie, J. P., Kanost, M. R., and Trenczek, T. (1997). Biological mediators of insect immunity. *Ann. Rev. Entomol.* **42:**611–643.

Godfray, H. C. J. (1994). *Parasitoids: Behavioral and evolutionary biology.* Princeton: Princeton University Press.

Goulson, D., and Cory, J. S. (1995). Sublethal effects of baculovirus in the cabbage moth, *Mamestra brassicae. Biol. Cont.* **5:**361–367.

Grossniklaus-Buergin, D., Connat, J. L., and Lanzrein, B. (1989). Ecdysone metabolism in the host-parasitoid system *Trichoplusia ni/Chelonus* sp. *Arch. Insect Biochem. Physiol.* **11:**79–92.

Grossniklaus-Buergin, C., Wyler, T., Pfister-Wilhelm, R., and Lanzrein, B. (1994). Biology and morphology of the parasitoid *Chelonus inanitus* (Braconidae, Hymenoptera) and effects on the development of its host *Spodoptera littoralis* (Noctuidae, Lepidoptera). *Inv. Repro. Develop.* **25:**143–158.

Gwadz, W. R., Kaslow, D., Lee, J., Maloy, W. L., Zaloff, M., and Miller, L. (1989). Effects of magainins and cecropins on the sporogonic development of malaria parasites in mosquitoes. *Infect. Immun.* **57:**2628–2633.

Hajek, A. E., and St. Leger, R. J. (1994). Interactions between fungal pathogens and insect hosts. *Ann. Rev. Entomol.* **39:**293–322.

Hall, J. C. (1994). The mating of a fly. *Science* **264:**1702–1714.

———. (1995). Tripping along the trail to the molecular mechanisms of biological clocks. *Trends Neurosci.* **18:**230–240.

Hayakawa, Y. (1995). Growth-blocking peptide: An insect biogenic peptide that prevents onset of metamphosis. *J. Insect Physiol.* **41:**1–6.

Hayakawa, Y., and Yasuhara, Y. (1993). Growth-blocking peptide or polydnavirus effects on the last instar larvae of some insect species. *J. Insect Physiol.* **23:**225–231.

Hoover, K., Schultz, C. M., Lane, S. S., Bonning, B. C., Duffey, S. S., McCutchen, B. F., and Hammock, B. D. (1995). Reduction in damage to cotton plants by a recombinant baculovirus that knocks moribund larvae of *Heliothis virescens* off the plant. *Biol. Cont.* **5:**419–426.

Horseman, G., Harmann, R., Virant-Doberlet, M., Loher, W., and Huber, F. (1994). Nervous control of juvenile hormone biosynthesis in *Locusta migratoria. Proc. Natl. Acad. Sci.* **91:**2960–2964.

Horton, D., and Moore, J. (1993). Behavioral effects of parasites and pathogens in insect hosts. In *Parasites and pathogens of insects* (N. E. Beckage, S. N. Thompson, and B. A. Federici, eds.), vol. 1, pp. 107–124. New York: Academic Press.

Hoy, M. A. (1994). *Insect molecular genetics: An introduction to principles and applications.* San Diego: Academic Press.

Jones, D. (1985). Parasite regulation of host metamorphosis: A new form of regulation of pseudoparasitized larvae of *Trichoplusia ni. J. Comp. Physiol.* **B155:**583–590.

Jones, D., Gelman, D., and Loeb, M. (1992). Hemolymph concentrations of host ecdysteroids are strongly suppressed in precocious prepupae of *Trichoplusia ni* parasitized and pseudoparasitized by *Chelonus* near *curvimaculatus. Arch. Insect Biochem. Physiol.* **21:**155–165.

Jones, G. (1995). Molecular mechanisms of action of juvenile hormone. *Ann. Rev. Entomol.* **40:**147–169.

Kaatz, H., Eichmuller, S., and Kreissl, S. (1994). Stimulatory effect of octopamine on juvenile hormone biosynthesis in honeybees *(Apis mellifera):* Physiological and immunocytochemical evidence. *J. Insect Physiol.* **10:**865–872.

Kallapur, V. L., Majumber, C., and Roe, R. M. (1996). *In vivo* and *in vitro* tissue specific metabolism of juvenile hormone during the last stadium of the cabbage looper, *Trichoplusia ni. J. Insect Physiol.* **42:**181–190.

Kavaliers, M., and Colwell, D. D. (1992). Parasitism, opioid systems and host behaviour. *Adv. Neuroimmunol.* **2:**287–295.

Kelly, T. J., Gelman, D. B., Reed, D. A., and Beckage, N. E. (1997). Brain ecdysiotropic hormone activity in host larvae of the tobacco hornworm, *Manduca sexta,* parasitized by the braconid wasp, *Cotesia congregata. J. Insect Physiol.* (submitted).

Krasnoff, S. B., Watson, D. W., Gibson, D. M., and Kwan, E. C. (1995). Behavioral effects of the entomopathogenic fungus, *Entomophthora muscae,* on its host *Musca domestica:* Postural changes in dying hosts and gated pattern of mortality. *J. Insect Physiol.* **41:**895–903.

Lanzrein, B., and Hammock, B. D. (1995). Degradation of juvenile hormone III *in vitro* by non-parasitized and parasitized *Spodoptera exigua* (Noctuidae) and by the endoparasitoid *Chelonus inanitus* (Braconidae). *J. Insect Physiol.* **41:**993–1000.

Lavine, M. D., and Beckage, N. E. (1995). Polydnaviruses: Potent mediators of host insect immune dysfunction. *Parasitol. Today* **11:**368–378.

Lawrence, P. O. (1986a). Host-parasite hormonal interactions: An overview. *J. Insect Physiol.* **32:**295–298.

———. (1986b). The role of 20-hydroxyecdysone in the moulting of *Biosteres longicaudatus*, a parasite of the Caribbean fruit fly, *Anastrepha suspensa. J. Insect Physiol.* **32:**329–337.

Lawrence, P. O., Baker, F. C., Tsai, L. W., Miller, C. A., Schooley, D. A., and Geddes, L. G. (1990). JH III levels in larvae and pharate pupae of *Anastrepha suspensa* (Diptera:

Tepritidae) and in larvae of the parasitic wasp *Biosteres longicaudatus* (Hymenoptera: Braconidae). *Arch. Insect Biochem.* **13L:**53–62.

Lawrence, P. O., and Lanzrein, B. (1993). Hormonal interactions between endoparasites and their host insects. In *Parasites and pathogens of insects* (N. E. Beckage, S. N. Thompson, and B. A. Federici, eds.), vol. 1, pp. 59–86. New York: Academic Press.

Lonard, D. M., Bhaskaran, G., and Dahm, K. H. (1996). Control of prothoracic gland activity by juvenile hormone in fourth instar *Manduca sexta* larvae. *J. Insect Physiol.* **42:**205–213.

Matsubara, T., Beeman, R. W., Shike, H., Besansky, N. J., Mukabayire, O., Higgs, S., James, A. A., and Burns, J. C. (1996). Pantropic retroviral vectors integrate and express in cells of the malaria mosquito, *Anopheles gambiae. Proc. Natl. Acad. Sci.* **93:**6181–6185.

Mattiacci, L., and Dicke, M. (1995). Host-age discrimination during host location by *Cotesia glomerata,* a larval parasitoid of *Pieris brassicae. Entomol. Exp. Appl.* **76:**37–48.

McCutchen, B. F., Herrman, R., Heinz, K. M., Parrella, M. P., and Hammock, B. D. (1996). Effects of recombinant baculoviruses on a nontarget endoparasitoid of *Heliothis virescens. Biol. Cont.* **6:**45–50.

Meinertzhagen, I. A., and Pyza, E. (1996). Daily rhythms in cells of the fly's optic lobe: Taking time out from the circadian clock. *Trends Neurosci.* **19:**285–291.

Møller, A. P. (1993). A fungus infecting domestic flies manipulates sexual behaviour of its host. *Behav. Ecol. Sociobiol.* **33:**403–407.

Moore, J. (1993). Parasites and the behavior of biting flies. *J. Parasitol.* **79:**1–16.

———. (1995). The behavior of parasitized animals. *Bioscience* **45:**89–96.

Mullens, B. A., Rodriguez, J. L., and Meyer, J. L. (1987). An epizotological study of *Entomophthora muscae* in muscoid fly populations in southern California poultry facilities, with emphasis on *Musca domestica. Hilgardia* **55:**1–41.

Mullens, B. A., and Rutz, D. A. (1982). Mermithid parasitism in *Culicoides variipennis* in New York state. *Mosq. News* **22:**231–235.

Mullens, B. A., and Velten, R. K. (1994). Laboratory culture and life history of *Heleidomermis magnapapula* in its host, *Culicoides variipennis* (Diptera: Ceratopogonidae). *J. Nematol.* **26:**1–10.

Neckameyer, W. S., (1996). Multiple roles for dopamine in *Drosophila* development. *Develop. Biol.* **176:**209–219.

Noguchi, H., Hayakawa, Y., and Downer, R. G. H. (1995). Elevation of dopamine levels in parasitized insect larvae. *Insect Biochem. Molec. Biol.* **25:**197–201.

O'Reilly, D. R. (1995). Baculovirus-encoded ecdysteroid UDP-glucosyltransferases. *Insect Biochem. Molec. Biol.* **25:**541–550.

O'Reilly, D. R., Brown, M. R., and Miller, L. K. (1992). Alteration of ecdysteroid metabolism due to baculovirus infection of the fall armyworm *Spodoptera frugiperda:* Host ecdysteroids are conjugated with galactose. *Insect Biochem. Molec. Biol.* **22:**313–320.

O'Reilly, D. R., Kelley, T. J., Masler, E P., Thyagaraja, B. S., Robson, R. M., Shaw, T. C., and Miller, L. K. (1995). Overexpression of *Bombyx mori* prothoracicotropic hormone using baculovirus vectors. *Insect Biochem. Molec. Biol.* **25:**475–485.

Oro, A. E., McKeown, M., and Evans, R. M. (1992). The *Drosophila* RXR homolog *ultraspiracle* functions in both female reproduction and eye morphogenesis. *Development* **115:**449–462.

Park, Y. C., Tesch, M. J., Toong, Y. C., and Goodman, W. G. (1993). Affinity purification and binding analysis of the hemolymph juvenile hormone binding protein from *Manduca sexta. Biochemistry* **32:**79-9–79-15.

Pennacchio, F., Digilio, M. C., and Tremblay, E. (1995). Biochemical and metabolic alterations in *Acyrthosiphon pisum* parasitized by *Aphidius ervi. Arch. Insect Biochem. Physiol.* **30:**351–367.

Pennachio, F., Vinson, S. B., Tremblay, E., and Ostuni, A. (1994). Alteration of ecdysone metabolism in *Heliothis virescens* (F.) (Lepidoptera: Noctuidae) larvae induced by *Cardiochiles nigriceps* Viereck (Hymenoptera: Braconidae) teratocytes. *Insect Biochem. Molec. Biol.* **24:**383–394.

Pfeifer, T. A., and Grigliatti, T. A. (1996). Future perspectives on insect pest management: Engineering the pest. *J. Inv. Pathol.* **67:**109–119.

Presnail, J. K., and Hoy, M. A. (1992). Stable genetic transformation of a beneficial arthropod, *Metaseiulus occidentalis* (Acari: Phytoseiidae), by a microinjection technique. *Proc. Natl. Acad. Sci.* **89:**7732–7736.

———. (1994). Transmission of injected DNA sequences to multiple eggs of *Metaseiulus occidentalis* and *Amblyseius finlandicus* (Acari: Phytoseiidae) following maternal microinjection. *Exp. Appl. Acarol.* **18:**319–330.

———. (1996). Maternal injection of the endoparasitoid *Cardiochiles diaphaniae* (Hymenoptera: Braconidae). *Ann. Entomol. Soc. Amer.* **89:**576–580.

Rachinsky, A. (1994). Octopamine and serotonin influence on corpora allata activity in honeybee *(Apis mellifera)* larvae. *J. Insect Physiol.* **40:**549–554.

Ransohoff, R. M., and Benveniste, E. N., eds. (1996). *Cytokines and the CNS.* Boca Raton, Fl: CRC Press.

Reed, D. A., and Beckage, N. E. (1997). Inhibition of testicular growth and development in *Manduca sexta* larvae parasitized by the braconid wasp *Cotesia congregata. J. Insect Physiol.* **43**(1):29–38.

Riddiford, L. M. (1996). Molecular aspects of juvenile hormone action in insect metamorphosis. In *Metamorphosis: Postembryonic reprogramming of gene expression in amphibian and insect cells* (L. I. Gilbert, J. R. Tata, and B. G. Atkinson, eds.), pp. 223–251. San Diego: Academic Press.

Rodriguez, M. del C., Zamudio, F., Torres, J. A., Gonzalez-Ceron, L., Possani, L. D., and Rodriguez, M. H. (1995). Effect of a cecropin-like synthetic peptide (*Shiva*-3) on the sporogonic development of *Plasmodium berghei. Exp. Parasitol.* **80:**596–604.

Rothman, L. D., and Myers, J. H. (1996). Debilitating effects of viral diseases on host Lepidoptera. *J. Inv. Pathol.* **67:**1–10.

Rybczynski, R., Mizoguchi, A., and Gilbert, L. A. (1996). *Bombyx* and *Manduca* prothoracicotropic hormones: An immunologic test for relatedness. *Gen. Comp. Endocrinol.* **102:**247–254.

Sait, S. M., Begon, M., and Thomson, D. J. (1994a). The effects of a sublethal infection in the Indian meal moth, *Plodia interpunctella. J. Anim. Ecol.* **63:**541–550.

———. (1994b). Long-term population dynamics of the Indian meal moth *Plodia interpunctella* and its granulosis virus. *J. Anim. Ecol.* **63:**861–870.

Sapp, J. (1994). *Evolution by association: A history of symbiosis.* Oxford: Oxford University Press.

Sayeed, O., and Benzer, S. (1996). Behavioral genetics of thermosensation and hygrosensation in *Drosophila. Proc. Natl. Acad. Sci.* **93:**6079–6084.

Schopf, A., Nussbaumer, C., Rembold, H., and Hammock, B. D. (1996). Influence of the braconid *Glyptapanteles liparidis* on the juvenile hormone titer of its larval host, the gypsy moth, *Lymantria dispar. Arch. Insect Biochem. Physiol.* **31:**337–351.

Shelby, K. S., and Webb, A. A. (1994). Polydnavirus infection inhibits synthesis of an insect plasma protein, arylphorin. *J. Gen. Virol.* **75:**2285–2292.

Shelby, K. S., and Webb, B. A. (1997). Polydnavirus infection inhibits translation of specific growth-associated host proteins. *Insect Biochem. Molec. Biol.* **27:**263–270.

Shimizu, T., Shiotsuki, T., Tankaka, Y., and Takeda, S. (1995). Larval development of a larval parasitoid *(Apanteles kariyai)* forced to oviposit in a pupal host *(Pseudaletia separata). Entomol. Exp. Appl.* **76:**109–112.

Six, D. L., and Mullens, B. A. (1996). Distance of conidial discharge of *Entomophthora muscae* and *Entomophthora schizophorae* (Zygomycotina: Entomophthorales). *J. Invert. Pathol.* **67:**253–258.

Smith, H. (1995). The revival of interest in mechanisms of bacterial pathogenicity. *Biol. Rev.* **70:**277–316.

Søller, M., and Lanzrein, B. (1996). Polydnavirus and venom of the egg-larval parasitoid *chelonus inanitus* (Braconidae) induce developmental arrest in the prepupa of its host *Spodoptera littoralis. J. Insect Physiol.* **42:**471–481.

Stairs, G. R. (1965). The effect of metamorphosis on nuclear polyhedrosis infecting certain Lepidoptera. *Can. J. Microbiol.* **11:**509–512.

———. (1970). The development of nuclear polyderosis virus in ligated larvae of the greater wax moth *Galleria mellonella. J. Inv. Pathol.* **15:**60–62.

Stanley-Samuelson, D. W., and Pedibhotla, V. K. (1996). What can we learn from prostaglandins and related eicosanoids in insects? *Insect Biochem. Molec. Biol.* **26:**223–234.

Strand, M. R., and Pech, L. L. (1995). Immunological basis for compatibility in parasitoid-host relationships. *Ann. Rev. Entomol.* **40:**31–56.

Szmidt-Adjide, V., Rondelaud, D., Dreyfuss, G., and Cabaret, J. (1996). The effect of parasitism by *Fasciola hepatica* and *Muellerius capillaris* on the nerve ganglia of *Lymnaea truncatula. J. Inv. Pathol.* **67:**300–305.

Tanada, Y., and Kaya, H. K. (1993). *Insect pathology.* New York: Academic Press.

Tanaka, T. (1987). Calyx and venom fluids of *Apanteles kariyai* (Hymenoptera: Braconidae) as factors which prolong the larval period of the host, *Pseudaletia separata* (Lepidoptera: Noctuidae). *Ann. Entomol. Soc. Am.* **80:**530–533.

Tanaka, T., Agui, N., and Hiruma, K. (1986). The parasitoid *Apanteles kariyae* inhibits pupation of its host, *Pseudaletia separata,* via disruption of prothoracicotrpic hormone release. *Gen. Comp. Endocrinol.* **67:**364–374.

Tanaka, Y., Asahina, M., and Takeda, S. (1994). Induction of ultranumerary larval ecdyses by ecdysone does not require an active prothoracic gland in the silkworm, *Bombyx mori. J. Insect Physiol.* **40:**753–757.

Thompson, S. N. (1993). Redirection of host metabolism and effects on parasite nutrition. In *Parasites and pathogens of insects* (N. E. Beckage, S. N. Thompson, and B. A. Federici, eds.), vol. 1, pp. 125–144. New York: Academic Press.

Vance, S. A. (1996). Morphological and behavioural sex reversal in mermithid-infected mayflies. *Proc. Roy. Soc. Biol. Sci. series B* (in press).

Vance, S. A., and Peckarsky, B. L. (1996). The infection of *Baetis bicaudatus* (Ephemeroptera) by the mermithid nematode *Gasteromermis* sp. *Ecol. Entomol.* (in press).

Van Driesche, R. G., and Bellows, T. S. (1995). *Biological control.* New York: Chapman & Hall.

Volkman, L. E. (1997). Nucleopolyhedrosis virus interactions with their insect hosts. *Adv. Virus Res.* (in press).

Washburn, J. O., Kirkpatrick, B. A., and Volkman, L. E. (1996). Insect protection against viruses. Nature **383:**767.

Watson, D. W., Mullens, B. A., and Petersen, J. J. (1993). Behavioral fever response of *Musca domestica* (Diptera: Muscidae) to infection by *Entomophthora muscae* (Zygomycetes: Entomophthorales). *J. Invert. Path.* **61:**10–16.

Watson, D. W., and Peterson, J. J. (1993). Sexual activity of male *Musca domestica* (Diptera: Muscidae) infected with *Entomophthora muscae* (Entomophthoraceae: Entomophthorales). *Biol. Cont.* **3:**22–26.

Webb, B. A., and Summers, M. D. (1992). Stimulation of polydnavirus replication by 20-hydroxyecdysone. *Experientia* **48:**1018–1022.

Whitfield, J. B. (1994). Mutualistic viruses and the evolution of host ranges in endoparasitoid Hymenoptera. In *Parasitoid community ecology* (B. Hawkins, and W. Sheehan, eds.), Oxford: Oxford University Press.

Woodring, J. K., and Hoffmann, K. H. (1994). The effects of octopamine, dopamine and serotonin on juvenile hormone synthesis, *in vitro*, in the cricket, *Gyllus bimaculatus. J. Insect Physiol.* **40:**797–802.

Wolff, K. M., and Scott, A. L. (1995). *Brugia malayi:* Retinoic acid uptake and localization. *Exp. Parasitol.* **80:**282–290.

Wulker, W. (1975). Parasite-induced castration and intersexuality in insects. In *Intersexuality in the animal kingdom* (R. Reinboth, ed.), pp. 121–134. New York: Springer-Verlag.

Zhang, D., Dahlman, D. L., and Gelman, D. (1992). Juvenile hormone esterase activity and ecdysteroid titer in *Heliothis virescens* larvae injected with *Microplitis croceipes* teratocytes. *Arch. Insect Biochem. Physiol.* **20:**231–242.

Zitnan, D., Kingan, T. G., and Beckage, N. E. (1995a). Parasitism-induced accumulation of FMRF-amide-like peptides in the gut innervation and endocrine cells of *Manduca sexta. Insect Biochem. Molec. Biol.* **25:**669–678.

Zitnan, D., Kingan, T. G., Kramer, S. J., and Beckage, N. E. (1995b). Accumulation of neuropeptides in the cerebral neurosecretory system of *Manduca sexta* larvae parasitized by the braconid wasp *Cotesia congregata. J. Comp. Neurol.* **356:**83–100.

Zuk, M., Simmons, L. W., and Cupp, L. (1993). Calling characteristics of parasitized and unparasitized populations of the field cricket *Teleogryllus oceanicus. Behav. Ecol. Sociobiol.* **33:**339–343.

Zuk, M., Simmons, L. W., and Rotenberry, J. T. (1995). Acoustically-orienting parasitoids in calling and silent males of the field cricket *Teleogryllus oceanicus. Ecol. Entomol.* **20:**380–383.

2

The Life History and Development of Polyembryonic Parasitoids

Michael R. Strand and Miodrag Grbic'

I. The Life History of Parasitoids

Of increasing interest in biology is understanding how evolutionary processes act on development to generate variation in morphology, reproduction, and adaptive radiation of new species (Buss, 1987; Barinaga, 1994). Some of the most extraordinary examples of speciation are found among parasites where specialization in large measure reflects adaptations for life within another organism, the host. A recurring theme in parasite life history is that host patches are often ephemeral and the probability of locating and successfully colonizing patches is low (Price, 1980). Parasites that locate hosts may also not encounter another conspecific during their reproductive life. From a developmental perspective these circumstances will favor characteristics that promote rapid exploitation of the host before it becomes unsuitable. They also will favor modes of reproduction that facilitate location of mates or that eliminate the need for mating. These could include mating between siblings before dispersal, parthenogenesis, hermaphroditic development, or clonal propagation.

One of the most diverse groups of parasites occur in the insect order Hymenoptera. Parasitic wasps (usually referred to as parasitoids) lay their eggs in or on the bodies of other arthropods, which serve as food for their developing offspring. After completing development, parasitoids emerge from their host as free-living adults that disperse to seek mates or new hosts to parasitize. Most parasitoids exploit patchily distributed resources, and examples of parthenogenesis or mating before dispersal are well documented (Godfray, 1994). Indeed, most Hymenoptera are facultatively parthenogenetic (haplodiploid) with males developing from unfertilized eggs and females from fertilized eggs. This allows unmated individuals to produce male offspring and also provides a mechanism by which mated individuals can bias their sex ratios toward females under conditions that favor sib mating. This chapter focuses on the life history of a unique group of parasitoids that have combined haplodiploidy with a clonal form of development called polyembryony. Each egg laid by a polyembryonic wasp divides to produce a brood of genetically identical siblings that feed together inside the host. More unusual still, some species exhibit a caste system whereby certain embryos develop into larvae that form adult wasps while others develop into larvae that have only worker functions. Here we summarize where polyembryony has arisen in the Hymenoptera, the developmental interactions between polyembryonic wasps and their hosts, and the role each caste plays in successful parasitism.

II. Polyembryony: An Overview

Polyembryony is defined as the formation of multiple embryos from a single egg. Sporadic polyembryony such as identical twinning occurs in most animal taxa. Obligate polyembryony, by contrast, is relatively rare, occurring primarily in certain parasites (some cestodes, trematodes and insects) and colonial, aquatic invertebrates (oligochaetes, bryzoans) (summarized by Bell, 1982; Hughes and Cancino, 1985). Polyembryony was first reported in insects by Marchal (1898) and has since been documented in two orders, the Hymenoptera and the exclusively parasitic Strepsiptera (Ivanova-Kasas, 1972). Because so little is known about polyembryony in the Strepsiptera, we will not discuss this group further here (see Askew, 1970, and Ivanova-Kasas, 1972 for general summaries). In the Hymenoptera, polyembryony occurs in selected genera from four families: the Platygasteridae, Braconidae, Encyrtidae, and Dryinidae (Strand, Wong, and Baehrecke, 1991). Polyembryony in these groups clearly arose from monoembryonic ancestors since most species in these large families are monoembryonic. The phylogenetic distance between these families (LaSalle and Gauld, 1991) also indicates that polyembryony arose independently in each group.

A. Polyembryony in the Braconidae, Platygasteridae and Dryinidae

Relatively little is known about polyembryony in these taxa. Most studies were conducted prior to 1940 with little information reported on the specific interactions that occur between these parasites and their hosts. In the braconids, polyembryony has been reported only in the genus *Macrocentrus*. This group consists of both mono- and polyembryonic species that lay their eggs into the larval stage of concealed Lepidoptera (moths and butterflies). Studies with *Macrocentrus grandii* Goidanich and *M. ancylivorus* Rohwer indicate that wasps usually oviposit a single egg into the host's hemocoel (Parker, 1931; Daniel, 1932). Within 24 h of oviposition, the wasp egg forms a single embryo enveloped by a multinucleated extraembryonic membrane (trophamnion) of unknown origin. As development continues, the number of embryonic cells in the original embryo increases and these cells then become partitioned by the extraembryonic membrane into additional embryonic masses. In addition, small spheres of extraembryonic membrane that do not surround any embryonic cells often form which Daniel (1932) called pseudogerms. All of the embryos and pseudogerms ultimately become separated from one another and distributed throughout the host hemocoel. From the published literature it is difficult to determine the total developmental time for these events, but the observations of Parker (1931) and Dittrick and Chang (1982) suggest that development varies with the timing of parasitization. In contrast, initiation of morphogenesis is synchronous, beginning during the host's final larval instar. In *M. grandii* an average of 25 larvae are produced per host (Wishart, 1946), and while multiple embryos initiate morphogenesis in *M. ancylivorus,* only one individual survives to adulthood (Daniel, 1932). As soon as one larva hatches, development of the others ceases. In *M. grandii,* the larvae feed internally through the third instar, but when the host initiates cocoon spinning in preparation for metamorphosis, the wasp larvae molt to the fourth instar and emerge (Strand, unpublished). They then consume the remainder of the host by feeding as external parasites, spin a cocoon, and pupate.

Platygasterids in the genus *Platygaster* range from being exclusively monoembryonic to facultatively or obligately polyembryonic (Marchal, 1904; Silvestri, 1921; Leiby and Hill, 1923; Leiby, 1924; Leiby and Hill, 1924; Hill and Emery, 1937). All species lay their eggs into the egg stage of gall midges (Diptera: Cecidomyiidae) and their progeny complete development in the host's larval stage (i.e. egg-larval parasitoids). Females lay one or more eggs per host, and the site of development is often very specific. For example, the eggs of *Platygaster vernalis* (Myers) develop in the midgut of the host *Phytophaga destructor* Say (Leiby and Hill, 1924), whereas the eggs of *Platygaster hiemalis* Forbes develop in the hemocoel enveloped by fat body tissue (Leiby and Hill, 1923). During the host egg stage, the wasp egg appears to undergo complete cleavage, forming a single morula stage embryo by the time the host egg hatches. In monoembryonic species this morula develops into a blastula stage embryo during the early stages of host larval development, and morphogenesis begins during the last instar of the host (Hill and Emery, 1937). In obligately polyembryonic species, the embryonic cells of the morula increase in number and become subdivided into several morulae during early larval development of the host. Morphogenesis, however, remains associated with the host's final instar. The host forms a puparium but is consumed by the parasitoid larva or larvae before pupating. An average of 8–13 parasitoids are produced per host by *Platygaster vernalis* and *Platygaster felti* Fouts (Leiby and Hill, 1924; Patterson, 1921), but only one or two parasitoids are produced per host by *Platygaster hiemalis* (Hill, 1926).

The dryinids are an unusual group of aculeate Hymenoptera that parasitize Homoptera. Little is known about the life history or habits of these wasps, yet one species, *Aphelopus theliae* (Gahan), is reported to be polyembryonic in its membracid host, *Thelia bimaculata* F. (Kornhouser, 1919). No information is available on development of *A. theliae*, but females are reported to oviposit into first or second instar nymphs. Approximately 50 wasp larvae are produced per host, and they consume the host in its final instar.

B. Polyembryony in the Encyrtidae

The most extreme form of polyembryony for any metazoan occurs in encyrtids from the tribe Copidosomatini. All are egg-larval parasitoids of Lepidoptera, with several species producing broods of more than 1,000 offspring per host (Marchal, 1904; Silvestri, 1906; Martin, 1914; Patterson, 1921; Leiby, 1922; Koscielski et al., 1978). Development in this group is best studied in *Copidosoma floridanum* (Ashmead) and its host, *Trichoplusia ni* (Hubner) (Strand, 1989a; Baehrecke and Strand, 1990; Baehrecke, Grbic', and Strand, 1992; Baehrecke, Strand, et al., 1992; Baehrecke et al., 1993; Grbic' et al., 1992; Grbic' et al., 1996a, b; Grbic' et al., 1997) (see Fig. 2-1). After oviposition into a host egg, the egg(s) of *C. floridanum* forms a morula stage embryo surrounded by an enveloping membrane of polar body origin. This stage is referred to as the primary morula. Late in host embryogenesis and during the first through fourth instars of the host larva, the embryonic cells of the primary morula increase in number and become subdivided into increasing numbers of proliferating morulae. Each proliferating morula is enveloped by a serosal membrane of embryonic cell origin and groups of proliferating morulae are surrounded by an extraembryonic membrane to form collectively a polymorula. Light and confocal microscopy studies indicate that the polymorula comprises two types of proliferating morulae: small proliferating morulae, 50–75 μm along their long axis and large proliferating morulae that are spherical and 90–110 μm in diameter. The number of proliferating morulae per host increases in synchrony with the host molting cycle, with a total of ca. 1,000–1,500 being produced by the host's fifth instar.

Certain proliferating morulae in the polymorula initiate morphogenesis during the host first-fourth instars forming what we call precocious larvae (Fig. 2-1). Within each host instar, morphogenesis of embryos that develop into precocious larvae is always synchronized with host critical period (i.e., initiation of the host molting cycle), as is proliferation of the polymorula. This results in bursts of precocious larvae developing and being released into the host hemocoel at discrete periods of each host instar, creating an age structure and a progressive increase in the number of precocious larvae per host with time. The number of precocious larvae produced per host instar ranges from 1 to 15, and a total of up to 200 precocious larvae can be present in the host's final instar. In contrast, morphogenesis of embryos that develop into reproductive larvae always begins shortly after critical period of the host fourth (penultimate) instar (Fig. 2-1). Reproductive larvae eclose at 60 h of the host fifth instar, consume the host and pupate within the remnant host cuticle, forming a "mummy". All precocious larvae die after the host has been consumed. Approximately 1,200 adult wasps emerge from the mummy seven days later.

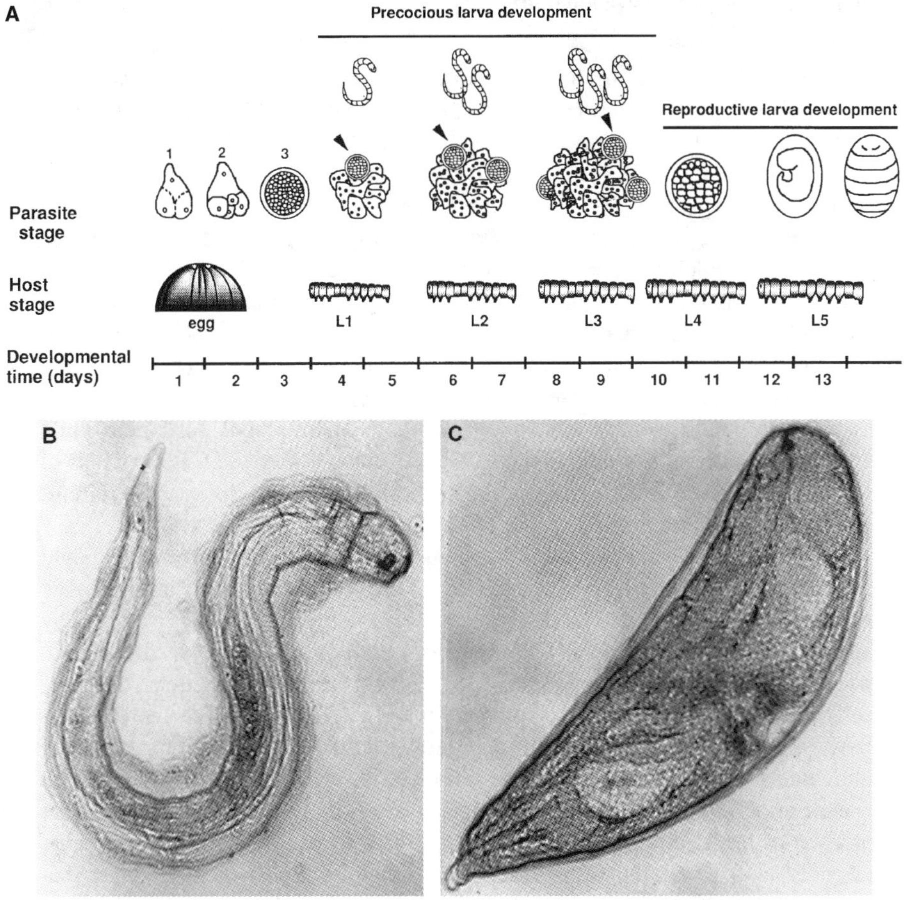

Figure 2-1 (*A*) Life cycle of the parasitic wasp *Copidosoma floridanum* in its host, *Trichoplusia ni*. The top line diagrams the developmental stages of the parasitic wasp while the middle line diagrams the corresponding stages of the host. Total developmental time (days) for both parasite and host is plotted at the bottom. The life cycle begins when the adult wasp deposits her egg into the host's egg. The first cleavage (1) of the wasp egg forms three cells: two equal-sized blastomeres and a large polar cell. The second cleavage (2) is unequal: one blastomere forms one small and one large blastomere, while the other blastomere forms two equal-sized cells. Further cleavages are asynchronous. After 24 hours a primary morula (3) forms. The primary morula then proliferates and subdivides, creating many morulae (polymorula stage) during the host first–fourth stadia (L1–L4). Selected morulae (arrows) initiate morphogenesis during this period, forming precocious larvae. During the penultimate host instar (L4), the remaining proliferating morulae synchronously form individual embryonic primordia. During the early fifth instar of the host (L5), the wasp embryos undergo germband extension, segmentation and eclose as reproductive larvae. (*B*) Phase contrast micrograph of a precocious larva (150×). (*C*) Phase-contrast micrograph of a reproductive larva (100×).

Silvestri (1906) was the first to report the formation of precocious larvae in *C. floridanum* (as *Litomastix truncatellus*; (see Noyes, 1988)), and they have since been described in several other species (Patterson, 1921; Parker and Thompson, 1928; Doutt, 1947; Fitzgerald and Simeone, 1971; Cruz, 1981). In a few studies, however, they are not reported (Martin, 1914; Leiby, 1922; Nenon, 1978), suggesting that precocious larvae are not produced by all polyembryonic encyrtids. Morphological characters that differ between precocious and reproductive larvae of *C. floridanum* include the following: (1) mandible and head capsule sizes differ between precocious larvae and reproductive larvae, (2) the width of thoracic and abdominal segments is greater (eight cells wide/segment) in precocious larvae than reproductive larvae (four cells wide), (3) lateral spines are associated with all thoracic and abdominal segments of precocious larvae while no spines are present on reproductive larvae, and (4) the overall body width of reproductive larvae is about double the body width of precocious larvae (Grbic' et al., 1992; Grbic' and Strand, unpublished) (Fig. 2-1).

Early workers proposed several hypotheses for how two larval morphs could develop from a single egg. Silvestri (1906, 1937) hypothesized that larval fate was mediated by a "germ cell determinant". This determinant was visualized by histological staining and appeared as cytoplasmic granules that localized to a single blastomere in the four cell stage of cleavage. He hypothesized that daughter cells inheriting the determinant form a lineage that develops into reproductive larvae while cells not inheriting the determinant form a precocious lineage. Other individuals suggested that precocious larvae arise aberrantly from "errors" that occur during polyembryony. For example, Patterson (1917, 1921) proposed that precocious larvae arise from embryonic cells that lose an X chromosome during the course of proliferation. Doutt (1952) noted in *Copidosoma koehleri* that precocious larvae develop primarily or exclusively from fertilized (female) eggs and suggested that entry of a sperm in some way stimulated formation of precocious larvae.

III. Regulation of Polyembronic Development

None of the aforementioned ideas on regulation of polyembryony were tested experimentally by their proponents, yet the observations of these workers influence two key questions in the contemporary developmental biology literature. Namely, how does a single cell give rise to the diversity of cell types and complexity of pattern manifested in multiple embryos, and how can different larval phenotypes arise from an identical genetic background? Patterning in insects is thought to be regulated endogenously by the coordinated expression of maternal and zygotic genes (St. Johnston and Nuslein-Volhard, 1992). Cell lineage can also play a significant role in development of metazoans with a strong tendency in lower animal phyla (annelids, molluscs, echinoderms, ascidians) toward autonomous specification of lineages while in higher phyla (amphibians, fish, mammals) lineages are strongly influenced by positional information and extracellular signalling molecules (reviewed by Davidson, 1991, but see Strehlow and Gilbert, 1993). In contrast, caste formation and other polyphenisms exhibited by insects are thought to be regulated by the interactions between environmental (photoperiod, crowding, pheromones, nutrition) and endocrine factors (Kumaran, 1983; de Wilde and Beetsma, 1982; Robinson, 1992; Nijhout, 1994). Within this framework, three factors could potentially influence patterning and caste determination in polyembryonic wasps: environmental or endocrine factors related to the host, endogenous factors associated with cell lineage, or molecular elements controlling pattern formation. Below we summarize our understanding of how environmental and endogenous events mediate polyembryonic development, emphasizing our recent experimental studies with *Copidosoma floridanum* in *Trichoplusia ni*.

A. *The Host Environment*

1. *Host Age and Sexual Asymmetries of Wasps*

Polyembryonic wasps produce broods composed of only males, only females (single-sex broods) or both sexes (mixed broods) (Patterson, 1918; 1921; Leiby 1926; Strand, 1989b, c; Strand and Ode, 1990). Single-sex and mixed broods arise from a female laying a specific number of eggs, yet the total number of adult wasps produced per host and the sex ratio of mixed broods are highly variable both within and between species (Leiby, 1926; Walter and Clarke, 1992; Hardy et al., 1993). In *C. floridanum*, for example, from 300 to 2,800 wasps can emerge per host, and sex ratios for mixed broods can range from strongly female biased to male biased (Strand, 1989b). In a few instances we and others have even documented the emergence of more than 2,000 females and one male wasp from a host (Patterson, 1919; Strand, 1989c; Ode and Strand, 1995). The number of precocious larvae that develop per host is also variable, ranging from zero in some host larvae to more than 100 in others (Strand, Johnson, and Culin, 1990). This variation has been enigmatic to many workers, and in large measure led Patterson and Doutt (see above) to conclude that precocious larvae arise from some random mechanism. It also prompted Patterson (1919) to suggest that fertilized eggs produce both sexes via somatic nondisjunction.

Our studies indicate that this variation is not random but is instead due to two factors: asymmetric development of precocious larvae from male and female eggs (Grbic' et al., 1992), and asymmetric responses by male and female embryos to subtle variations in the host environment (Ode and Strand, 1995). Cytogenetic studies confirm that *C. floridanum* is haplodiploid, with single-sex and mixed broods arising solely from a female laying one (male or female) or two eggs (male and female) per host. By monitoring the sex of the egg(s) laid into a host, we also observed that through the fourth instar, precocious larvae are always present in female and mixed broods but are usually absent in male broods. The significance of this asymmetry is that female precocious larvae attack and kill male embryos, resulting in strongly female biased sex ratios for mixed broods (Grbic' et al., 1992). How such siblicidal behavior could have evolved is discussed below.

The second factor affecting brood size and sex ratio is the age of the host egg when parasitized by the female wasp (Ode and Strand, 1995). Under standard conditions, *T. ni* eggs require 72 h to complete embryogenesis. Female wasps successfully oviposit in all embryonic stages of the host, but the number of progeny that ultimately develop varies with age (i.e., developmental stage) of the host egg and sex of the brood. The number of wasps that emerge from female broods decreases significantly with host egg age, whereas the number of wasps produced from male broods does not. Mixed broods also decrease in size with host egg age while sex ratio (proportion males) increases. This is because the number of precocious larvae produced by the female egg also declines with host egg age, resulting in a reduction in the number of male embryos killed and a concomitant increase in sex ratio. Thus, there is little variation in brood sizes and sex ratios within defined age classes of host eggs, yet there is large variation in these characters between age classes of host eggs. Because female wasps prefer to oviposit in younger (pregastrulation) host eggs, most mixed broods collected from the field or produced in the laboratory exhibit strongly female-biased sex ratios (Hardy et al., 1993; Ode and Strand, 1995). We do not understand why small differences in host egg age so profoundly affect the number of embryos produced by female eggs. One possibility is that, because proliferation is an exponential event, initiating development in a young host egg (i.e., 12 h old) results in significantly more proliferating morulae by the host's fourth instar than initiating development in an old host egg (72 h old).

2. *Host Endocrine Physiology*

Regardless of when a host egg is parasitized, proliferation and morphogenetic events in male, female, and mixed broods are synchronized with the molting cycle of the host larva (see Fig. 2-1). This circumstantially suggests that host endocrine physiology also influences development. Molting and metamorphosis of insects is mediated primarily by the ecdysteroid, 20-hydroxyecdysone (20-E) and the sesquiterpenoid juvenile hormones (JHs) (Riddiford, 1985). Development of *T. ni* parasitized by *C. floridanum* is very similar to that of unparasitized larvae with the exception of the fifth instar, during which parasitized larvae attain a higher final weight and initiate wandering in preparation for metamorphosis 24 h later than unparasitized larvae. Hemolymph ecdysteroid and JH titers of the host fluctuate in a manner consistent with these observations (Strand, Johnson, and Dover, 1990; Strand, Wong, and Baehrecke, 1991; Strand, Goodman, and Baehrecke, 1991). Ecdysteroid titers in parasitized larvae are similar to those of unparasitized larvae with the exception that the titer rises in the fifth instar 24 h later in parasitized larvae than in unparasitized larvae. Similarly, JH titers in parasitized larvae are similar to those of unparasitized larvae except in the fifth instar, during which the titer declines in parasitized larvae 24 h later than in unparasitized larvae. Comparing these data to *C. floridanum* development (see Fig. 2-1) indicates that morphogenesis of embryos that form reproductive larvae is synchronized with the fall in JH titer and rise in ecdysteroid titer during the host fourth instar. Morphogenesis of embryos forming precocious larvae is similar, albeit iterative, occurring cyclically in the host first–fourth instar (instars with an elevated JH titer).

Three hypotheses can be proposed for the timing of morphogenetic events in *C. floridanum*. First, proliferation and morphogenesis of embryos forming precocious larvae might occur in response to ecdysteroids and a high JH titer, but individuals forming reproductive larvae may initiate morphogenesis in response to ecdysteroids and a low JH titer. This is analogous to the proposed regulation of larval-pupal development by JH and ecdysone (Riddiford, 1985). Second, morphogenesis may be induced by ecdysone, but the synchrony of the event with the molting cycle may be due to embryos becoming competent to respond to an ecdysone stimulus. According to this model, competence is regulated endogenously, and embryos do not initiate morphogenesis in earlier instars because they either cannot perceive or cannot respond to this hormonal signal. This is similar to morphogenesis of imaginal discs where ecdysteroid induction is associated with acquisition of competence (Mindek, 1972; Oberlander, 1985; Bryant and Schmidt, 1990). The third possibility is that host hormones are not involved in induction of morphogenesis.

Results to date support the second hypothesis (Strand, Goodman, and Baehrecke, 1991; Baehrecke et al. 1993; Grbic' et al., 1997). If the fall in host JH titer is responsible for induction of morphogenesis of the reproductive morph, an elevated titer should maintain embryos in a proliferative state, promote continued morphogenesis of the precocious morph, yet inhibit morphogenesis of the reproductive morph. However, topical application of the analog methoprene or JH II does not have this effect. Parasitoid embryos do not continue to proliferate, additional precocious larvae are not formed and while morphogenesis of the reproductive caste is delayed 48 h, it is not arrested (Strand, Goodman, and Baehrecke, 1991; Baehrecke, Grbic', and Strand, 1992; Grbic' et al., 1997). Moreover, maintainance of an elevated JH titer in the host does not alter the proportion of embryos that develop into precocious and reproductive larvae. This, combined with the fact that all embryos in a polymorula develop in an identical environment (the host hemocoel), strongly suggests that caste formation is not determined solely by interactions between environmental factors and endocrine state.

In contrast, transplantation experiments clearly support a role for ecdysteroids and competency in morphogenesis. Temporally speaking, *C. floridanum* embryos that develop into the reproductive morph are 8–9 days old when the host molts from the fourth to the fifth instar (Baehrecke et al., 1993). If embryos must attain this age to initiate morphogenesis in response to a rise in the ecdysteroid titer of the host, transplantation of embryos to a novel host stage, such as pupae, should still induce morphogenesis provided embryos are at least nine days old prior to a rise in the endogenous ecdysteroid titer of the recipient host. In contrast, embryos transplanted into a noninductive host stage, such as moths, should not initiate morphogenesis at any age, but should be rescued by back transplantation into host larvae or pupae if ≥ 9 days old.

This is precisely what occurs (Strand, Wong, and Baehrecke, 1991; Baehrecke et al., 1993; Rivers, Grbic' et al., 1997). Ecdysteroid titers in *T. ni* pupae peak at 60 h coincident with the morphological character of tarsal tanning. Transplanted embryos proliferate but do not initiate morphogenesis if they are ≤ 7 days old when the recipient host pupa reaches the tarsal tanning stage. In contrast, embryos initiate morphogenesis, forming reproductive larvae, but do not proliferate if they are ≥ 9 days old when the recipient pupa initiates tarsal tanning. Morphogenesis of ≥ 9 day old embryos is arrested if the source of ecdysone in recipient pupae is ablated but is rescued dose-dependently by 20-HE. Parallel experiments indicate that proliferation occurs if embryos ≥ 9 days old are transplanted into recipient pupae. These results indicate that the timing of morphogenesis by embryos that develop into reproductive larvae is due to an endogenous competency effect rather than some unique feature of the host fourth instar. We have not defined competency periods for embryos that develop into precocious larvae, but *in vitro* studies confirm that morphogenesis is an ecdysteroid-inducible event (Grbic' et al., 1997). Both proliferation and morphogenesis of the reproductive morph is associated with the coordinated expression of a putative ecdysone receptor gene or related gene in the steroid receptor family (Baehrecke et al., 1993; Strand, Rivers, and Johnson, unpublished).

B. Endogenous Factors

The above discussion indicates that host endocrine factors play an important role in synchronizing developmental transitions of *C. floridanum* but are not the sole mediators of caste fate. In contrast, endogenous developmental events could influence caste fate at two levels. The first possibility is that caste-specific development is lineage-dependent with the progeny of one blastomere or a restricted group of blastomeres forming one morph, and the progeny of another blastomere forming the other morph. Alternatively, all early blastomeres could participate in formation of both morphs. By this we mean that there is no lineage restriction during early cleavage, and that caste fate is determined later in proliferation via localized cell-cell interactions.

1. Early Cleavage and Cell Lineages

The role of lineage in the early stages of insect development is not well understood because in most species early cleavage takes place in a syncytial environment. In *Drosophila melanogaster*, for example, 13 rounds of nuclear division occur before nuclei migrate to the periphery of the egg, cellularize (ca. 5,000 cells), and initiate blastoderm formation (Schweisguth et al., 1991; Schejter and Wieschaus, 1993). Polyembryonic wasps differ markedly from this situation in that their eggs lack yolk and appear to undergo holoblastic cleavage (Ivanova-Kasas, 1972). This was recently confirmed in *C. floridanum* (Grbic' et al., 1996a). Injection of individual blastomeres with flourescein (FDA) or tetramethylrhodamine dextran amine indicated that cellularization is complete at the four cell stage. Dye also remains confined to the injected cell in the primary morula, polymorula or embryonic primordium stages, indicating that embryogenesis proceeds almost entirely in a cellularized environment. Parallel studies further reveal that the cytoplasmic granules observed by Silvestri indeed localize to a single blastomere at the four cell stage (Strand and Grbic', 1997), suggesting that some lineage restriction occurs early in development. We do not know whether localization of these granules is involved in caste formation. However, the results of transplanting large and small proliferating morulae from the polymorula suggest these masses represent different castes. Large proliferating morulae explanted from second or third stadium hosts and implanted into host pupae always form precocious larvae. In contrast, small proliferating morulae almost always form reproductive larvae.

2. *Pattern Formation*

The second class of endogenous factors that could influence polyembryony are the molecular elements mediating pattern formation. Our understanding of patterning in insects again stems primarily from the genetic study of *D. melanogaster*. As mentioned previously, egg cleavage in *D. melanogaster* begins in a syncytial environment that is considered essential for establishing egg polarity. Morphogen gradients from maternal coordinate genes establish large domains that activate the gradients of zygotic gap genes. Gap gene products in turn activate pair-rule genes at cellularization to establish parasegmental periodicity (Ingham, 1988; Akam, 1987). These genes then activate segment polarity genes which specify segmental periodicity, and finally proteins of the gap, pair rule and segment polarity genes interact to regulate the homeotic selector genes whose transcripts influence developmental fate of each segment (Sommer and Tautz, 1991; Patel, 1994). Thus, for *D. melanogaster* and many other insects, the entire body plan is established by the blastoderm stage. This cascade of events clearly must differ in the cellurized environment of a *C. floridanum* egg. Moreover, it is not intuitively obvious how the patterning cascade might operate when the original anterior-posterior axis of the egg is lost during proliferation and must be reestablished within each embryo at the onset of morphogenesis. To begin addressing these questions, we analyzed expression patterns in *C. floridanum* of the Even-skipped (*Eve*), Engrailed (*En*) and Ultrabithorax/abdominal-A (*Ubx/*

abd-A) proteins, whose cognates play major roles in *D. melanogaster* segmentation (Grbic' et al., 1996a,b).

Emphasis was first placed on embryos that form reproductive larvae. The first indication of segmental periodicity in *D. melanogaster* occurs when pair rule genes such as *Eve* are activated in a pattern of seven stripes, in alternating segments. The *Eve* expression starts as a broad domain, missing only from the terminal regions of the blastoderm and resolves into the pair rule stripes (one stripe for every two future segments) by repression of the inter stripes. The *D. melanogaster Eve* expression later evolves from a pair rule pattern to a segmentally reiterated pattern (Frasch et al., 1988; Lawrence et al., 1987). We surveyed *C. floridanum* embryos over the entire course of embryogenesis for expression of the *Eve* antigen. We did not detect any *Eve* expression in eggs, primary morulas or proliferating morulas up to the host fourth instar (day 8). However, *Eve* expression does occur in individual embryonic primordia at the end of the host fourth instar (day 9). At this time, expression extends from the posterior end to 55 percent of the length of the newly formed primordium, resembling the early *D. melanogaster Eve* syncytial blastoderm pattern. Unlike *D. melanogaster*, however, this broad domain does not enter a pair rule phase but instead resolves directly into segmental stripes of *Eve* expression. Later, *Eve* was detected in several paired neuroblasts in a pattern conserved in all insects examined to date (Patel, 1994).

Despite the change in pair rule expression, subsequent tiers of the *D. melanogaster* segmentation hierarchy are conserved. In *D. melanogaster* the segment polarity gene *En* is regulated by *Eve* (DiNardo and O'Farrell, 1987): with *En* stripes appearing in rapid anteroposterior progression and even-numbered stripes appearing before odd-numbered stripes. In *C. floridanum*, however, there is no suggestion of pair rule regulation of the *En* antigen. Expression occurs in 15 stripes with a slight lag in the anterior to posterior direction. In fully segmented embryos, *En* marks the posterior of every segment as it does in all arthropods examined. To examine whether other, downstream elements of the *D. melanogaster* segmentation cascade were conserved, we used a monoclonal antibody that recognizes the expression pattern of both the Ultrabithorax (*Ubx*) and Abdominal-A (*abd-A*) proteins (referred to as *Ubx/abd-A*). The *Ubx* and *abd-A* genes are members of the homeotic selector group in *D. melanogaster*, responsible for the specification of metathoracic and abdominal segment identity (Lewis, 1978; Sanchez-Herrero et al. 1985). The earliest expression of *C. floridanum Ubx/ abd-A* occurs following gastrulation, in a broad domain, from the metathoracic *En* stripe to the penultimate *En* stripe with a parasegmental modulation. This is also consistent with *Ubx/abd-A* expression in *D. melanogaster* and its role in abdomen and last thoracic segment specification.

From the perspective of insect development generally, these results indicate that neither a syncytial environment nor formation of double segments is essential for segmental patterning. That none of the examined antigens are expressed during the proliferative period also indicates that these genes were not co-opted for novel functions during a phase of embryogenesis in polyembryony that has no counterpart in monoembryonic species. It is likely that changes in the genetic program that led to polyembryony occurred upstream of the pair rule genes, perhaps in programs that regulate the fate of germ cell determinants or proliferation of embryonic cells. How might these patterning events relate to caste formation? One possibility is that fine grained differences in the way cells or genes interact can produce morpologically distinct outcomes. For example, precise boundaries and expression levels of homeotic selector genes have been implicated in defining differences in abdominal segmentation and wing orientation (Tear et al., 1990). Similarly, small differences in expression of patterning genes may be significant in the formation of precocious and reproductive larvae.

C. Conclusions

The picture that emerges from these studies is that both host environment and endogenous events influence development of *C. floridanum*. First, early cleavage is total and thereafter the polymorula becomes divided into large and small proliferating morulae. Second, both proliferation and morphogenesis are induced by 20-hydroxyecdysone, resulting in synchronization of parasite development with the molting cycle of its larval host. Based upon our transplantation and hormonal manipulation experiments, however, we conclude that host endocrine state is not the primary factor mediating caste fate. This leaves us with the alternative that caste formation is mediated endogenously, possibly by cell lineage. We hypothesize, therefore, that separate precocious and reproductive lineages are established early in embryogenesis, but that the absolute number of individuals in each caste that develops is affected by age of the host egg when parasitized and the endocrine environment of the host larva. Under this model, precocious and reproductive lineages proliferate in response to a rise in the host ecdysteroid titer, but timing of morphogenesis, competency, and axial patterning differs between the two lineages. Our current studies focus on testing these predictions.

IV. The Role of Precocious and Reproductive Larvae in Parasitism

Reproductive larvae form adult wasps that seek mates and/or new hosts to parasitize (female). Precocious larvae have at least two functions: interspecific defense and, as alluded to earlier, manipulation of brood sex ratios (Cruz, 1981; Strand, Johnson, and Dover, 1990; Grbic' and Strand, 1991; Grbic' et al., 1992). Most insects are parasitized by more than one species of parasitoid, resulting in frequent opportunities for interspecific competition to occur. The possibility of a host being attacked by another species is perhaps heightened by the long life cycle of polyembryonic wasps relative to many other larval endoparasites. Precocious larvae attack interspecific competitors with their mandibles, protecting brood mates from resource competition. A more complex and potentially significant function, however, is the killing of male embryos. Manipulation experiments confirm that female precocious larvae attack brother embryos but not sisters *in vivo* and *in vitro*. As mentioned previously, the siblicidal behavior of precocious larvae explains, mechanistically, how mixed broods become so female biased. But how could such a phenomenon evolve? Grbic' et al. (1992) suggest it arose as a consquence of genetic conflict between siblings. Female-biased sex ratios are predicted under conditions of local mate competition (Hamilton, 1967) where female wasps mate with their brothers at the site of emergence, and mating opportunities for males are insignificant away from the brood. Under such conditions no genetic conflict exists between parent, sons, and daughters, and sex ratios are predicted to be strongly female biased. However, if males can obtain a significant number of matings away from the natal brood, the male clone would favor a higher sex ratio to capitalize on opportunities to mate with other females, but the female clone would favor a lower sex ratio to reduce competition for the limited food resources of the host. Any difference in evolutionary optima would be increased by the genetic asymmetries between males and females due to haplodiploidy. Field data suggest that males do obtain additional matings away from the natal host, and that the resulting conflict between siblings has been resolved in favor of sisters through the actions of precocious larvae.

V. The Evolution of Polyembryony in the Hymenoptera

What kinds of factors have favored the evolution of polyembryony and sterile castes in parasitic wasps? From an ecological perspective, sterile castes are associated almost exclusively with eusociality, a condition that has arisen in at least two groups of insects, the Isoptera (termites) and Hymenoptera (bees, ants), and once in mammals (naked mole rats) (Alexander et al., 1991). Sterile castes are also reported in certain aphids (Homoptera) and thrips (Thysanoptera) where again, sterile individuals assist reproductives (Aoki, 1987; Crespi, 1992). Ecological conditions suggested to favor formation of sterile castes and parent-offspring groups include nesting sites that are safe, defensible, and that contain abundant food. Clonal development has also evolved most often under conditions where high quality patches suitable for rapid exploitation exist (Bell, 1982). Location of a patch or host by clonal organisms is often hazardous. Once located, however, opportunities for monopolizing resources and cooperative defense between members of the clone arise, since evolution of cooperative behavior generally revolves around securing resources and defense for the benefit of related individuals (Alexander, 1974). Thus, we suggest that parent-offspring and clonal groups that occupy protected, rich habitats both exist in conditions where older or more developed individuals could help younger or less developed relatives. The soldier castes described by Aoki (1987) and Crespi (1992) arose in clonally developing aphids and thrips that exploit discrete nest sites (galls) similar to those of sexually reproducing eusocial species like ants, termites and naked mole rats. Hosts provide a similar environment for polyembryonic wasps.

From a developmental perspective, Ivanova-Kasas (1972) noted that eggs of polyembryonic species are distinguished by their small size, lack of yolk and thin chorion. She further suggested that holoblastic cleavage and the formation of an enveloping membrane of polar body origin were also unique. Yet the eggs of many monoembryonic parasitoids are as small as those of polyembryonic species, and polar-body-derived membranes are also reported in monoembryonic encyrtids, platygasterids, braconids, and scelionids (Silvestri, 1921; Tremblay and Calvert, 1972; Tremblay and Caltagirone, 1973; Strand et al., 1986). These features, therefore, are not associated solely with polyembryonic wasps, but they could be important preadaptations that favored the evolution of polyembryonic development. We also note that all polyembryonic wasps develop endoparasitically in an aquatic environment in which an exogenous food source (the host) meets their energy needs. Such conditions differ substantially from the terrestrial environment in which most insect eggs are placed, but are similar to the conditions in which mammalian embryos develop. That early cleavage in polyembryonic wasps resembles mammalian embryogenesis more than that of most arthropods may thus reflect congruent adaptations to circumstances in which a yolk source is not required (Strand and Grbic', 1997). Finally, morphogenesis in all polyembryonic wasps occurs synchronously with their host's final immature stage. This trend reflects a pattern observed commonly in organisms that exhibit both asexual and sexual phases of development; namely, asexual propagation occurs when resources are abundant and stable, but shifts to sexual reproduction occur when resource quality declines and dispersal becomes necessary. Although minor modifications in developmental mechanisms seem to be the normal mode of evolution, radical changes in development occasionally arise with shifts in life history strategies (Wray, 1995). Parasitism and clonality are two of the most common life history transformations in which punctuated shifts in developmental mechanisms might be predicted to arise. Polyembryonic wasps embrace both events, and, as illustrated here, offer an array of opportunities for studying the relationship between development and life history.

VI. Acknowledgments

Many individuals contributed to the work presented in this review. Most importantly, we acknowledge the efforts of E. H. Baehrecke, J. A. Johnson, L. Nagy, P. J. Ode, and D. Rivers. Funding for our studies on polyembryonic parasitoids was provided by grants from the U.S. Department of Agriculture, National Science Foundation, National Institutes of Health, Fulbright Foundation, and Sigma Xi.

VII. References

Akam, M. (1987). The molecular basis for metameric pattern in the *Drosophila* embryo. *Development* **101:**1–22.

Alexander, R. D. (1974). The evolution of social behavior. *Annu. Rev. Ecol. Syst.* **5:**325–383.

Alexander, R. D., Noonan, K. M., Crespi, B. J. (1991). The evolution of eusociality. In *The biology of the naked mole-rat* (P. W. Sherman, J. U. M. Jarvis and R. D. Alexander, eds.), pp. 3–44. Princeton, NJ: Princeton Univ. Press.

Aoki, S. (1987). Evolution of sterile soldiers in aphids. In *Animal societies: Theories and facts* (Y. Ito, J. L. Brown, and J. Kikkawa, eds.), pp. 53–65. Tokyo: Japan Science Press.

Askew, R. R. (1970). *Parasitic insects.* New York: Elsevier.

Baehrecke, E. H., Aiken, J. M., Dover, B. A., and Strand, M. R. (1993). Ecdysteroid induction of embryonic morphogenesis in a parasitic wasp. *Dev. Biol.* **158:**275–287.

Baehrecke, E. H., Grbic', M., and Strand, M. R. (1992). Serosa ontogeny in two embryonic morphs of *Copidosoma floridanum*, the influence of host hormones. *J. Exp. Zool.* **262:**30-39.

Baehrecke, E. H., and Strand, M. R. (1990). Embryonic morphology and growth of the polyembryonic parasitoid *Copidosoma floridanum* (Ashmead) (Hymenoptera: Encyrtidae). *Int. J. Insect Morphol. Embryol.* **19:**165-175.

Baehrecke, E. H., Strand, M. R., Williamson, J. L., and Aiken, J. M. (1992). Stage-specific protein and mRNA synthesis during morphogenesis of the polyembryonic parasitoid *Copidosoma floridanum* (Ashmead) (Hymenoptera: Encyrtidae). *Arch. Insect Biochem. Physiol.* **19:**81–92.

Barinaga, M. (1994). Looking to development's future. *Science* **266:**561–564.

Bell, G. (1982). *The masterpiece of nature: The evolution and genetics of sexuality.* London: Croom Helm.

Bryant, P. J., and Schmidt, O. (1990). The genetic control of cell proliferation in *Drosophila* imaginal discs. *J. Cell Sci. Suppl.* **13:**169–189.

Buss, L. W. (1987). *The evolution of individuality.* Princeton, NJ: Princeton University Press.

Crespi, B. J. (1992). Eusociality in Australian gall thrips. *Nature* **359:**724–726.

Cruz, Y. P. (1981). A sterile defender morph in a polyembryonic hymenopterous parasite. *Nature* **295:**446–447.

Daniel, D. M. (1932). *Macrocentrus ancylivorus* Rohwer, a polyembryonic braconid parasite of the oriental fruit moth. *NY State Agric. Exp. Stn. Tech. Bul.* **187:**101pp.

Davidson, E. H. (1991). Spatial mechanisms of gene regulation in metazoan embryos. *Development* **113:**1–26.

DiNardo, S., and O'Farrell, P. (1987). Establishment and refinement of segmental pattern in the *Drosophila* embryo: Spatial control of engrailed expression by pair-rule genes. *Genes & Dev.* **1:**1212–1225.

Dittrick, L. E., and Chiang, H. C. (1982). Developmental characteristics of *Macrocentrus grandii* as influenced by temperature and instar of its host, the European corn borer. *J. Insect Physiol.* **28:**47–52.

Doutt, R. L. (1947). Polyembryony in *Copidosoma koehleri* Blanchard. *Am. Nat.* **81:**435–453.

Doutt, R. L. (1952). The teratoid larva of polyembryonic encryrtids (Hymenoptera). *Can. Entomol.* **84:**247–250.

Fitzgerald, T. D., and Simeone, J. B. (1971). Polyembryony in *Paralerocerus bicoloripes. Ann. Entomol. Soc. Am.* **64:**774–777.

Frasch, M., Warrior, R., Tugwood, J., and Levine, M. (1988). Molecular analysis of *even-skipped* mutants in *Drosophila* development. *Genes & Dev.* **2:**1824–1838.

Godfray, H. C. J. (1994). *Parasitoids: Behavioral and evolutionary ecology.* Princeton, NJ: Princeton University Press.

Grbic', M., Nagy, L., Carroll, S. B., and Strand, M. R. (1996a). Polyembryonic development: Insect pattern formation in a cellurized environment. *Development* **122:**795–804.

Grbic', M., Ode, P. J., and Strand, M. R. (1992). Sibling rivalry and brood sex ratios in polyembryonic wasps. *Nature* **360:**254–256.

Gbric', M., and Strand, M. R. (1991). Intersexual variation in the development of precocious larvae from the parasitoid Copidosoma floridanum. In *Proceedings of the III. International Congress on* Trichogramma *and Other Egg Parasitoids* (W. C. Nettles, ed.), pp. 25–28. Paris: INRA.

Grbic', M., Nagy L. M., and Strand, M. (1996b). Pattern duplication in the polyembryonic wasp *Copidosoma floridanum. Dev. Genes Evol.* **206:**281–287.

Grbic', M., Rivers, D. and Strand, M. R. (1997). Caste formation in the polyembryonic wasp *Copidosoma floridanum: in vivo* and *in vitro* analysis. *J. Physiol.* in press.

Hamilton, W. D. (1967). Extraordinary sex ratios. *Science* **156:**477–488.

Hardy, I. C. W., Ode, P. J., and Strand, M. R. (1993). Factors influencing brood sex ratios in polyembryonic Hymenoptera. *Oecologia* **93:**343–348.

Hill, C. C. (1926). *Platygaster hiemalis* Forbes, a parasite of the Hessian fly. *J. Agric. Res.* **22:**261–265.

Hill, C. C., and Emery, W. T. (1937). The biology of *Platygaster herrickii,* a parasite of the Hessian fly. *J. Agric. Res.* **55:**199–213.

Hughes, R. N., and Cancino, J. M. (1985). An ecological overview of cloning in metazoa. In *Population biology and evolution of clonal organisms* (J. B. C. Jackson, L. W. Buss, and R. E. Cook, eds.), pp. 153–186. New Haven, Conn.: Yale University Press.

Ingham, P. W. (1988). The molecular genetics of embryonic pattern formation in *Drosophila. Nature* **335:**25–33.

Ivanova-Kasas, O. M. (1972). Polyembryony in insects. In *Developmental systems,* vol. 1., *Insects* (S. J. Counce, and C. H. Waddington, eds.), pp. 243–271. New York: Academic Press.

Kornhauser, S. I. (1919). The sexual characteristics of the membracid, *Thelia bimaculata*

(Fabr.). I. External changes induced by *Aphelopus theliae* (Gahan). *J. Morphol.* **32:**531–635.

Koscielski, B., Koscielska, M. K., and Szroeder, J. (1978). Ultrastructure of the polygerm of *Ageniaspis fuscicollis* Dalm. (Chalcidoidea, Hymenoptera). *Zoomorphologie* **89:**279–288.

Kumaran, A. K. (1983). Introduction: Evolution of regulatory controls in insect life cycles. In *Endocrinology of insects,* vol. 1 (R. G. H. Downer and H. Laufer, eds.), pp. 333–336. New York: Liss.

LaSalle, J., and Gauld, I. D. (1991). Parasitic Hymenoptera and the biodiversity crisis. *Redia* **74:**540–544.

Lawrence, P., Johnston, P., Macdonald, P., and Struhl, G. (1987). Borders of parasegments in *Drosophila* embryos are delimited by the *fushi-tarazu* and *even-skipped* genes. *Nature* **328:**440–442.

Leiby, R. W. (1922). The polyembryonic development of *Copidosoma gelechiae* with notes on its biology *J. Morphol.* **37:**195–285.

Leiby, R. W. (1924). The polyembryonic development of *Platygaster vernalis. J. Agric. Res.* **28:**829–839.

Leiby, R. W. (1926). The origin of mixed broods in polyembryonic Hymenoptera. *Ann. Entomol. Soc. Am.* **19:**290–299.

Leiby, R. W., and Hill, C. C. (1923). The twinning and monembryonic development of *Platygaster hiemalis,* a parasite of the Hessian fly. *J. Agric. Res.* **19:**337–350.

Leiby, R. W., and Hill, C. C. (1924). The polyembryonic development of *Platygaster vernalis. J. Agric. Res.* **28:**829–839.

Lewis, E. B. (1978). A gene complex controlling segmentation in *Drosophila. Nature* **276:**565–570.

Marchal, P. (1898). La dissociation de l'oeuf en un cycle evolutif chez l'*Encyrtus fuscicollis* (Hymenoptère). *Comp. Rend. Acad. Sci.* **126:**662–664.

Marchal, P. (1904). Recherches sur la biologie et le développement des hyménoptères parasites. I. La polyembryonie spécifique ou germinogonie. *Arch. de Zool. et Gen.* **2:**257–335.

Martin, F. (1914). Zur entwicklungsgeschichte des polyembryonalen chalcidiers *Ageniaspis* (Encyrtus) *fuscicollis. Z. Wiss. Zool.* **110:**419–479.

Mindek, G. (1972). Metamorphosis of imaginal discs of *Drosophila melanogaster. Roux's Arch. Entw-Mech. Org.* **169:**353–356.

Nenon, J. P. (1978). La polyembryonie de *Ageniaspis fuscicollis* Thoms. (Hyménoptère, Chalcidien, Encyrtidae). *Bull. Biol. Fr. Belg.* **112:**13–107.

Nijhout, F. (1994). *Insect endocrinology.* Princeton, NJ: Princeton Univ. Press.

Noyes, J. S. (1988). *Copidosoma truncatellum* (Dalman) and *C. floridanum* (Ashmead) (Hymenoptera, Encyrtidae), two frequently misidentified polyembryonic parasitoids of caterpillars (Lepidoptera). *Syst. Entomol.* **13:**197–204.

Oberlander, H. (1985). The imaginal discs. In *Comprehensive insect physiology, biochemistry and pharmacology,* vol. 2, *Postembryonic development* (L. I. Gilbert and G. A. Kerkut, eds.), pp. 151–182. Oxford: Pergamon Press.

Ode, P. J., and Strand, M. R. (1995). Progeny and sex allocation decisions of the polyembryonic wasp *Copidosoma floridanum. J. Anim. Ecol.* **64:**213–224.

Parker, H. L. (1931). *Macrocentrus gifuensis* Ashmead, a polyembryonic braconid parasite of the European corn borer. *U. S. Dept. Agric. Tech. Bull.* **230:**62pp.

Parker, H. L., and Thompson, W. R. (1928). Contribution a la biologie des chalcidiens entomophages. *Annals Soc. Entomol. Fr.* **97:**425–465.

Patel, N. H. (1994). Developmental evolution: Insights from studies of insect segmentation. *Science* **266:**581–590.

Patterson, J. T. (1917). Studies on the biology of *Paracopidosomopsis.* I. Data on the sexes. *Biol. Bull.* **32:**291–305.

Patterson, J. T. (1918). Studies on the biology of *Paracopidosomopsis* III. Maturation and fertilization. *Biol. Bull.* **35:**362–377.

Patterson, J. T. (1919). Polyembryony and sex. *J. Hered.* **10:**344–352.

Patterson, J. T. (1921). The development of *Paracopidosomopsis. J. Morphol.* **36:**1–69.

Price, P. W. (1980). *Evolutionary biology of parasites.* Princeton, NJ: Princeton University Press.

Riddiford, L. M. (1985). Hormone action at the cellular level. In *Comprehensive insect physiology, biochemistry, and pharmacology,* vol. 8 (G. A. Kerkut and L. I. Gilbert, eds.), pp. 37–84. Oxford: Pergamon Press.

Robinson, G. E. (1992). Regulation of division of labor in insect societies. *Annu. Rev. Entomol.* **37:**637–665.

Sanchez-Herrero, E., Vernos, I., Marco, R., and Morata, G. (1985). Genetic organisation of the *Drosophila* bithorax complex. *Nature* **313:**108–113.

Schejter, E., and Weischaus, E. (1993). Functional elements of the cytoskeleton in the early *Drosophila* embryo. *Annu. Rev. Cell Biol.* **9:**67–99.

Schweisguth, F., Vincent, A., and Lepesant, J. A. (1991). Genetic analysis of the cellularization of the *Drosophila* embryo. *Biol. Cell* **72:**15–23.

Silvestri, F. (1906). Contribuzioni alla conoscenza biologica degli imenotteri parassiti. Biologia del *Litomastix truncatellus* (Dalm.) (2 nota preliminare). *Ann. Regia Sc. Super. Agric. Portici.* **6:**3–51.

Silvestri, F. (1921). Contribuzioni alla conoscenza biologica degli imenotteri parassiti. V. Sviluppo del *Platygaster dryomyiae* Silv. *Ann. Regia Sc. Super. Agric. Portici.* **11:**299–326.

Silvestri, F. (1937). Insect polyembryony and its general biological aspects. *Bull. Mus. Comp. Zool. Harv. Univ.* **81:**468–496.

Sommer, R., and Tautz, D. (1991). Segmentation gene expression in the housefly *Musca domestica. Development* **113:**419–430.

St. Johnston, D., and Nüsslein-Volhard, C. (1992). The origin of pattern and polarity in the *Drosophila* embryo. *Cell* **68:**201–219.

Strand, M. R. (1989a). Development of the polyembryonic parasitoid *Copidosoma floridanum* in *Trichoplusia ni. Entomol. Exp. Appl.* **50:**37–46.

———. (1989b). Oviposition behavior and progeny allocation by the polyembryonic wasp *Copidosoma floridanum. J. Insect Behav.* **2:**355–369.

———. (1989c). Clutch size, sex ratio and mating by the polyembryonic encyrtid *Copidosoma floridanum. Fla. Entomol.* **72:**32–42.

Strand, M. R., Goodman, W. G., and Baehrecke, E. H. (1991). The juvenile hormone titer of *Trichoplusia ni* and its potential role in embryogenesis of the polyembryonic wasp *Copidosoma floridanum. Insect Biochem.* **21:**205–214.

Strand, M. R., Johnson, J. A., and Culin, J. D. (1990). Intrinsic interspecific competition between the polyembryonic parasitoid *Copidosoma floridanum* and solitary endoparasitoid *Microplitis demolitor* in *Pseudoplusia includens. Entomol. Exp. Appl.* **55:**275–284.

Strand, M. R., Johnson, J. A., and Dover, B. A. (1990). Ecdysteroid and juvenile hormone esterase profiles of *Trichoplusia ni* parasitized by the polyembryonic wasp *Copidosoma floridanum. Arch. Insect Biochem. Physiol.* **13:**41–51.

Strand, M. R., Meola, S. M., and Vinson, S. B. (1986). Correlating pathological symptoms in *Heliothis virescens* eggs with development of the parasitoid *Telenomus heliothidis. J. Insect Physiol.* **32:**389–402.

Strand, M. R., and Ode, P. J. (1990). Chromosome number of the polyembryonic parasitoid *Copidosoma floridanum* (Ashmead) (Hymenoptera: Encyrtidae). *Ann. Entomol. Soc. Am.* **83:**834–837.

Strand, M. R., Wong, E. A., and Baehrecke, E. H. (1991). The role of host endocrine factors in the development of polyembryonic parasitoids. *Biol. Control* **1:**144–152.

Strand, M. R., and Grbic', M. (1997). Development and evolution of polyembryonic insects. *Curr. Topics Devp. Biol.* in press.

Strehlow, D., and Gilbert, W. (1993). A fate map for the first cleavages of the zebrafish. *Nature* **361:**451–453.

Tear, G., Akam, M., and Martines-Arias, A. (1990). Isolation of an *abdominal-A* gene from the locust *Shistocerca gregaria* and its expression during early embryogenesis. *Development* **110:**915–925.

Tremblay, E., and Caltagirone, L. E. (1973). Fate of polar bodies in insects. *Annu. Rev. Entomol.* **18:**421–444.

Tremblay, E., and Calvert, D. (1972). New cases of polar nuclei utilization in insects. *Ann. Soc. Entomol. Fr.* **8:**495–498.

Wishart, G. (1946). Laboratory rearing of *Macrocentrus gifuensis* Ashmead, a parasite of the European corn borer. *Can. Entomol.* **78:**78–82.

de Wilde, J., and Beetsma J. (1982). The physiology of caste development in social insects. *Adv. Insect Physiol.* **16:**167–246.

Walter, G. H., and Clarke, A. R. (1992). Unisexual broods and sex ratios in a polyembryonic encyrtid parasitoid (*Copidosoma* sp.: Hymenoptera). *Oecologia* **89:**147–149.

Wray, G. A. (1995). Punctuated evolution of embryos. *Science* **267:**1115–1116.

3

Schistosome Parasites Induce Physiological Changes in their Snail Hosts by Interfering with Two Regulatory Systems, the Internal Defense System and the Neuroendocrine System

Marijke de Jong-Brink, Robert M. Hoek, Wessel Lageweg, and August B. Smit

I. Introduction

The complex life cycle of trematodes involves a remarkable and highly characteristic alternation of asexual and sexual reproductive phases in molluscan and vertebrate hosts, respectively. The parasitic worms that cause schistosomiasis ("bilharzia") in humans belong to the trematode genus *Schistosoma*. The avian schistosome *Trichobilharzia ocellata* can be studied as a model for the interactions between schistosomes and their host. This avian schistosome is also an important parasite as it causes cercarial dermatitis, and is becoming a serious problem in some freshwater areas. The life cycle of *T. ocellata* involves two hosts: a duck as the definitive host (*Anas platyrhynchos*) and a freshwater snail as the intermediate host (*Lymnaea stagnalis*). Eggs produced by the adult female parasite are shed with the feces of the duck, hatch in water, and produce a ciliated larva ("miracidium"). The miracidia can penetrate the skin of the mantle or the head-foot of a snail host to develop into a primary or mother sporocyst near the site of penetration only during a limited period of time. Daughter sporocysts develop in the mother sporocyst and emerge 3–4 weeks postinfection (Amen and Meuleman, 1992), after which they migrate to the hind part of the snail. In the digestive gland/ovotestis area they grow and give rise to the final larval stage, the cercariae. The production of cercariae may continue for the rest of the life of the snail. Upon receipt of an as yet unidentified stimulus, the cercariae leave the snail and must find a suitable definitive host within a few hours.

We are interested in the strategies this schistosome parasite uses to manipulate its snail host so that the parasite is able to survive, multiply, and complete its life cycle. Our studies indicate that *T. ocellata* interferes with two regulatory systems in the host, the internal defense system (IDS) and the neuroendocrine system (NES), and that the endocrine and immune systems of the host are physiologically linked.

II. Effects of the Parasite on the Internal Defense System of the Snail Host

When schistosome miracidia penetrate a snail host, the parasites have to circumvent the snail's internal defense system (IDS). This IDS comprises both cellular and humoral elements that cooperate in the process of discrimination between self and non-self, resulting in the killing of non-self invaders (Bayne, 1983). Hemocytes are the primary effector cells in the internal defense system and these freely moving cells are able to recognize non-self antigens, synthesize lectins (Van der Knaap et al., 1981, 1983), to encapsulate microbes or multicellular parasites (e.g., Sminia et al., 1974) and to phagocytose (e.g., Van der Knaap and Loker, 1990). Furthermore, hemocyte-mediated cytotoxicity is responsible for the killing and digestion of phagocytozed small organisms or encapsulated larger organisms (Bayne et al., 1980a,b). Besides nonoxidative killing, which is effected intracellularly by lysosomal enzymes (Cheng and Dougherty, 1989) and extracellularly by the release of lysosomal enzymes and bactericidins (Cushing et al., 1971; Adema et al., 1991a), oxidative killing mechanisms are involved (Dikkeboom et al., 1987; Adema et al., 1991b). In incompatible parasite-snail combinations the responses described above will result in elimination of the parasite, as is the case when *Schistosoma mansoni* miracidia enter the resistant 10-R2 strain of *Biomphalaria glabrata* (Loker et al., 1982). In compatible combinations, on the other hand, host hemocytes fail to respond and the parasite survives. The parasite is capable of circumventing or suppressing the host's defense system via several possible mechanisms (Lie, 1982; Bayne and Yoshino 1989; Van der Knaap and Loker, 1990; De Jong-Brink, 1995).

The cytotoxic activities of hemocytes are influenced by multiple plasma factors. Injection of plasma from *S. mansoni*-resistant strains of *B. glabrata* into susceptible strains transfers resistance (Granath and Yoshino, 1984). In addition, hemocytes from susceptible snails are capable of the *in vitro* killing of sporocysts when incubated with plasma from resistant snails (Bayne et al., 1980b). Plasma factors alone, however, are incapable of direct killing of trematode larvae (Bayne et al., 1980a; Dikkeboom et al., 1988). These data indicate that humoral factors in the plasma of a resistant host enhance the killing capacity of hemocytes from susceptible snails. Hemocytes from resistant snails, however, can kill parasites in the absence of humoral factors (see Bayne et al., 1980a). These humoral factors, with the properties of lectins, are often found in circulating hemolymph and have frequently been shown to agglutinate foreign particles. They also function as opsonins and facilitate phagocytosis by molluscan hemocytes (Bayne, 1990). Interestingly, the opsonizing activity present in hemolymph of strains of *B. glabrata* resistant to *S. mansoni* is absent in a susceptible strain (Fryer et al.,

1989). However, hemocytes can be engaged in direct immunorecognition of foreign particles without the involvement of humoral factors since they possess surface molecules that have lectin activity (Richards and Renwrantz, 1991).

III. Parasite Excretory/Secretory (E/S) Products Affect the Internal Defense System of the Host

Their central role in defense makes hemocytes the targets of parasite-mediated interference with the host's internal defense system. *In vivo* studies suggest that the immunosuppressive effect is strongest immediately adjacent to the trematode larva and weaker at a distance, raising the possibility that larvae release substances that interfere with hemocyte activities (Lie, 1982). This hypothesis is supported by the results of *in vitro* experiments in which *T. ocellata* was cultured for 96 h and the media were changed after 33 and 72 h (see below; De Jong-Brink, 1995; Núñez and De Jong-Brink, 1995). The interference of parasitic larval stages or their excretory-secretory (E/S) products includes hemocyte aggregation (Noda and Loker, 1989), induction of lysosomal enzyme activity (Granath and Yoshino, 1983), polypeptide synthesis and release (Yoshino and Lodes, 1988; Lodes et al., 1991), superoxide production (Connors and Yoshino, 1990), motility of hemocytes (Lodes and Yoshino, 1990) and modulation of phagocytic activity (Abdul Salam and Michelson, 1980a,b; Van der Knaap et al., 1987; Noda and Loker, 1989; Connors and Yoshino, 1990; Fryer and Bayne, 1990; Núñez et al., 1994).

The components of the E/S products of *S. mansoni* responsible for the modulatory effects on hemocyte activity are still unknown although some preliminary information is available. E/S products of 1 day primary cultures of *S. mansoni*, which can modulate the protein metabolism of *B. glabrata* hemocytes, are mainly concentrated in a heat-stable fraction composed of molecules of > 30 kDa together with a smaller fraction with a molecular weight of < 10 kDa (Yoshino and Lodes, 1988). Both silver staining and fluorography of SDS-PAGE-separated E/S products revealed a wide variety of glycoproteins ranging in molecular weight from 13 to > 200 kDa (Lodes and Yoshino, 1989). The secretion of resistant hemocytes was not affected; inhibition of polypeptide secretion by susceptible hemocytes could be attributed to a high molecular protein aggregate consisting of subunits of 22–24 kDa in the E/S products of *S. mansoni*. This differential effect on hemocytes was only observed in the presence of homologous plasma (Lodes et al., 1991). This E/S factor differs from a 108 kDa molecule found by Lodes and Yoshino (1990) which differentially affects motility of hemocytes from susceptible and resistant strains. These results concur with those of findings of Loker and colleagues (1992), who reported on E/S products of *Echinostoma paraensei* sporocysts interfering with *B. glabrata* hemocyte functions. The active E/S components were > 100 kDa in their native configuration and heat- and trypsin-labile (Loker et al., 1992).

To study the effects of E/S products of *T. ocellata* on the internal defense of its susceptible host *L. stagnalis,* a bacterial clearance assay (BCA), which detects changes in the capacity of hemocytes to eliminate bacteria, was used as a parameter for modulation of hemocyte activity (Núñez et al., 1994). Two main fractions of E/S products were released by *T. ocellata in vitro*, which had direct but antagonistic effects on the killing activity of hemocytes (Núñez and De Jong-Brink, 1995). A low molecular weight (MW) fraction of approximately 2 kDa, as determined by means of high performance gel-permeation column chromatography (HPGPC), actually activated the hemocytes. Release of this low MW fraction was optimal during the first 33 h of the *in vitro* culture of miracidia, which corresponds to 1.5–6 h *in vivo* (Amen et al., 1992a). E/S products released between 33–72 h and separated by means of HPGPC showed a prominent high MW fraction of about 40 kDa that inhibited the bacterial killing activity of hemocytes from *L. stagnalis* (Núñez and De Jong-Brink, 1996). These findings are in agreement with the *in vivo* situation in which hemocytes from snails that had been exposed to the trematode for 1.5 h had an enhanced clearance capacity, whereas cells obtained from snails at 24–96 h postinfection showed decreased clearance of bacteria, indicating suppression of hemocyte activity by the parasite (Núñez et al., 1994).

These observations suggest that, in a compatible situation, the suppressive E/S fraction acts directly on hemocytes to protect the parasite from being attacked by the host defense system. Indeed, hemocytes from an incompatible host for *T. ocellata*, *Planorbarius corneus*, were not affected by the suppressive factor. On the other hand, the activating factor was able to increase bacterial killing activity of hemocytes from both *P. corneus* and *T. ocellata* (De Jong-Brink, 1995). The fact that a parasite-derived suppressive factor is not recognized by hemocytes of incompatible hosts might explain why parasites in an incompatible combination are encapsulated and eliminated.

These results are consistent with those of Lodes and colleagues (1991), Lodes and Yoshino (1990), and Fryer and Bayne (1990), who isolated E/S fractions with an inhibitory effect on motility and/or phagocytotic activity of susceptible hemocytes that left hemocytes from resistant strains unaffected. These observations could be related to the presence or absence of particular lectins that recognize specific parasitic E/S products in resistant and susceptible hosts, respectively. Summarizing, the results obtained for both *T. ocellata* and *S. mansoni* clearly show that the parasite E/S products play a crucial role in determining compatibility and susceptibility as opposed to incompatibility and resistance characteristics.

Recently, Crews-Oyen and Yoshino (1995) described the temporal synthesis and release of *S. mansoni* E/S proteins by daughter sporocysts isolated from mother sporocysts. They speculated that daughter sporocyst E/S products play a role in disrupting host immune defense since completion of the parasite life cycle relies on the parasite's continued ability to successfully escape detection by the host's immune system (Crews-Oyen and Yoshino, 1995). Since these daughter sporocysts had been removed from the mother sporocysts, this might not reflect the *in vivo* situation. Indeed, the suggestion that parasites have to suppress the host's internal defense throughout their development in the snail is not in agreement with results described by Amen and coworkers (1992a), who studied the phagocytotic activity in (cross-) combinations of hemocytes and plasma of later stages of infected and noninfected snails. At 2, 4, and 6 weeks after *L. stagnalis* had been exposed to miracidia of *T. ocellata* no significant differences were found in the phagocytotic activity of hemocytes of infected and noninfected snails in the presence of plasma from both infected and noninfected snails. A higher percentage of phagocytosing hemocytes, however, was found in snails that had been infected for 8 weeks. Activation of the defense system coincided with the presence of fully differentiated cercariae within and emerging from daughter sporocysts, and eventually from the snail. The mechanical and lytic tissue damage caused by migrating cercariae, which themselves may be protected by masking snail antigens (Van der Knaap et al., 1985; De Jong-Brink, 1995), may be responsible for this phenomenon.

Whether the E/S products of parasites only affect hemocytes or also affect other tissue systems in the host is not known. For instance, it might be possible that such an effect is indirect: upon infection, hemocytes may initiate synthesis of factors that influence other physiological processes. This situation would be comparable to that in vertebrates where activated macrophages are known to produce and release cytokines, which are able to elicit physiological changes by acting on the neuroendocrine system (NES). Since *S. mansoni* E/S products have different effects on the secretion of metabolically labeled polypeptides from hemocytes in resistant and susceptible *B. glabrata* strains (Lodes et al., 1991), investigation of whether these peptides play a role in the regulation of the NES might yield critical new information.

IV. Effects of the Parasite on the Neuroendocrine System of the Host

After having invaded their snail host and escaped from the initial attack of the IDS of the host, the parasites migrate to tissues with high concentrations of nutrients such as the connective tissue bathed in hemolymph of the open circulatory system in the snail host. In meeting their energy demands, parasites often affect their host in a much more complex way than merely by competition for nutrients. Some schistosomes (e.g., *T. ocellata*) reduce or completely stop reproduction of the snail host ("parasitic castration") and also enhance body growth, inducing gigantism. In this way the parasites claim both energy and space, especially in the stage during which the development and release (shedding) of enormous numbers of cercariae take place in/from the daughter sporocysts of *T. ocellata*. Previous work (see De Jong-Brink, 1995) has demonstrated that the effects of the parasite on the physiology of the snail host are established by humoral interference with the neuroendocrine system regulating these processes. This hypothesis has been verified for *T. ocellata* in the *Lymnaea* host. In the hemolymph of infected snails approaching or during patency a humoral factor interferes both with the action of snail hormones upon their target organs (peripheral effects) and with electrical activity of neurosecretory cells and hence with release of neuropeptides (central effects).

A. Peripheral Effects of the Parasite on Target Organs

Bioassays for three hormones involved in regulation of female reproduction were used for demonstrating the presence of a factor in the hemolymph of parasitized snails that inhibits the hormone responses in their respective target cells. Caudodorsal cell hormone (CDCH), inducing ovulation and egg laying, calfluxin (CaFl), stimulating perivitellin fluid release from the secretory cells of the albumen gland, one of the female accessory sex glands, and dorsal body hormone (DBH), regulating differentiation and growth of the female reproductive tract and growth and maturation of oocytes (for more details see De Jong-Brink, 1995), were assayed. The neuropeptides CDCH and CaFl are derived from the same precursor in the neurosecretory caudodorsal cells (CDCs) located in the cerebral ganglia of the central nervous system (CNS), whereas DBH is produced by the endocrine dorsal bodies (DB) attached to the cerebral ganglia (Vreugdenhil et al., 1988; Geraerts et al., 1991). The bioassays for CaFl and DBH measure the induced changes in the second messenger cascade in their respective target cells. As a bioassay for CDCH, the number of snails responding with ovulation upon an injection with synthetic CDCH (2 pmol per snail) was counted. The bioactivity of these three female gonadotropic hormones appeared to be inhibited in the presence of hemolymph from infected snails as soon as differentiating cercariae were present in the daughter sporocysts. This inhibiting effect of hemolymph from parasitized snails was ascribed to schistosomin, a protein consisting of 79 amino acids with 4 intramolecular disulphide bridges (De Jong-Brink et al., 1986, 1988; Joosse et al., 1988; Hordijk et al., 1991a,b). Schistosomin antagonizes the effects of structurally different hormones and likely binds its own receptor and interacts with the action of other hormones in their intracellular signal transduction pathway (see De Jong-Brink, 1995).

Appropriate bioassays to study the biological activities of the hormones involved in regulation of metabolism and body growth in *Lymnaea* are still lacking. Data describing the effects of schistosomin on the central part of the NES indicate that the effects on growth and metabolism can also be ascribed to parasitic interference with the NES.

B. Central Effects of the Parasite on Neuroendocrine Cells

1. Electrical Activity of Neurosecretory Cells

The electrical properties of two types of neuroendocrine cells, the already mentioned CDCs and the light green cells (LGCs) involved in regulation of growth, change considerably due to parasitism. The excitability of these two cell types was studied *in vitro* in the presence of hemolymph from infected and noninfected snails (Hordijk et al., 1992). CDCs can display three different states of excitability, the refractory or inhibited state, the resting state, and the afterdischarge, which are closely related to different stages in the egg-laying cycle (Kits, 1980). During an afterdischarge, massive release of neuropeptides occurs in the neurohemal area of the CDCs, the cerebral commissure, followed by ovulation and egg laying. In the resting state CDCs are silent but excitable: prolonged stimulation gives an afterdischarge, whereas it is impossible to elicit an afterdischarge in CDCs in the refractory or inhibited state. However, after replacing the normal hemolymph with hemolymph from parasitized snails the cells in a resting state cannot generate an afterdischarge. Similar effects were observed when the cells were incubated in saline containing purified 10^{-8} M schistosomin (Hordijk et al., 1992).

LGCs *in situ* in an isolated CNS kept in hemolymph from non-parasitized snails were silent and responded with a single action potential upon suprathreshold stimulation. When the hemolymph was replaced by hemolymph from parasitized snails, the excitability of the cells increased. This last effect was also observed with purified schistosomin from parasitized snails. The same phenomena were observed using freshly dissociated LGCs in a primary culture. This shows that schistosomin acts directly upon LGCs and that the response is not mediated by interneurons, which implies that LGCs have receptors for schistosomin (Hordijk et al., 1992). This is supported by the observation that increased excitability of the LGCs caused by schistosomin coincides with activation of adenylate cyclase resulting in the production of cAMP (De Jong-Brink et al., 1992). The evidence that schistosomin affects the LGCs, in addition to the CDCs, suggests that it is not only involved in establishing the effects on reproduction, but also on growth and metabolism in parasitized snails. Onset of enhanced growth and glycogen depletion coincides with the appearance of schistosomin in hemolymph of infected snails (Schallig et al., 1991b).

2. *Changes in Gene Expression in the CNS*

An interesting question is whether parasites might affect gene expression in neuroendocrine systems in their snail host, which may result in a changed physiology enabling the parasites to survive and continue their development. To test this hypothesis a cDNA library of the CNS of patently infected snails was constructed and differentially screened using probes made of mRNA extracted from CNS of parasitized and nonparasitized animals. In this procedure 50,000 individual clones were screened and plaques corresponding to lambda phages that showed a differential hybridization signal to either probe were selected, leading to the isolation of 148 cDNA clones corresponding to mRNAs that are differentially—either up or down—regulated in the CNS of parasitized snails. These clones were grouped on the basis of nucleotide sequence identity by cross-hybridization, leading to the characterization of 22 groups that ranged in size from 63 clones to individual clones. From the six groups comprising more than one clone, the longest cDNA was selected for sequence analysis. In addition to these genes, one individual clone was sequenced. Northern blot analysis confirmed that all clones contained full-length cDNA sequences, since the estimated lengths of the transcripts were in good agreement with the insert sizes of the clones.

Sequence analysis and *in situ* hybridization have shown that four of these seven groups of clones encode neuropeptides. A group of five identical clones appeared to encode the precursor containing the neuropeptide FMRFa, which had already been isolated and sequenced from several molluscan species and is probably present in all protostomes (see Price and Greenberg, 1995). The largest remaining group of 4 identical clones encoded a multipeptide precursor containing 16 copies of a neuropeptide homologous to the *Aplysia* pedal peptide (Lloyd and Conolly, 1989). Furthermore, a group of three identical clones was shown to encode the structural homologue of the neuropeptide Y (NPY) precursor. One individual clone contained the sequence of a set of novel, hitherto unknown neuropeptides each containing the C-terminal sequence Arg-Phe-amide (Hoek et al., pers. comm.).

Of the three remaining groups of clones not encoding neuropeptides, one appeared to encode a novel member of the immunoglobulin (Ig) superfamily. Within the predicted protein, 10 cysteine residues participate in the formation of 5 intra-Ig domain disulphide bridges (Fig. 3-1; Hoek et al., 1996). Primary sequence identity of this molecule was found with the insect hemolins, which play a role as an agglutination factor in the defense reaction to bacterial infection (Sun et al., 1990). Furthermore, two clones encoding mitochondrial products have been characterized. One encodes cytochrome c, an electron carrier participating in the respiratory chain, and the other represents 16S mitochondrial ribosomal RNA.

Of the seven characterized clones, four appeared to encode neuropeptide precursors, one a mitochondrial protein, one the 16S mitochondrial rRNA sequence, and one a novel member of the Ig superfamily.

The results obtained with this differential screening were confirmed by data obtained by reverse Northern blot analysis using different cDNA clones. In this

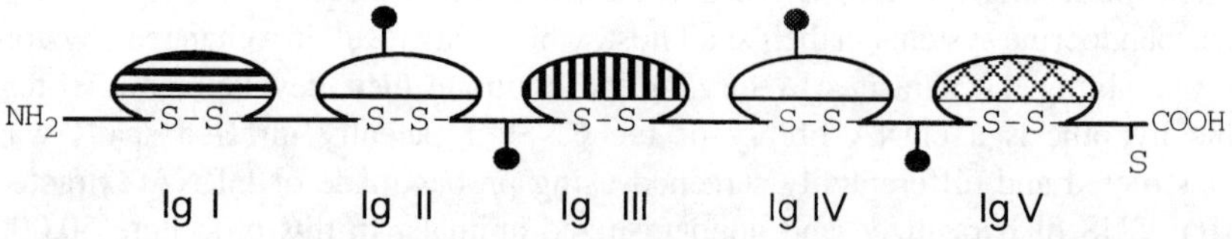

Figure 3-1. Schematic representation of the novel type of protein, which is a member of the Ig superfamily. Differences in shading indicate different chemical characteristics; putative glycosylation sites are indicated by ball-and-stick; S–S indicate intrachain disulphide bridges. The free cysteine residues at the C-terminus are indicated by S (Hoek et al., 1996).

semiquantitative analysis it is possible to analyze in more detail the changes in steady-state levels of previously identified CNS-specific mRNAs during parasitic infection. For this purpose cDNA clones that had been characterized in the differential screening (see Table 3-1), and a number of previously identified cDNA clones from the CNS were chosen. The last group comprised the specific messengers for (1) CDCH, involved in the regulation of ovulation/egg laying; (2) the tetrapeptide Ala-Pro-Gly-Trp-amide (APGWa; Croll et al., 1991) and the vasopressin-related peptide conopressin (Van Kesteren et al., 1995), which are both synthesized by neurons belonging to the neuronal circuitry that controls male copulation behaviour; and (3) the MIPs, molluscan insulin-related peptides, synthesized in the LGCs, involved in regulation of growth (Smit et al., 1988). In addition, cDNA of some clones encoding structural proteins or enzymes were tested. Slot blots of the entire set of cDNA clones were made and hybridized to a cDNA probe of either the CNS of parasitized or of nonparasitized animals. The autoradiographs were analyzed by densitometry and the results obtained were statistically analyzed (Table 3-1).

Table 3-1. Reverse Northern Blot Analysis of 10 Transcripts Shows Differential Regulation in the CNS of Parasitized Snails versus Nonparasitized Snails

Encoded precursor	Relative change ± SEM in mRNA levels in P
FMRFa[1]	184 ± 33*
Pedal peptide[1]	252 ± 19**
NPY[1]	162 ± 19**
xLFRFa[1]	190 ± 7**
Cyt-C[1]	256 ± 5**
16S mt rRNA[1]	142 ± 6**
MIg[1]	41 ± 2**
CDCH[2]	34 ± 2**
MIPs[2]	96 ± 4
APGWa[2]	106 ± 7
Conopressin	130 ± 10

Indicated is the mean percentage change in parasitized animals, taking the nonparasitized animals as 100% (± SEM). 1: isolated from differential screening; 2: previously characterized transcripts; *$p < 0.08$ unpaired *t*-test; **$p < 0.05$ unpaired *t*-test, $n = 3$.

The seven clones detected with the differential screening strategy also showed significant changes in the corresponding mRNA levels. These observations validate the differential screening strategy. The levels of the mRNAs encoding the four neuropeptides—pedal peptide, FMRFa, -RFa, and NPY—appeared to be higher in parasitized snails, indicating that up-regulation of these genes occurs.

The mRNA levels of neuropeptides involved in regulation of reproduction and growth showed that CDCH mRNA had decreased significantly in parasitized snails, which does not necessarily mean that protein synthesis completely stopped in these neurosecretory cells. Electronmicroscopy (Fig. 3-2a) of CDCs from parasitized snails shows Golgi bodies that are still active in synthesizing secretory granules. Also, high amounts of secretory material are stored in the cell bodies as well as in their axon terminals in the neurohemal area, the cerebral commissure (Fig. 3-2b). The CDCs are apparently prevented from an afterdischarge so that massive release of their content cannot occur. Schistosomin and/or possibly higher order neurons/neurosecretory cells may be involved in this inhibition. The mRNA for cytochrome P450 (Teunissen et al., 1992), which is abundantly present in the endocrine DBs of non-parasitized snails, did not change significantly in parasitized snails. Electronmicrographs of DB cells from parasitized snails, on the other hand, have been interpreted as showing hyperactive secretory cells (Sluiters, 1981), suggesting that cytochrome P450 is not particularly involved in this elevation of synthetic activity.

The levels of mRNAs encoding the neuropeptides APGWa and conopressin did not change significantly in parasitized snails. These neuropeptides are synthesized in neurons of the male copulation network in the anterior lobe of the right cerebral ganglion whose axons synapse with muscle cells in the penial complex. Surprisingly, the levels of mRNAs encoding the neuropeptides NPY and PP, which are also produced in these neurons, show an increase in parasitized snails. As these two neuropeptides are also synthesized in many other neurons throughout the snail's CNS, this upregulation may also reflect other hitherto unknown functions of these peptides.

The mRNA levels of MIP 1, 3, and 7 did not significantly change. Thus, these MIPs, which may be involved in regulation of growth, are not influenced by the parasite at the level of transcription. However, at least two other known MIPs, MIP 2 and 5, have not yet been studied. For all five MIPs it remains possible that parasites affect transcript translation and/or MIP release.

The mRNA levels of the other peptides not representing neuropeptides, cytochrome c, and 16S RNA, also significantly increased in parasitized snails compared to the non parasitized ones. This may reflect a higher metabolic activity in parasitized snails. For the novel molecule belonging to the Ig superfamily, which is synthesized in the granular cells, a connective tissue cell type, a significant decrease was detected. The functional significance of this observation is not yet clear.

In summary, these data show that the schistosome parasites are able to influence gene expression in their snail host, in particular a number of genes encoding neuropeptides involved in regulation of physiological processes. Whether these changes in neuropeptide gene expression in parasitized snails result from the physiological changes evoked in these snails or from a direct, specific action of parasites on these genes is not clear. There is evidence that E/S products of

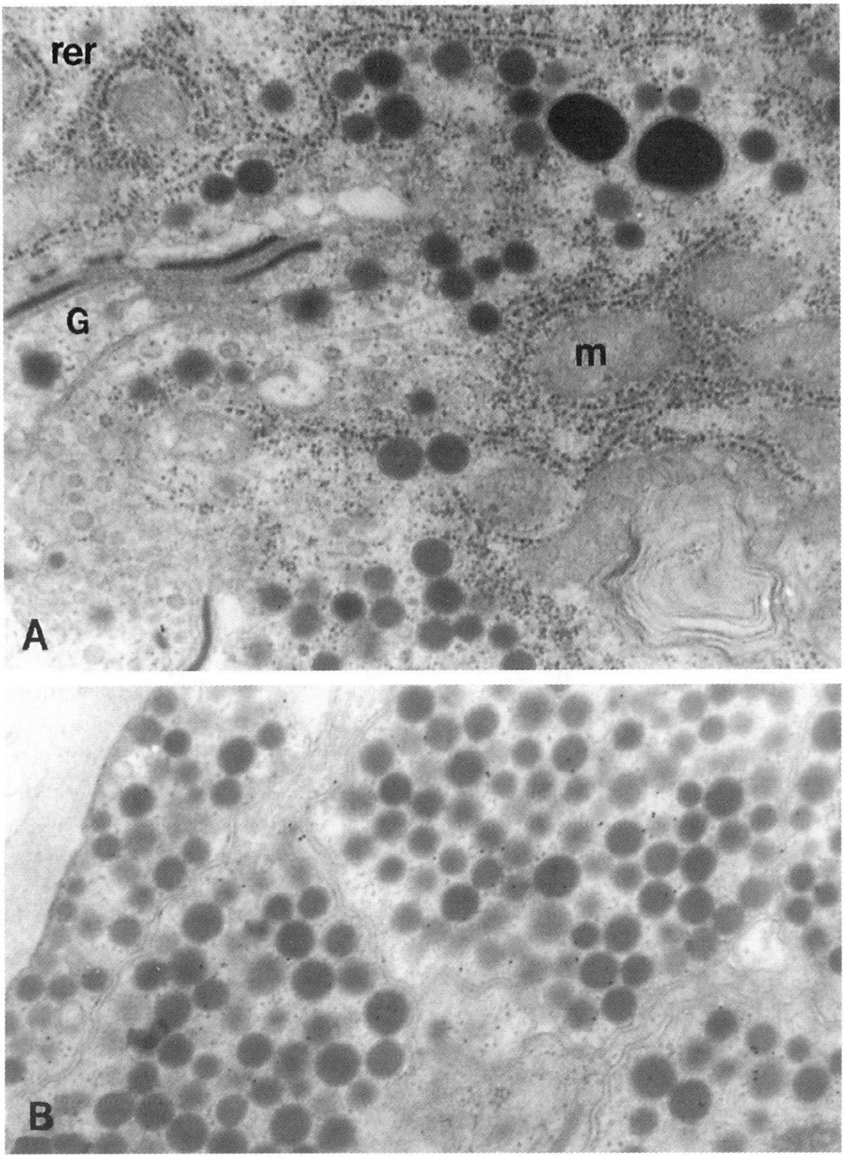

Figure 3-2. Electronmicrographs of caudodorsal cells (CDCs) from a patently parasitized snail. The sections were stained with immuno-gold labelled anti-α CDCP, one of the CDC peptides synthesized in these cells. (× 40,000). (A) The cell body of a CDC. The Golgi bodies (*G*) are still synthesizing secretory granules (two types). *m*, mitochondria; *rer*, rough endoplasmic reticulum. (B) The cerebral commissure (COM), the neurohemal area of the CDCs, from the same snail. The axons are filled with the small type of secretory granules. As in nonparasitized snails, the large electron-dense type of granules remain in the cell bodies.

parasites are involved in acting on hemocytes and/or on other cell types of the IDS in the snail host. Factors derived from cells belonging to the IDS, for instance cytokine-like factors, may then act on gene expression in cells belonging to the NES. Alternatively, parasite-derived factors may directly induce changes in gene expression in the CNS of the host.

V. Interaction between the Internal Defense and Neuroendocrine Systems

Clearly, *T. ocellata* interferes with the two regulatory systems in its snail host, the IDS and the NES. In studying the strategies parasites employ to interfere with these systems we must recognize that in vertebrates these systems intercommunicate. An increasing number of neuropeptides derived from the NES affect immunoreactive cells with neuropeptide receptors displayed on their surface (e.g., Carr, 1992). Neuropeptides are also produced by immunoreactive cells acting as paracrine agents (see e.g., Smith and Johnson, 1991). In addition, factors previously supposed to be exclusively produced by and acting on cells of the immune system, namely cytokines, also exert effects on the NES and are in some cases even produced by it (see Blalock, 1992).

Data on interference between these two systems in invertebrates and in particular in molluscs are still based on circumstantial evidence as the molecules and/or their receptors involved have not yet been identified. Nevertheless, any findings on the communication between the two systems would have consequences for the unravelling of the mechanisms underlying the physiological changes induced by parasites in their host. Changes in one of the two systems must be considered in relation to the activities of the other, which requires extensive knowledge of both systems in the host. Changes in expression of genes encoding certain neuropeptides may be reflected by changed activity of the IDS. In this context the up-regulation of the gene encoding NPY in parasitized snails might be interesting since this neuropeptide has a clear immunosuppressive effect in vertebrates (Elitsur et al., 1994). With this in mind it is very interesting to investigate the origin of schistosomin, which circulates in the hemolymph of parasitized snails as soon as differentiating cercariae are present in the sporocysts (Schallig et al., 1991b), and which has clear effects on the NES. It is also intriguing to investigate how *T. ocellata* can induce its synthesis and release, and in particular which humoral factors are involved.

Origin and Induction of Schistosomin, and Its Interaction with the NES

Schistosomin appeared to be a host-derived factor as it could not be extracted from any of the developmental stages of the parasites (Schallig et al., 1991a). Based on the fact that it could be extracted from CNS preparations (of noninfected snails) it was first supposed to be a neuropeptide. However, it is more likely that schistosomin is synthesized in cells of the connective tissue, for example, in the teloglial cells, which contain secretory granules showing immunostaining with a polyclonal antiserum against native schistosomin (see De Jong-Brink, 1995). This would explain how this factor can be extracted from all snail tissues (see De Jong-Brink, 1995). However, efforts to establish the production sites of schistosomin by *in situ* hybridization techniques have failed because the amount of mRNA per cell is below detection level. Thus indirect evidence indicates that schistosomin is derived from cells in the connective tissue; whether the cells producing schistosomin belong to the IDS of the snail host remains to be determined.

In the connective tissue of *Lymnaea* different cell types have been characterized (Sminia, 1972). Besides fibroblasts, glycogen-storing vesicular connective tissue cells (VCTCs) and calcium cells regulating the acid-base balance in hemolymph, as well as cells that function in the internal defense of the snail are present. These are (1) tissue-dwelling hemocytes, which can move freely from hemolymph to connective tissue and vice versa, because snails have an open circulatory system; (2) endocytosing cells, the reticulum cells lining the blood spaces; and (3) pore cells, which selectively endocytose proteins smaller in size than the pores in their cell membrane. The role(s) of the granular cells in which mRNA of the new molecule of the Ig superfamily has been recently demonstrated (Hoek et al., 1996) remain unknown. In addition, a gene encoding a peptide, granularin, with a high homology to (parts of) molecules involved in protein-protein interaction, e.g., the Von Willebrand factor, has also been localized in these granular cells (Smit et al., 1995). The function of the already mentioned teloglial cells is probably the production of schistosomin. So, besides hemocytes, at least two types of cells in the connective tissue are known to be part of the internal defense system of the animal, and such a role seems very likely for two other types.

In *Lymnaea*, hemocytes occur as "round cells," also named "hyalinocytes," or as "spreading cells" ("granulocytes"; see Van der Knaap and Loker, 1990). In contrast to previously published data on the same snail species, many efforts to distinguish functional subclasses of hemocytes with either morphological or immunocytochemical techniques, using monoclonal antibodies against surface epitopes, were unsuccesful (see Amen et al., 1992b). Therefore, we assume that snails have only one type of hemocyte. The observed heterogeneity has no functional significance other than reflecting differences in development or activity. Since it is not likely that only one cell type can exert all the activities of the IDS, we suggest that a number of functions of the internal defense in snails are exerted by cells in the connective tissue. A recent paper on hemocytes in *Manduca sexta*, an insect that has an open circulatory system like *Lymnaea*, demonstrated that four types of hemocytes could be distinguished not only on the basis of their functional morphology but also on that of the composition of their cell surface (Willott et al., 1994). Thus, in addition to the fat body, which is the main source of defense proteins, the distinguished types of hemocytes are also important for the internal defense in insects. The connective tissue of insects, which mainly consists of fat-storing cells, probably does not play a role in the internal defense. Surprisingly, in the caterpillar *Calpodes ethlius* hemocytes have appeared to be involved in synthesizing peptides targeting the integument (Sass et al., 1994), which also contributes to successful defense (Brey et al., 1993). Hemocytes in molluscs are not known to play a role in the production of the mucus covering their body surface. This mucus also represents part of the defense system containing antibacterial factors (e.g., Otsuka-Fuchino et al., 1992).

The stimulation of the production and release of schistosomin is probably caused by a humoral factor derived from cercariae (Schallig et al., 1992a). In this way E/S products of cercariae are responsible for affecting the reproduction of the host. Earlier studies from Crews and Yoshino (1990) also established the role of E/S products of *S. mansoni* daughter sporocysts in the modulation of a specific aspect of the reproductive physiology of the molluscan host. We suppose that a similar situation occurs in an early stage of *T. ocellata* infected *Lymnaea*. Mother sporocyst E/S products are probably involved in establishing the inhibiting effects on the development of the reproductive tract of the snail host (Sluiter, 1981; De Jong-Brink, 1992). Whether this effect is caused by a direct action of the mother sporocyst E/S products on the NES or is mediated via the IDS–NES route is presently under investigation.

VI. References

Abdul-Salam, J. M., and Michelson, E. H. (1980a). *Biomphalaria glabrata* amoebocytes: Effect on *in vitro* phagocytosis. *J. Invertebr. Pathol.* **35:**241–248.

———. (1980b). *Biomphalaria glabrata* amoebocytes: Assays of factors influencing *in vitro* phagocytosis. *J. Invertebr. Pathol.* **36:**52–59.

Adema, C. M., Van der Knaap, W. P. W. and Sminia, T. (1991a). Molluscan hemocyte-mediated cytotoxicity: The role of reactive oxygen intermediates. *Reviews in Aquatic Sciences* **4:**201–223.

Adema, C. M., Van Deutekom-Mulder, E. C., Van der Knaap, W. P. W., Meuleman, E. A. and Sminia, T. (1991b). Generation of oxygen radicals in hemocytes of the snail *Lymnaea stagnalis* in relation to the rate of phagocytosis. *Dev. Comp. Immunol.* **15:**17–26.

Amen, R. I., Aten, J. A., Baggen, J. M. C., Meuleman, E. A., de Lange-de Klerk, E. S. M. and Sminia, T. (1992b). *Trichobilharzia ocellata* in *Lymnaea stagnalis:* A flow cytometric approach to study its effects on hemocytes. *J. Invertebr. Pathol.* **59:**97–98.

Amen, R. I., Baggen, J. M. C., Bezemer, P. D. and De Jong-Brink, M. (1992a). Modulation of the activity of the internal defence system of the pond snail *Lymnaea stagnalis* by the avian schistosome *Trichobilharzia ocellata. Parasitol.* **104:**33–40.

Amen, R. I., and Meuleman, E. A. (1992). Isolation of mother and daughter sporocysts of *Trichobilharzia ocellata* from *Lymnaea stagnalis. Parasitol. Res.* **78:**265–266.

Bayne, C. J. (1983). Molluscan immunobiology. In *The mollusca* (A. S. M. Salenddin and K. M. Wilbur, eds.), vol. 5, part 2, pp. 407–486. New York: Academic Press.

Bayne, C. J. (1990). Phagocytosis and non-self recognition in invertebrates. *Biosci.* **40:**723–731.

Bayne, C. J., Buckley, P. M. and Dewan, P. C. (1980a). Macrophage-like hemocytes of resistant *Biomphalaria glabrata* are cytotoxic for sporocysts of *Schistosoma mansoni in vitro. J. Parasitol.* **66:**413–419.

———. (1980b). *Schistosoma mansoni:* Cytotoxicity of hemocytes from susceptible snails for sporocysts in plasma from resistant *Biomphalaria glabrata. Exp. Parasitol.* **50:**409–416.

Bayne, C. J. and Yoshino, T. P. (1989). Determinants of compatibility in mollusc-trematode parasitism. *Am. Zool.* **29:**399–407.

Blalock, J. E. (1992). Production of peptide hormones and neurotransmitters by the immune system. In *Neuroimmunoendocrinology* (J. E. Blalock, ed.), Vol. 52, pp. 1–24. Basel: Karger.

Brey, P. T., Lee, W. J., Yamakawa, M., Koizumi, Y., Perrot, S., Francois, M. and Ashida, M. (1993). Role of the integument in insect immunity: Epicuticular abrasion and induction of cecropin synthesis in cuticular epithelial cells. *Procs. Natl. Acad. Sci. USA* **90:**6275–6279.

Carr, D. J. J. (1992). Neuroendocrine peptide receptors on cells of the immune system. In *Neuroimmunoendocrinology* (J. E. Blalock, ed.), vol. 52, pp. 84–105. Basel: Karger.

Cheng, T. C. and Dougherty, W. J. (1989). Ultrastructural evidence for the destruction of *Schistosoma mansoni* sporocysts associated with elevated lysosomal enzyme levels in *Biomphalaria glabrata. J. Parasitol.* **75:**928–941.

Connors, V. A. and Yoshino, T. P. (1990). *In vitro* effect of larval *Schistosoma mansoni* excretory-secretory products on phagocytosis-stimulated superoxide production in haemocytes from *Biomphalaria glabrata. J. Parasitol.* **76:**895–902.

Crews, A. E. and Yoshino, T. P. (1990). Influence of larval schistosomes on polysaccharide synthesis in albumen glands of *Biomphalaria glabrata. Parasitol.* **101:**351–359.

Crews-Oyen, A. E. and Yoshino, T. P. (1995). *Schistosoma mansoni:* Characterization of excretory-secretory polypeptides synthesized *in vitro* by daughter sporocysts. *Exp. Parasitol.* **80:**27–35.

Croll, R. P., Minnen, J. van, Smit, A. B. and Kits, K. S. (1991). APGWamide: Molecular, histological and physiological examination of a novel neuropeptide involved with reproduction in the snail *Lymnaea stagnalis.* In *Molluscan neurobiology* (K. S. Kits, H. H. Boer, and J. Joosse, eds.), pp. 248–254. Amsterdam: North Holland Publishing Comp.

Cushing, J. E., Evans, E. E. and Evans, M. L. (1971). Induced bacterial responses of abalones. *J. Invertebr. Pathol.* **17:**446.

De Jong-Brink, M. (1992). Neuroendocrine mechanisms underlying the effects of schistosome parasites on their intermediate snail host. *J. Invertebr. Repr. Dev.* **22:**127–138.

De Jong-Brink, M. (1995). How schistosomes profit from the stress responses they elicit in their hosts. *Adv. Parasitol.* **35:**177–256.

De Jong-Brink, M., Elsaadany, M. M. and Boer, H. H. (1988). *Trichobilharzia ocellata:* Interference with the endocrine control of female reproduction of its host *Lymnaea stagnalis. Exp. Parasitol.* **65:**91–100.

De Jong-Brink, M., Elsaadany, M. M., Boer, H. H. and Joosse, J. (1986). Influence of trematode parasites upon reproduction activity of their intermediate hosts, freshwater snails. In *Advances in invertebrate reproduction 4* (M. Porchet, J. C. Andries, and A. Dhainaut), pp. 163–172. Amsterdam: Elsevier Science Publishers B. V. (Biomedical Division).

De Jong-Brink, M., Hordijk, P. L., Vergeest, D. P. E. J., Schallig, H. D. F. H., Kits, K. S. and Maat, A. ter (1992). The anti-gonadotropic neuropeptide schistosomin interferes with peripheral and central neuroendocrine mechanisms involved in the regulation of reproduction and growth in the schistosome-infected snail *Lymnaea stagnalis.* In *The peptidergic neuron* (J. Joosse, R. M. Buijs, and F. J. H. Tilders, eds.). *Progress in Brain Research,* Vol. 92, pp. 385–396. Amsterdam: Elsevier Science Publishers (Biomedical Division).

Dikkeboom, R., Bayne, C. J., Van der Knaap, W. P. W. and Tijnagel, J. M. G. H. (1988). Hemocytes of the pond snail *Lymnaea stagnalis* generate reactive forms of oxygen. *Parasitol. Res.* **75:**148–154.

Dikkeboom, R., Tijnagel, J. M. G. H., Mulder, E. C. and Van der Knaap, W. P. W. (1987). Hemocytes of the pond snail *Lymnaea stagnalis* generate reactive forms of oxygen. *J. Invertebr. Pathol.* **49:**321–331.

Elitsur, Y., Luk, G. D., Colberg, M., Gesell, M. S., Dosescu, J. and Moshier, J. A. (1994). Neuropeptide Y (NPY) enhances proliferation of human colonic lamina propria lymphocytes. *Neuropeptides* **26:**289–295.

Fryer, S. E. and Bayne, C. J. (1990). *Schistosoma mansoni* modulation of phagocytosis in *Biomphalaria glabrata. J. Parasitol.* **76:**45–52.

Fryer, S. E., Hull, C. J. and Bayne, C. J. (1989). Phagocytosis of yeast by *Biomphalaria glabrata:* Carbohydrate specificity of hemocyte receptors and a plasma opsonin. *Dev. Comp. Immunol.* **13:**9–16.

Geraerts, W. P. M., Smit, A. B., Li, K. W., Hordijk, P. L. and Joosse, J. (1991). Molecular biology of hormones involved in the regulation of reproduction and growth in molluscs. *Bull. Inst. Zool., Academia Sinica Monograph* **16:**387–440.

Granath, W. O. and Yoshino, T. P. (1983). Lysosomal enzyme activities in susceptible and refractory strains of *Biomphalaria glabrata* during the course of infection with *Schistosoma mansoni. J. Parasitol.* **69:**1018–1026.

———. (1984). *Schistosoma mansoni:* Passive transfer of resistance by serum in the vector snail *Biomphalaria glabrata. Exp. Parasitol.* **58:**188–193.

Hoek, R. M., Smit, A. B., Frings, H., Vink, J. M., De Jong-Brink, M. and Geraerts, W. P. M. (1996). A new Ig superfamily member, molluscan defence molecule (MDM) from *Lymnaea stagnalis* is down-regulated during parasitosis. *Eur. J. Immunol.* **26:**939–944.

Hordijk, P. L., De Jong-Brink, M., Lodder, J. C., Pieneman, A. B., Ter Maat, A. and Kits, K. S. (1992). The neuropeptide schistosomin and hemolymph from parasitized snails induce similar changes in excitability in neuroendocrine cells controlling reproduction and growth in a freshwater snail. *Neurosci. Lett.* **136:**193–197.

Hordijk, P. L., Ebberink, R. H. M., De Jong-Brink, M. and Joosse, J. (1991a). Isolation of schistosomin, a neuropeptide which antagonizes gonadotropic hormones in a freshwater snail. *Eur. J. Biochem.* **195:**131–136.

Hordijk, P. L., Schallig, H. D. F. H., Ebberink, R. H. M., De Jong-Brink, M. and Joosse, J. (1991b). Primary structure and origin of schistosomin, an anti-gonadotropic neuropeptide of the pond snail *Lymnaea stagnalis. Biochem. J.* **279:**837–842.

Joosse, J., Van Elk, R., Mosselman, H., Wortelboer, H. and Van Diepen, J. C. E. (1988). Schistosomin: a pronase-sensitive agent in the hemolymph of *Trichobilharzia ocellata*-infected *Lymnaea stagnalis* inhibits the activity of albumen glands *in vitro. Parasitol. Res.* **74:**228–234.

Kits, K. S. (1980). States of excitability in ovulation hormone–producing neuroendocrine cells of *Lymnaea stagnalis* (Gastropoda) and their relation to the egg-laying cycle. *J. Neurobiol.* **11:**397–410.

Lie, K. J. (1982). Survival of *Schistosoma mansoni* and other trematode larvae in the snail *Biomphalaria glabrata.* A discussion of the interference theory. *Trop. Geogr. Med.* **34:**111–122.

Lloyd, P. E. and Connolly, C. B. (1989). Sequence of pedal peptide: A novel neuropeptide from the central nervous system of *Aplysia. J. Neurosci.* **9:**312–317.

Lodes, M. J., Connors, V. A. and Yoshino, T. P. (1991). Isolation and functional characterization of snail hemocyte–modulating polypeptide from primary sporocysts of *Schistosoma mansoni. Mol. Biochem. Parasitol.* **49:**1–10.

Lodes, M. J. and Yoshino, T. P. (1989). Characterization of excretory-secretory proteins synthesized *in vitro* by *Schistosoma mansoni* primary sporocysts. *J. Parasitol.* **75;**6:853–862.

———. (1990). The effect of schistosome excretory-secretory products on *Biomphalaria glabrata* haemocyte motility. *J. Invertebr. Pathol.* **56:**75–85.

Loker, E. S., Bayne, C. J., Buckley, P. M. and Kruse, K. T. (1982). Ultrastructure of encapsulation of *Schistosoma mansoni* mother sporocysts by hemocytes of juveniles of the 10-R2 strain of *Biomphalaria glabrata. J. Parasitol.* **68:**84–94.

Loker, E. S., Cimino, D. F. and Hertel, L. A. (1992). Excretory-secretory products of *Echinostoma paraensei* sporocysts mediate interference with *Biomphalaria glabrata* hemocyte functions. *J. Parasitol.* **78:**(1):104–115.

Noda, S. and Loker, E. S. (1989). Phagocytotic activity of hemocytes of M-line *Biomphalaria glabrata* snails: Effect of exposure to the trematode *Echinostoma paraensei. J. Parasitol.* **75:**261–269.

Núñez, P. E., Adema, C. M. and De Jong-Brink, M. (1994). Modulation of the bacterial clearance activity of hemocytes from the freshwater mollusc, *Lymnaea stagnalis,* by the avian schistosome, *Trichobilharzia ocellata. Parasitol.* **109:**299–310.

Núñez, P. E., Molenaar, M. J., Lageweg, W., Li, K. W., and Jong-Brink, M. de (1996). Excretory-secretory products of *Trichobilharzia ocellata* and their modulating effects on the internal defence system of *Lymnaea stagnalis.* Parasitology 114.

Otsuka-Fuchino, H., Watanabe, Y., Hirakawa, C., Timiya, T., Matsumoto, J. J. and Tsuchiya, T. (1992). Bactericidal action of a glycoprotein from the body surface mucus of giant African snail. *Comp. Biochem. Physiol.* **101C:**607–613.

Price, D. A. and Greenberg, M. J. (1995). Comparative aspects of FMRFamide gene organization in molluscs. *Neth. J. Zool.* **44**(3–4):421–431.

Richards, E. H. and Renwrantz, L. R. (1991). Two lectins on the surface of *Helix pomatia* hemocytes: A Ca^{2}+-dependent, GalNac-specific lectin and a Ca^{2+}-independent, mannose 6-phosphate-specific lectin which recognizes activated homologous opsonins. *J. Comp. Physiol.* **161:**43–54.

Sass, M., Kiss, A. and Locke, M. (1994). Integument and hemocyte peptides. *J. Insect Physiol.* **40**(5):407–421.

Schallig, H. D. F. H., Hordijk, P. L., Oosthoek, P. W. and De Jong-Brink, M. (1991a). Schistosomin, a peptide present in the hemolymph of *Lymnaea stagnalis* infected with *Trichobilharzia ocellata,* is only produced in the snail's central nervous system. *Parasitol. Res.* **77:**152–156.

Schallig, H. D. F. H., Sassen, M. J. M., Hordijk, P. L. and De Jong-Brink, M. (1991b). *Trichobilharzia ocellata* infection influence on the fecundity of its intermediate snail host *Lymnaea stagnalis* and cercarial induction of the release of schistosomin, a snail neuropeptide antagonizing female gonadotropic hormones. *Parasitology* **102:**85–91.

Sluiters, J. F. (1981). Development of *Trichobilharzia ocellata* in *Lymnaea stagnalis* and the effect of infection on the reproductive system of the host. *Z. Parasitenkd.* **64:**303–319.

Sminia, T. S. (1972). Structure and function of blood and connective tissue cells of the freshwater pulmonate *Lymnaea stagnalis* studied by electron microscopy and enzyme histochemistry. *Z. Zellforsch.* **130:**497–526.

Sminia, T. S., Borghart-Reinders, E. and Van der Linde, A. W. (1974). Encapsulation of foreign material experimentally introduced into the freshwater snail *Lymnaea stagnalis. Cell Tiss. Res.* **153:**307–326.

Smit, A. B., Vreugdenhil, E., Ebberink, R. H. M., Geraerts, W. P. M., Klootwijk, J. and Joosse, J. (1988). Growth-controlling molluscan neurons contain the precursor of molluscan insulin-related peptide. *Nature* **331:**535–538.

Smit, A. B., Li, K. W., van Elk, R. and Geraerts, W. P. M. (1995). Granularin is a peptide from the connective tissue of the central nervous system of *Lymnaea stagnalis* that is structurally related to protein domains of cellular adhesion factors. Submitted.

Smith, E. M. and Johnson, E. W. (1991). Neuropeptides and neuropeptide receptors in the immune system. In *Studies in neuroscience. Comparative aspects of neuropeptide function* (E. Florey, and G. B. Stefano, eds.), pp. 313–316.

Sun, S. C., Lindström, L., Boman, H. G., Faye, I. and Schmidt, O. (1990). Hemolin: An insect immune protein belonging to the immunoglobulin superfamily. *Science* **250:**1729–1732.

Teunissen, Y., Geraerts, W. P. M., Heerikhuizen, H. van, Planta, R. J. and Joosse, J. (1992). Molecular cloning of a cDNA encoding a novel member of the cytochrome P450 family in the mollusc *Lymnaea stagnalis. J. Biochem.* **112:**249–252.

Van der Knaap, W. P. W., Boots, A. M. H., Meuleman, E. A. and Sminia, T. (1985). Search for shared antigens in the schistosome–snail combination *Trichobilharzia ocellata–Lymnaea stagnalis. Z. Parasitenkd.* **71:**219–226.

Van der Knaap, W. P. W. and Loker, E. S. (1990). Immune mechanisms in trematode–snail interactions. *Parasit. Today* **6:**175–182.

Van der Knaap, W. P. W., Sminia, T., Kroesse, F. G. M. and Dikkeboom, R. (1981). Elimination of bacteria from the circulation of the pond snail *Lymnaea stagnalis. Dev. Comp. Immunol.* **5:**21–32.

Van der Knaap, W. P. W., Sminia, T., Schutte, R. and Boerrigter-Barendsen, L. H. (1983). Cytophilic receptors for foreignness and some factors which influence phagocytosis by invertebrate leukocytes: *In vitro* phagocytosis by amoebocytes of the snail *Lymnaea stagnalis. Immunol.* **48:**377–383.

Van Kesteren, R. E., Smit, A. B., Lange, R. P. J. de, Kits, K. S., Golen, F. A. van, Schors, R. C. van der, With, N. D., Burke, J. F. and Geraerts, W. P. M. (1995). Structural and functional evolution of the vasopressin/oxytocin superfamily: Vasopressin-related conopressin is the only member present in *Lymnaea,* and is involved in the control of sexual behaviour. *J. Neurosci.* **15:**5989–5998.

Vreugdenhil, E., Jackson, J. F., Bouwmeester, T., Smit, A. B., Van Minnen, J., Van Heerikhuizen, H., Klootwijk, J. and Joosse, J. (1988). Isolation, characterization and evolutionary aspects of a cDNA clone encoding multiple neuropeptides involved in the stereotyped egg-laying behavior of the freshwater snail *Lymnaea stagnalis. J. Neurosci.* **8:**4184–4191.

Willott, E., Trenczek, T., Thrower, L. W. and Kanost, M. R. (1994). Immunocytochemial identification of insect hemocyte populations: Monoclonal antibodies distinguish four major hemocyte types in *Manduca sexta. Eur. J. Cell Biol.* **65:**417–423.

Yoshino, T. P. and Lodes, M. J. (1988). Secretory protein biosynthesis in snail hemocytes: *In vitro* modulation by larval schistosome excretory-secretory products. *J. Parasitol.* **74:**538–547.

4

Infection with *Echinostoma paraensei* (Digenea) Induces Parasite-Reactive Polypeptides in the Hemolymph of the Gastropod Host *Biomphalaria glabrata*

Coen M. Adema, Lynn A. Hertel, and Eric S. Loker

I. Overview

In addition to affecting the endocrine systems and behavior of their hosts, parasites also have a profound impact on host internal defense systems. We have examined alterations in circulating plasma (cell-free hemolymph) polypeptides of the host snail *Biomphalaria glabrata* following exposure to the digenetic trematode parasites *Echinostoma paraensei* and *Schistosoma mansoni*. Such humoral components are of interest because they may influence the outcome of host-parasite associations and they serve to illuminate the general nature of internal defense responses of invertebrates. Infection of juvenile *B. glabrata* with *E. paraensei* resulted in increased abundance of plasma polypeptides that precipitate parasite antigens present in excretory/secretory (E/S) products derived from cultured sporocysts of *E. paraensei*. When added to 70 μl of plasma from snails infected for 4–14 days, 1 μl of E/S products ("low dose") precipitated a protein complex consisting of two previously described groups of polypeptides referred to as G1M and G2M. Addition of 10 μl of E/S products ("high dose") to 70 μl of plasma from snails infected for 1–8 days precipitated a second complex comprised of covalently and noncovalently associated subunits of 65–70 kDa. The 65–70 kDa complex was also shown to bind to cilia from miracidia of *E. paraensei*. This complex serves as a marker for exposure to *E. paraensei*; it also precipitated from plasma of parasite-exposed snails that did not develop viable infections, such as refractory adults or juveniles exposed to noninfective aged miracidia. The G1M/G2M complex only precipitated from plasma of snails harboring viable *E. paraensei* infections. Plasma from unexposed snails or snails exposed to *S. mansoni* did not form precipitates upon exposure to E/S products, suggesting that production of these protein complexes is specifically provoked by *E. paraensei*.

II. Introduction

Colonization of a host by a parasite places several demands on both symbiotic partners, and this volume specifically focuses on how parasites influence host behavior and endocrinology. It may be advantageous for the parasite to modify the host's energy budget, developmental program, or behavior (Beckage, 1991, 1993; Combes, 1991; Mouritsen and Jensen, 1994; Moore, 1995), yet it must do so without inciting an effective attack by the host's internal defense system (IDS). Thus several parasite-induced changes in host physiology might be simultaneously provoked and it is often challenging to ascertain the causes and purposes of such alterations. This is particularly true of invertebrate hosts in which the underlying homeostatic mechanisms may be less well understood than in their mammalian counterparts.

This chapter examines parasite-induced alterations that are interpreted to represent immunological phenomena. Specifically, we have examined changes that occur in the hemolymph proteins of the freshwater snail *Biomphalaria glabrata* following exposure to infection with the digenetic trematodes *Echinostoma paraensei* and *Schistosoma mansoni*. Although one motivation is to learn more about the specific factors dictating success or failure of larval trematode development in snails, we also hope to reveal basic properties of the molluscan IDS using this model system (Loker and Adema, 1995). Thus the parasites employed can be thought of as probes that facilitate detection and study of immunological phenomena that might otherwise remain cryptic.

The relationships between digeneans and snails are of considerable interest because some digeneans such as *S. mansoni* are widespread human pathogens and novel means to control their impact might be directed at disrupting their ability to establish infection in the molluscan host (Lim and Heyneman, 1972; Woodruff, 1978). Furthermore, digeneans undergo an interesting and complex pattern of development within the snail host, one that features intimate and prolonged contact with host tissues and extensive asexual reproduction (Brown, 1978; Jourdane and Theron, 1987; Lie et al., 1987).

The stages of a typical digenean life cycle are briefly outlined below, using *E. paraensei* as a specific example (Lie and Basch, 1967; see Roberts and Janovy, 1996 for a more general discussion). A free-swimming ciliated miracidium stage hatched from an egg locates a snail host, the freshwater planorbid snail *B. glabrata* in this particular case, and burrows into the head-foot of the snail. The miracidium sheds its ciliated epithelial plates and develops a syncytial tegumental surface. At this point it is referred to as a sporocyst. Sporocysts migrate through the arterial system to reach the host's heart by 2 days postinfection (dpi). Sporocysts attach to the ventricle wall, increase in size and by 8 dpi have produced and released mother rediae. Mother rediae colonize other parts of the snail and by 14 dpi release the next life cycle stage, secondary or daughter rediae. These are typically found in the digestive gland and the ovotestis where they give rise to cercariae that leave the snail by 4 weeks postinfection. Cercariae encyst as metacercariae in another snail and a vertebrate final host (hamsters are used experimentally) becomes infected by ingesting metacercariae which develop directly into hermaphroditic adult worms. These worms produce eggs which are passed in the feces of the definitive host.

One of the typical consequences of larval digenean infection for the snail is partial or complete castration (Crews and Yoshino, 1991; Coustau et al., 1991; de Jong-Brink, 1992). Several profound endocrinological alterations might be expected (see Chapter 3) but the exact mechanisms underlying parasitic castration remain to be elucidated.

Finally such interactions are of interest because larval digeneans exhibit considerable specificity with respect to the snails they are able to colonize (Basch, 1976; Coustau et al., 1990; Preston and Southgate, 1994). Thus the two parasites mentioned above are unable to complete their life cycle in snails other than *Biomphalaria*. Compatibility, or lack thereof, depends on a dynamic immunobiological interaction between host and parasite (Bayne and Yoshino, 1989; Van der Knaap and Loker, 1990).

To the best of our current knowledge, the IDS of molluscs and other invertebrates lack the ability to produce rearranging members of the immunoglobulin superfamily as either humoral or cell surface–associated receptors of non-self (Marchalonis and Schluter, 1990; Millar and Ratcliffe, 1994). Self–non-self discrimination in molluscs nonetheless occurs but the underlying mechanism is still poorly understood. It remains a distinct possibility that molluscs may possess a system for recognition of self, such that parasites and other foreign entities failing to display self signals might be immediately subject to attack (Coombe et al., 1984). More conventionally, however, it is considered that lectins, nonenzymatic proteins with the ability to bind carbohydrates (Brossmer et al., 1992; Knibbs et al., 1993), mediate recognition of non-self in molluscs (Renwrantz, 1986; Olafsen, 1986). Molluscan lectins may circulate as humoral factors or be associated with the surfaces of multipurpose, motile macrophage-like cells called hemocytes (Richards and Renwrantz, 1991). Cellular and humoral components of the molluscan IDS are clearly able to interact collaboratively (Renwrantz and Stahmer, 1983; Yang and Yoshino, 1990a,b). Thus parasites might be bound by humoral lectins that function as opsonins and that increase the rate at which phagocytosis or encapsulation of the parasite by hemocytes occurs. There is evidence that binding of a humoral lectin to a foreign particle induces conformational changes in the lectin that make it more interactive with hemocytes (Fryer et al., 1989; Richards and Renwrantz, 1991). Hemocytes possess batteries of lytic enzymes (Granath and Yoshino, 1983) and are able to generate toxic oxygen radicals as a possible means of killing parasites (Adema et al., 1991; Anderson, 1994).

The full spectrum of humoral factors functioning in IDS in molluscs is not known. Agglutinins have been reported from many molluscs and their ability to clump target cells or to precipitate soluble antigens may also facilitate clearance from the hemolymph by fixed or circulating phagocytic cells (Bayne, 1983, 1990). Humoral factors able to lyse various target cells are also known and may play a role in internal defense (Leippe and Renwrantz, 1988). Additionally, some humoral factors are capable of modulating hemocyte activities (e.g., Granath et al., 1994; Connors et al., 1995). Thus far there has been no convincing demonstration that molluscs are capable of producing suites of inducible microbicidal peptides comparable to those induced in arthropods (Hultmark, 1993; Cociancich et al., 1994; Boman, 1995; Ham et al., 1995) following immunological challenge. Although some molluscs have been shown to possess phenoloxidase activity (Smith and Soderhall, 1991; Coles and Pipe, 1994), convincing evidence that the phenoloxidase pathway plays a substantial role in defense from parasites in this phylum is lacking. Molluscan hemolymph is also not conspicuous for its ability to coagulate, in contrast to the dramatic clotting processes observed in some other invertebrate groups (Iwanaga, 1993).

The roles of several other molluscan humoral factors in internal defense, including aminopeptidase (Cheng et al., 1978), lysozyme (Chu and LaPeyre, 1989), alpha-macroglobulin (Bayne et al., 1992), interleukin–1-like activity (Granath et al., 1994; Connors et al., 1995), tumor necrosis factor-alpha (Ouwemissioukemboyer et al., 1994), and fibronectin (Dirosa et al., 1994), await further study. Another molluscan feature worthy of additional study in the context of defense against parasitism is their copious body surface mucus, which is known to contain antibacterial peptides (Otsukafuchino et al., 1992).

With respect to molluscan defense from larval digeneans, studies using laboratory-derived strains of *B. glabrata* differing in susceptibility to *S. mansoni* (Richards et al., 1992) suggest that humoral factors can significantly influence susceptibility. Plasma derived from resistant strains of *B. glabrata* interacts with hemocytes from susceptible snails to significantly increase their cytotoxic potential for *S. mansoni* sporocysts (Bayne et al., 1980; Granath and Yoshino, 1984). Resistant strains have also been shown to possess agglutinating and opsonizing activities not present in plasma from susceptible snails (Loker et al., 1984; Fryer and Bayne, 1989) and some plasma components from resistant snails not found in susceptible snail plasma are able to bind sporocyst surfaces (Spray and Granath, 1990). Convincing evidence has not been forthcoming that host humoral factors are routinely directly toxic to digenean larvae, however (Fryer and Bayne, 1995).

It is also clear that hemocytes from resistant snails are qualitatively different from those of susceptible snails (Fryer and Bayne, 1989; Granath and Aspevig, 1993; Coustau and Yoshino, 1994; Agner and Granath, 1995) and possess the ability to kill parasites in the absence of host humoral factors (Loker and Bayne, 1982). Because of their obvious role in adhering to, encapsulating, and killing larval digeneans, attention has focused on hemocytes as the primary determinants of compatibility in snail-digenean systems. However, much more study of humoral factors is required to fully understand their role in defense. Of particular importance is the need to identify at the molecular level the specific humoral factors present in molluscs. In developing a more precise understanding of the molluscan IDS, it is also important to consider that the distinction between humoral and cellular components of the molluscan IDS is artificial in the sense that hemocytes are known to release polypeptides that logically would then circulate in the plasma. Hemocyte-secreted molecules include agglutinins and lysins (Leippe and Renwrantz, 1988) and polypeptides that can bind to parasite surfaces (Lodes and Yoshino, 1993). Production of hemocyte-secreted polypeptides could be directly modulated by exposure to parasite products, as shown by Yoshino and Lodes (1988).

The discussion above emphasizes the role of the host IDS in digenean-snail encounters, but a state of compatibility also depends critically on activities of the parasite. Parasites can produce host-like molecules that might prevent immunorecognition (Yoshino and Boswell, 1986; Dissous et al., 1990; Granath and Aspevig, 1993; Weston and Kemp, 1993; Weston et al., 1994). Or, larval digeneans can actively interfere with the function of the host's IDS, particularly hemocytes (Lie et al., 1987; Loker, 1994; Loker and Adema, 1995). These strategies are likely to be very specialized, thus accounting for the specificity of digenean-snail associations.

Regarding the particular model system discussed below, previous studies have shown that following infection with *E. paraensei* the protein content of the hemocyte-free hemolymph, or plasma, of *B. glabrata* increases significantly (Loker and Hertel, 1987). This increase is attributed to specific diffusely banded plasma proteins with the properties of lectins, previously designated as group 1 and group 2 molecules, or G1M and G2M (Uchikawa and Loker, 1992). Although other invertebrates, particularly insects, have been shown to be capable of generating complex, inducible humoral responses effective against a variety of infectious agents (see discussion above), comparable evidence for inducible humoral responses in molluscs has been slow to materialize.

As stated above, lectins have often been suggested as recognition factors in invertebrates, so the heightened levels of plasma lectins observed following *E. paraensei* infection might comprise an inducible defensive function in a mollusc. Both G1M and G2M have been shown to bind to foreign objects, including the surfaces of sporocysts and rediae of *E. paraensei,* further suggestive of a role in internal defense (Hertel et al., 1994).

To further investigate the humoral response of *B. glabrata* to trematode infection, we have examined the interaction between excretory/secretory (E/S) products collected from cultured sporocysts of *E. paraensei* and plasma from either unexposed control or trematode-infected specimens of *B. glabrata.* Such interactions result in the formation of extensive precipitate material (PM). Our analysis of PM has revealed the presence of previously observed humoral factors as well as newly identified plasma polypeptides that are reactive with *E. paraensei* antigens.

The composition of PM is monitored at different times postinfection and the production of PM is compared in juvenile or adult snails exposed to different doses of infection. Snail age is a relevant parameter because as *B. glabrata* grow and become adults, they become progressively more refractory to infection with *E. paraensei* (Loker et al., 1987). The exposure dose was also varied because previous studies indicated that the number of *E. paraensei* miracidia used to infect snails influenced agglutination titers in the plasma of *B. glabrata* (Loker et al., 1994). The age of the eggs used as a source of *E. paraensei* miracidia was also varied because the infectivity of miracidia is significantly lower when derived from aged eggs (DeGaffe, 1994). The PM responses observed with plasma derived from *E. paraensei*–infected snails are also compared with those obtained from snails exposed to *S. mansoni*. These studies complement other work suggesting the response of *B. glabrata* to these two parasites is very different (reviewed by Loker and Adema, 1995).

III. Material and Methods

A. Parasites and Hosts

Echinostoma paraensei was maintained in Syrian hamsters and in the M-line strain of *Biomphalaria glabrata* according to Loker and Hertel (1987). *Schistosoma mansoni* was maintained in mice and in the M-line strain of *Biomphalaria glabrata* as described by Stibbs and colleagues (1979).

Unless otherwise stated, the snails used for this investigation were juveniles (6–8 mm shell diameter) of the M-line strain. Juvenile M-line *B. glabrata* snails are susceptible to infection with both *E. paraensei* and *S. mansoni*. For some experiments, adult M-line snails of 12 mm or 20 mm shell diameter were exposed to *E. paraensei* infection; adult snails are relatively resistant to infection (Loker et al., 1987).

B. Infection of Snails

Individual snails were placed in wells of a 24-well tissue culture plate and exposed to recently hatched miracidia of either *E. paraensei* (50 miracidia) or *S. mansoni* (5 miracidia). *Echinostoma paraensei* miracidia were derived from eggs that had incubated for the minimum incubation period of 2 weeks at 24°C. Presence of developing sporocysts and/or rediae in the ventricles of exposed snails was used to confirm that exposed snails had actually become infected.

C. Collection of Excretory/Secretory Products (E/S)

E/S products were collected using general procedures and media previously described (Loker et al., 1992). A volume of 1 ml of culture medium (CM: Medium 199 [Sigma, St. Louis, MO], diluted 1:1 [v/v] with distilled water, 45 ug/ml gentamycin) conditioned by approximately 10,000 transforming sporocysts was collected after 24 h of culture, and centrifuged (16,000 × g for 3 min) through a Z-spin filter unit (Gelman Sciences; Ann Arbor, MI) to remove ciliated plates, cilia, and other particulate debris. Conditioned medium was then concentrated

20-fold in 10 kDa cut-off Centricon units (Amicon Inc.; Beverly, MA) to a volume of 50 μl and thereafter was referred to as E/S products.

D. Collection and Analysis of PM Obtained from Snail Plasma

Hemolymph was obtained from the desired group of snails by cardiac puncture. For this procedure the shell of the snail was cleaned and partially removed in the area over the heart. The heart was then punctured and hemolymph was collected with a drawn-out glass Pasteur pipette. The samples were pooled and held on ice until the required volume was obtained. Hemocytes and subsequently remaining extraneous particulate matter were removed by centrifugation (for 10 min at 110 × g and for 10 min at 16,000 × g, respectively) and the resulting supernatant was designated as "plasma."

E/S products and plasma were then mixed together to determine if PM would form. Initially the volumes of both plasma (from 50–200 μl) and E/S products (1–20 μl) mixed together were varied to determine conditions favoring formation of PM. Based on these trials, a standard protocol for collection of PM was devised as follows: 1 μl E/S products was first added to a 70 μl plasma sample for 1 h (25C) and any PM that formed ("low dose" PM) was removed by centrifugation (10 min at 16,000 × g). The supernatant received an additional volume of 10 μl of SEP and after an additional 1 h of incubation any resultant PM ("high dose" PM) was again collected by centrifugation. Using this protocol distinct plasma protein complexes could be monitored by their different preferences for binding E/S product components. Plasma obtained from unexposed control snails or from snails with infections of *E. paraensei* or *S. mansoni* were examined using this approach.

PM pellets were solubilized in SDS-PAGE sample buffer with or without beta-mercaptoethanol or with native sample buffer (BioRad, Hercules, CA) and visualized by silver staining following separation on 5–20 percent gradient gels (Loker and Hertel, 1987). Some PM samples were subjected to SDS-PAGE under reducing conditions, transferred to nitrocellulose (Biotrace NT; Gelman Sciences, Ann Arbor, MI). Blots were blocked with 3 percent (w/v) gelatin and probed with a polyclonal antiserum (1/100) raised in rabbits against E/S products obtained from cultured *E. paraensei* sporocysts. This antiserum recognizes several proteins from the E/S mixture as can be seen in Fig. 4-1. Recognized proteins were visualized using alkaline phosphatase labeled goat anti-rabbit IgG (Sigma; St. Louis, MO) as secondary antibody (1/1000) and NBT/BCIP as substrate (modified from Biorad Immuno-Blot Assay, Biorad; Hercules, CA; Loker et al., 1992).

E. PM Formation in Plasma Derived from Snails with E. paraensei *Infections of Different Duration*

Juvenile snails were exposed to *E. paraensei* as described above; infection was confirmed by visual inspection of the hearts of exposed snails. Pools of plasma

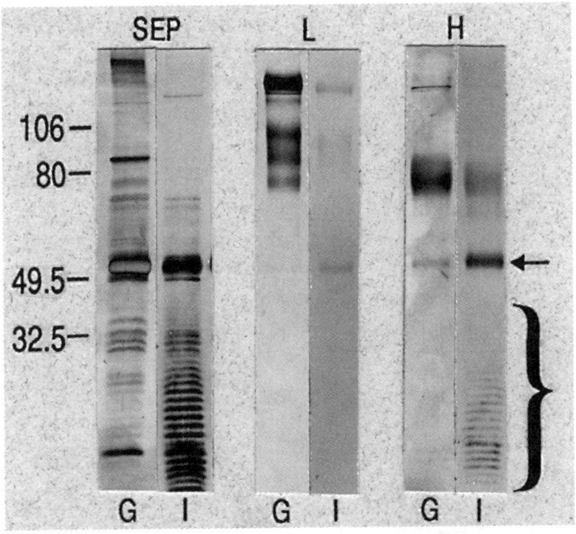

Figure 4-1. SDS-PAGE gel lanes (*G*) and corresponding immunoblots probed with rabbit anti-E/S products (*I*) of secretory-excretory products from *E. paraensei* sporocysts, low-dose (*L*) and high-dose (*H*) PM obtained from plasma of snails with 4-day-old *E. paraensei* infections. Note E/S components that also appear in PM: a 51 kDa band (*arrow*) and a group of regularly-spaced bands (*bracket*). Molecular weights indicated are in kDa.

were obtained from a minimum of three snails bled at 1, 4, 7, 14, 20, or 27 days postinfection and from unexposed controls. Plasma was mixed with low and high doses of SEP and precipitates collected and analyzed as described above.

F. PM Formation in Plasma Derived from E. paraensei*–Infected Snails of Different Sizes and with Different Infection Doses*

For this experiment the following groups of *E. paraensei*–exposed snails were used: juvenile snails individually exposed to 2, 10, or 50 miracidia; 12 mm or 20 mm shell diameter adult snails exposed individually to 50 miracidia; or juvenile snails exposed individually to 50 "aged" miracidia. Aged miracidia were derived from eggs held for more than 4 weeks prior to hatching; these miracidia swim normally and penetrate snails but do not colonize the ventricle and establish viable infections (DeGaffe, 1994). Prior to collection of plasma, snails from each group were examined. Ventricular sporocysts were noted in juvenile snails but not in adult snails exposed to normal sporocysts, as anticipated based on results obtained by Loker and coworkers (1987). Sporocysts were also not observed in juveniles exposed to aged miracidia. Pools of plasma derived from at least three snails in each group were collected at 3, 4, and 8 days following exposure to *E. paraensei*, mixed with E/S products, and precipitates collected and analyzed as described above.

G. Binding of the 65–70 kDa Complex to Cilia of Miracidia of E. paraensei

Cilia and ciliated plates from *E. paraensei* were used to investigate whether the 65–70 kDa complex also binds particulate antigens. When allowed to settle from suspension, transforming *E. paraensei* miracidia are effectively separated from shed cilia and ciliated plates since these have a lower sedimentation rate. The supernatant containing cilia and ciliated plates was washed three times in CM (4 min, 12,000 × g) to remove E/S components and incubated in a volume of 30 μl of plasma obtained from juvenile (6–8 mm) unexposed control snails or from snails infected with *E. paraensei* for 4 days. After 2 h they were centrifuged, washed three times in CM, transferred to clean tubes, solubilized, and analyzed by SDS-PAGE.

IV. Results

A. Formation and Structure of PM

PM did not form in plasma from unexposed control snails to which 1 μl and 10 μl aliquots of SEP were sequentially added (Fig. 4-2).

Addition of 1 μl of SEP to plasma from snails infected for 4 days with *E. paraensei* was followed within 1 h by the appearance of PM ("low dose"). Low-dose PM, when dissolved in SDS-PAGE sample buffer containing beta-

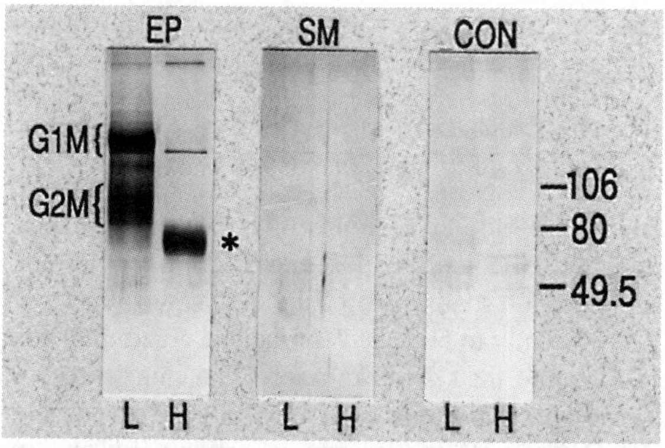

Figure 4-2. Separation on 5–20 percent gradient gels of solubilized low-dose (*L*) and high-dose (*H*) PM obtained from plasma of snails infected for 4 days with either *Echinostoma paraensei* (*EP*) or *Schistosoma mansoni* (*SM*) and from control snails (*CON*). See text for methods used to obtain PM. Molecular weight markers are indicated in kDa. For *E. paraensei*–infected snails, brackets indicate G1M and G2M components from low-dose PM, the 65–70 kDa band in high-dose PM is indicated with an asterisk (*). The focused band observed in the high-dose PM at 190 kDa represents hemoglobin, an abundant plasma protein present as a minor contaminant in PM.

mercaptoethanol, contained two previously recognized groups of diffusely banded plasma molecules, of 200 and 80–120 kDa, designated as G1M and G2M, respectively (Uchikawa and Loker, 1992; Figs. 4-2, 4-3). When placed in SDS-PAGE sample buffer lacking beta-mercaptoethanol, G1M again migrated as a separate 200 kDa band. Under similar conditions, the 80–120 kDa G2M band disappeared and was replaced by two > 200 kDa bands. Under native conditions, G1M and G2M were associated into a band of approximately 1,300 kDa, designated as the native G1M/G2M complex (see Fig. 4-2). The pattern obtained suggests that G2M are covalently associated with other G2M to form > 200 kDa molecules that interact noncovalently with G1M to form a complex with native molecular weight of 1,300 kDa.

Subsequent to the initial formation of low-dose PM, only trace amounts of the G1M/G2M complex could be precipitated from the same plasma sample by adding further 1 μl aliquots of E/S products; if increasing doses of E/S products were added, only the high-dose PM described below began to form. Examination of plasma supernatants following removal of low-dose PM revealed additional nonreactive G1M/G2M components to be present.

Following removal of low-dose PM by centrifugation, addition of a 10 μl aliquot of E/S products to the supernatant resulted in formation of a second precipitate ("high dose") that contained a prominent 65–70 kDa band on SDS-PAGE gels containing beta-mercaptoethanol (see Figs. 4-2, 4-3). In SDS-PAGE buffer lacking beta-mercaptoethanol, high-dose PM migrated as a band of > 200

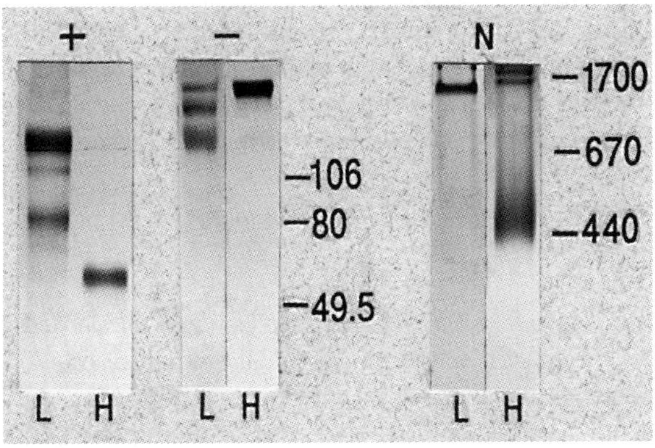

Figure 4-3. Separation on 5–20 percent gels of low-dose (*L*) and high-dose (*H*) PM obtained from plasma of snails infected with *E. paraensei* for 4 days. PM was dissolved in SDS-PAGE buffer with (+) or without beta-mercaptoethanol (−) added, or in native sample buffer (*N*). Molecular weight markers (kDa) are shown for each set of gels. The migration of *B. glabrata* hemoglobin was also used for calibration of native gels, assuming a native molecular weight of 1,700 kDa (similar to hemoglobin of the related planorbid species, *Helisoma trivolvis*; Terwilliger et al., 1976).

kDa (see Fig. 4-3). On native gels, a band of approximately 1,600 kDa was noted and was interpreted as incompletely dissolved high-dose PM.

Also evident on native gels was a broad smear centered at about 400 kDa, interpreted as dissolved high-dose PM (see Fig. 4-3). The pattern suggests that 65–70 kDa subunits are covalently associated, probably into hexamers, to form a molecule of circa 400 kDa. These hexamers are further associated noncovalently to form the 1,600 kDa complex.

Following removal of high-dose PM by centrifugation, only small amounts of the 65–70 kDa complex could still be harvested from plasma by adding additional aliquots of E/S products. Plasma supernatants remaining following removal of high-dose PM contained little 65–70 kDa material, suggesting it had been mostly removed by addition of E/S products.

Aliquots of 1 and 10 μl of *E. paraensei* sporocyst E/S products did not stimulate formation of PM when added to plasma from snails exposed to *S. mansoni* (see Fig. 4-2); this was true for snails exposed to *S. mansoni* for 1, 4, or 8 days.

Both low- and high-dose PM derived from plasma of *E. paraensei*–infected snails were probed on immunoblots with anti-E/S products to determine if E/S components were present within PM (Fig. 4-1). Both precipitates contained a band of 51 kDa that comigrated with a prominent component of E/S products. Although this band was typically not visible on silver-stained gel lanes of either precipitate (e.g., Figs 4-2, 4-3), it did appear in the lane loaded with high-dose PM in Figure 4-1. Also apparent in both high-dose PM and E/S products immunoblot lanes was a peculiar assemblage of regularly spaced, low molecular weight bands that appear only faintly or not at all on the corresponding silver-stained lanes. The anti-E/S products also bound to the 65–70 kDa band found in high-dose PM. This is interpreted as a cross-reaction because this band does not appear in the E/S lane; also, the massive amount of protein present in the 65 kDa band would favor nonspecific binding of the anti-E/S polyclonal antiserum.

B. *PM Formation in Plasma Derived from Snails with* E. paraensei *Infections of Different Durations*

Neither low- nor high-dose PM formed in plasma of unexposed control snails (Fig. 4-4). By 1 day postinfection (dpi) a small amount of 65–70 kDa complex was evident and was unusual for its precipitation following only 1 μl of E/S products. By 4 dpi, a prominent low-dose PM containing G1M/G2M and a high-dose PM containing the 65–70 kDa complex were present. By 7 dpi, only a low-dose PM containing G1M/G2M could be obtained from plasma. This persisted through 14 dpi and then disappeared (see Fig. 4-4). Other experiments have produced small amounts of low-dose PM containing G1M/G2M and abundant high-dose PM containing the 65–70 kDa complex at 3 dpi (Fig. 4-5a). Also, high-dose PM containing 65–70 kDa has been obtained from plasma taken from snails as late as 8 dpi (data not shown).

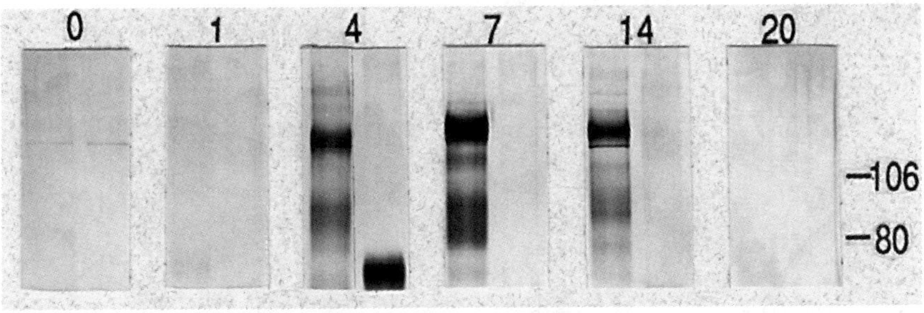

Figure 4-4. SDS-PAGE gel lanes loaded with PM obtained from plasma of snails infected with *E. paraensei* for 0 (unexposed controls), 1, 4, 7, 14, or 20 days. For each time point, the left lane indicates the low-dose PM and the right lane the high-dose PM. The kDa values of the molecular weight markers are shown.

C. PM Formation in Plasma Derived from Snails With E. paraensei *Infections of Snails of Different Sizes and Infection Doses*

At 3 days post exposure, plasma from most groups of snails showed only modest PM following addition of E/S products (see Fig. 4-5*a*). Prominent high-dose PM containing the 65–70 kDa complex formed in plasma from juvenile and 12 mm adult snails exposed to 50 miracidia each. A small amount of low-dose PM containing G1M was visible in some lanes. Also, it was unusual as compared to previous experiments to see small quantities of 65–70 kDa material appear in low-dose PM lanes and small quantities of G1M appear in high-dose PM lanes. Plasma from juvenile snails exposed to low infection doses of 2 or 10 miracidia, from 20 mm adults (more refractile than juveniles) exposed to 50 miracidia, and juveniles exposed to aged miracidia showed less PM response than the other groups of snails.

By 4 days post exposure, more prominent PM was noted in most lanes (Fig 4-5*b*). Of particular interest was the appearance of high-dose PM containing the 65–70 kDa complex in plasma from all the groups of exposed snails. Plasma from juveniles exposed to increasing numbers of miracidia had increasing amounts of low-dose PM containing G1M/G2M. A similar pattern of results was obtained for plasma derived from snails at 8 days post exposure (data not shown). At 4 dpi, only traces of low-dose PM could be found in plasma from exposed adult snails and none was recovered from plasma of juveniles exposed to aged miracidia.

D. Binding of the 65–70 kDa Complex to Cilia of Miracidia of E. paraensei

Cilia derived from *E. paraensei* miracidia, when incubated in plasma from *E. paraensei*–infected snails, were encased by large amounts of PM (Fig. 4-6*a*). When this PM was pelleted and analyzed by SDS-PAGE, the 65–70 kDa complex was found to be present (Fig. 6*b*). No PM formation occurred when cilia were placed in plasma from unexposed control snails.

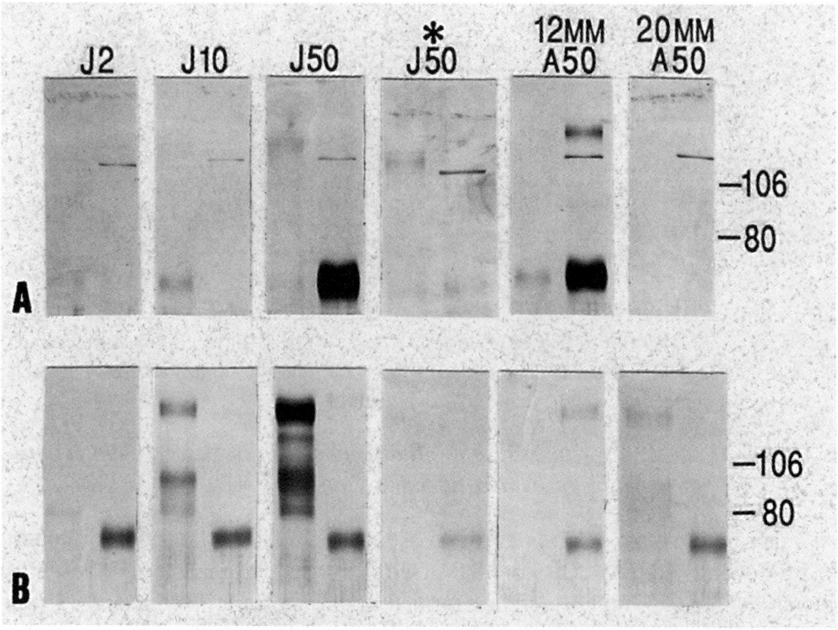

Figure 4-5. SDS-PAGE gel lanes loaded with PM obtained from plasma of 6 mm shell diameter juvenile (*J*) snails exposed to 2, 10, or 50 freshly hatched and infective miracidia, or to 50 aged miracidia (*) of *E. paraensei*. Also adult (*a*) snails with a shell diameter of either 12 mm or 20 mm were exposed to 50 freshly-hatched infective miracidia. (*a*) PM obtained from plasma of the different groups of snails at 3 days postexposure (dpe). (*b*) PM obtained from plasma of these groups of snails at 4 dpe. For each group of snails, the left lane indicates low-dose PM and the right lane high-dose PM. The molecular weight markers are in kDA.

V. Discussion

Secretory-excretory products derived from cultured *E. paraensei* sporocysts have been used as a stimulus to provoke precipitation reactions in the plasma of *B. glabrata*. This approach has identified two distinct reactive protein complexes in plasma from *E. paraensei*–infected snails that are either lacking or unresponsive in plasma from unexposed control snails or *S. mansoni*-infected snails. The 65–70 kDa protein complex is the first to appear following exposure to infection. Its constituent subunits are both covalently and noncovalently associated into high molecular weight complexes of around 1,600 kDa. Present within precipitates containing the 65–70 kDa complex are a prominent E/S polypeptide of 51 kDa and a peculiar ladder-like assemblage of bands noted in E/S products. That the 65–70 kDa complex usually precipitates from plasma only following addition of high doses of E/S products suggests that the ratio of the reactants must be

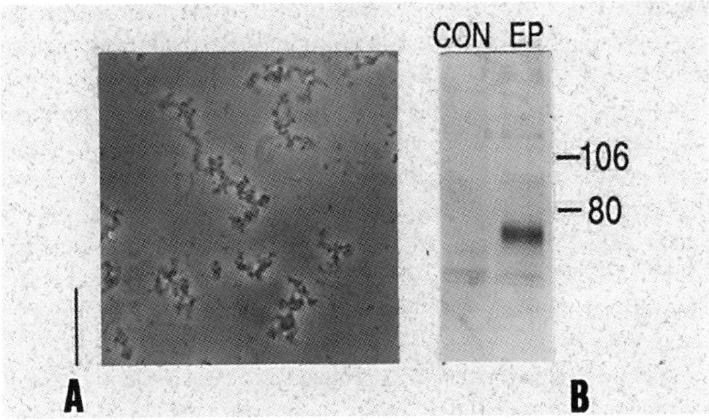

Figure 4-6. (*a*) Photograph showing PM formation in plasma from *E. paraensei*–infected snails (4 days postinfection) to which cilia from transforming *E. paraensei* miracidia have been added. Scalebar = 10 μm. (*b*) SDS-PAGE gel lanes of PM collected from plasma of control snails (*CON*) and *E. paraensei*–exposed snails (*EP*) to which cilia of *E. paraensei* miracidia have been added. Note the presence of a 65–70 kDa polypeptide in the latter lane. Positions of molecular weight (kDa) markers are shown.

optimized before precipitation occurs, implying that the precipitate formation is analogous to an antigen-antibody precipitation reaction. This is further supported by the demonstration that the 65–70 kDa complex is able to bind to particulate parasite antigens such as cilia from miracidia.

Previous studies to identify components within *B. glabrata* plasma that bind to monosaccharide affinity columns, latex beads, mammalian erythrocytes, or *E. paraensei* sporocysts and rediae failed to reveal the 65–70 kDa complex (Uchikawa and Loker, 1992; Hertel et al., 1994). Further studies are underway to characterize more precisely the nature of the 65–70 kDa subunits.

Appearance of this protein complex is associated with exposure to *E. paraensei*, regardless of whether the invading parasites succeed or fail in establishing a successful infection. Thus plasma from all groups of exposed M-line snails, including refractory adults and juveniles exposed to noninfective miracidia, contained precipitable 65–70 kDa complex material. The appearance of this band may therefore serve as a marker for exposure to *E. paraensei* infection.

By 4 dpi a second protein complex, the G1M/G2M complex, can be precipitated from plasma of infected snails by low doses of E/S products. This continues to be present until about 14 dpi. Composition of low-dose PM is very similar to that of particulate material that circulates in the hemolymph of *E. paraensei*–infected snails (Loker and Hertel, 1987); both G1M and G2M are prominent components of this in vivo precipitate. Low-dose PM also contains an E/S component of 51 kDa, again suggesting that the G1M/G2M is reactive with parasite antigens.

G1M and G2M have also been identified previously as monosaccharide-inhibitable plasma polypeptides that increase in abundance following exposure to *E. paraensei* (Monroy et al., 1992). Abundance of G1M and G2M in plasma following infection has been measured previously by monitoring agglutination titers for mammalian erythrocytes (Couch et al., 1990; Loker et al., 1994). Although the different binding targets used in these studies account for some discrepancies in timing, generally similar temporal patterns were observed; an observable increase in G1M/G2M occurred within a few days postinfection, peaked between 6 and 8 dpi and then declined by 16 dpi. Also, G1M and G2M have been shown to bind to various non-self entities including mammalian erythrocytes and *E. paraensei* sporocysts and rediae (Hertel et al., 1994).

G1M and G2M precipitated by E/S products are noncovalently associated into a high molecular weight complex. Other studies of G1M and G2M suggest they behave independently of one another as determined by their different binding preferences for various target particles (Hertel et al., 1994). G1M and G2M may become loosely associated following interaction with E/S products or they may exist in multiple forms in plasma, only some of which contain molecules of both groups. Heterogeneity among components of the G1M/G2M complex is implied by the observation that addition of repeated aliquots of E/S products to plasma is not able to deplete plasma of G1M or G2M.

The G1M/G2M complex did not precipitate detectably from plasma of exposed adult snails or juveniles exposed to aged miracidia. Such snails did not contain developing intraventricular parasites. Also, the amount of G1M/G2M precipitated from plasma of susceptible juvenile snails exposed to fully infective miracidia increased with dosage of miracidia. G1M/G2M abundance as measured by agglutination titer increased significantly in snails exposed to 10 or 100 miracidia of *E. paraensei* as compared to snails exposed to 1 miracidium (Loker et al., 1994).

The results of the present study suggest that the presence of G1M/G2M complex in E/S products-provoked precipitates is an indication that the snail harbors successfully developing intraventricular parasites. This observation is supported by previous studies in which titers of G1M- or G2M-mediated agglutination of mammalian erythrocytes were highest for snails with thriving infections of *E. paraensei* (Loker et al., 1994).

What is the functional significance of the polypeptides identified above to the interactions between *E. paraensei* and *B. glabrata*? The following possibilities can be considered.

First, the polypeptides may not be part of an immune response but are released into the host's hemolymph as a consequence of tissue damage or some other physiological disturbance resulting from *E. paraensei* infection. This possibility does not seem likely because they are found relatively early during the course of infection when the invading parasites are small and few in number. Tissue damage is minimal at this time. Abundance of these polypeptides wanes in later stages of infection when pathological effects resulting from the presence of numerous redial stages would be more likely or when physiological perturbations resulting from the presence of extensive parasite biomass would be more pronounced.

Second, they may represent host defense molecules whose production is stimulated by exposure to, and subsequent infection with, *E. paraensei*, yet are not part of an effective host response. As part of a deliberate strategy of immune evasion, the parasite might provoke the production of polypeptides that are incapable of adversely affecting parasite development. As noted by Ham and colleagues (1995), increased levels of humoral factors following infection could signal a lack of, or loss of, functional binding of that component to the parasite. By stimulating an irrelevant response, the parasite may effectively deny the host the resources to mount an appropriate response. Although this possibility can not be ruled out, it does not seem adaptive to either host or parasite. For example, in a host with an energy budget already severely compromised by the parasite's presence, an additional expenditure of energy on a superfluous IDS response would seem to have distinct costs for the parasite as well.

Third, the observed polypeptides might represent an effective and relevant response to the presence of *E. paraensei* because they are able to bind to and inactivate soluble parasite products such as factors that overtly interfere with hemocyte function, toxic metabolites or waste products, or factors that mediate parasitic castration. If this is the case, although it is possible that the induced host polypeptides might reduce the concentration of parasite-produced factors, it is also clear that they are not capable of preventing castration or impaired hemocyte function (Loker, 1994; Loker and Adema, 1995) or of eliminating infection. Also, one would have to postulate that the presence of induced polypeptides early in the course of infection implies that the putative parasite factors are produced predominantly at that time. This does not seem likely, particularly in the case of parasite waste products, which would be expected to increase in abundance as the infection progressed.

Fourth, these plasma polypeptides may be capable of binding to and inactivating a variety of opportunistic pathogens as a way to compensate for damage inflicted upon other defense components, particularly hemocytes (Lie, 1982), by *E. paraensei* infection (Loker, 1994). Again the evident production of these polypeptides early in the course of infection would seem to be perplexing but one possible explanation is that they are needed only during phases of the life cycle in which the sporocyst predominates. Later, as rediae are produced in massive numbers, polypeptide production wanes. Rediae possess a mouth and a gut and are known to be capable of attacking other parasites (Sousa, 1993). They may assume partial responsibility for defense of the host-parasite complex from opportunistic parasites (Loker, 1994). Also, hemocytes in infected hosts may regain some ability to spread normally and engage in defense activities (Noda and Loker, 1988). As argued by Loker and Adema (1995), the failure of *S. mansoni* infection to provoke comparable changes in plasma polypeptide composition may be related to the fact that this parasite apparently does not interfere as extensively with host hemocytes as does *E. paraensei* and may rely more on a strategy of disguise to evade host responses.

In conclusion, mixture of E/S products derived from *E. paraensei* sporocysts with plasma from *E. paraensei*–infected snails has revealed the presence of antigen-reactive plasma protein complexes that have escaped detection using other methods of analysis. Furthermore, because this method results in precipitation of these protein complexes, it provides a convenient way to harvest relatively pure preparations of these trematode-induced humoral factors from plasma. Further study is needed to reveal the functional significance to the snail of their elaboration. Although our working assumption is that the described snail plasma proteins serve the snail in an immunological context, it is important to allow for the possibility that they may function in some entirely unforeseen context pertaining to other facets of the host-parasite relationship.

We must also retain a flexible attitude about invertebrate immunobiology in general because for many phyla, including the Mollusca, key components of the IDS remain poorly known and molecular-level investigations simply have not been undertaken. For example, an immunoglobulin superfamily member has recently been identified from the snail *Lymnaea stagnalis* (see Chapter 3 of this volume), the first report of immunoglobulin-like molecules in molluscs. Also, it should be noted that recent studies discussed by Drickamer (1995) have identified immunoglobulin superfamily members with lectin activity (I lectins), thus obscuring a dichotomy between lectins and immunoglobulins previously emphasized in the invertebrate immunobiological literature. However, at this point in time, none of the described invertebrate immunoglobulin superfamily members is believed to function as a lectin in an immunological context. Equally startling discoveries pertaining to non-self recogniton and other fundamental aspects of invertebrate IDS undoubtedly lie ahead, particularly as the tools of molecular biology are brought to bear, and will facilitate interpretation of observations such as those described in this paper.

VI. Acknowledgments

The work presented in this study was supported by NIH grant AI 24340.

VII. References

Adema, C. M., van der Knaap, W. P. W. and Sminia, T. (1991). Molluscan hemocyte-mediated cytotoxicity: The role of reactive oxygen intermediates. *Crit. Rev. Aquat. Sci.* **4**:201–223.

Agner, A. E. and Granath, W. O. (1995). Hemocytes of schistosome-resistant and -susceptible *Biomphalaria glabrata* recognize different antigens on the surface of *Schistosoma mansoni* sporocysts. *J. Parasitol.* **81**:179–186.

Anderson, R. S. (1994). Hemocyte-derived reactive oxygen intermediate production in four bivalve mollusks. *Dev. Comp. Immunol.* **18:**89–96.

Basch, P. F. (1976). Intermediate host specificity in *Schistosoma mansoni. Exp. Parasitol.* **39:**150–169.

Bayne, C. J. (1983). Molluscan immunobiology. In *The mollusca.* (A. S. M. Saleuddin, and K. M. Wibur, eds.), Vol. 5, part 2, pp. 407–486. San Diego: Academic Press.

Bayne, C. J. (1990). Phagocytosis and non-self recognition in invertebrates. *Bioscience.* **40:**723–731.

Bayne, C. J., Bender, R. C. and Fryer, S. E. (1992). Proteinase inhibitory activity in plasma of a mollusc: Evidence for the presence of alpha-macroglobulin in *Biomphalaria glabrata. Comp. Biochem. Physiol.* **102B:**821–824.

Bayne, C. J., Buckley, P. M. and DeWan, P. C. (1980). *Schistosoma mansoni:* Cytotoxicity of hemocytes from susceptible snail hosts for sporocysts in plasma from resistant *Biomphalaria glabrata. Exp. Parasitol.* **50:**409–416.

Bayne, C. J. and Yoshino, T. P. (1989). Determinants of compatibility in mollusc–trematode parasitism. *Am. Zool.* **29:**399–407.

Beckage, N. E. (1991). Minireview: Host-parasite hormonal relationships: A common theme? *Exp. Parasitol.* **72:**332–338.

Beckage, N. E. (1993). Endocrine and neuroendocrine host-parasite relationships. *Receptor.* **3:**233–245.

Boman, H. G. (1995). Peptide antibiotics and their role in innate immunity. *Ann. Rev. Immunol.* **13:**61–92.

Brossmer, R., Wagner, M. and Fischer, E. (1992). Specificity of the sialic acid-binding lectin from the snail *Cepaea hortensis. J. Biol. Chem.* **267:**8752–8756.

Brown, D. (1978). Pulmonate molluscs as intermediate hosts for digenetic trematodes. In *The pulmonates,* 1st ed. (V. Fretter, and J. Peake, eds.), Vol. 2, pp. 289–333. London: Academic Press.

Cheng, T. C., Lie, K. J., Heyneman, D. and Richards, C. S. (1978). Elevation of aminopeptidase activity in *Biomphalaria glabrata* (Mollusca) parasitized by *Echinostoma lindoense* (Trematoda). *J. Invertebr. Pathol.* **31:**57–62.

Chu, F. E. and La Peyre, J. F. (1989). Effect of environmental factors and parasitism on hemolymph lysozyme and protein of American oysters *(Crassostrea virginica). J. Invertebr. Pathol.* **54:**224–232.

Cociancich, S., Bulet, P., Hetru, C. and Hoffman, J. A. (1994). The inducible antibacterial peptides of insects. *Parasitology Today* **10:**132–139.

Coles, J. A. and Pipe, R. K. (1994). Phenoloxidase activity in the hemolymph and hemocytes of the marine mussel *Mytilus edulis. Fish. Shellfish. Immunol.* **4:**337–352.

Combes, C. (1991). Ethological aspects of parasite transmission. *Amer. Naturalist.* **138:**866–880.

Connors, V. A., Deburon, I. and Granath, W. O. (1995). *Schistosoma mansoni:* Interleukin-1 increases phagocytosis and superoxide production by hemocytes and decreases output of cercariae in schistosome-susceptible *Biomphalaria glabrata. Exp. Parasitol.* **80:**139–148.

Coombe, D. R., Ey, P. L. and Jenkin, C. R. (1984). Self/non-self recognition in invertebrates. *Quart. Rev. Biol.* **59:**231–255.

Couch, L., Hertel, L. A. and Loker, E. S. (1990). Humoral response of the snail *Biomphalaria glabrata* to trematode infection: Observations on a circulating hemagglutinin. *J. Expt. Zool.* **255:**340–349.

Coustau, C., Combes, C., Maillard, C., Renaud, F. and Delay, B. (1990). *Prosorhynchus squamatus* (Trematoda) parasitosis in the *Mytilus edulis–Mystilus galloprovicialis* complex: Specificity and host–parasite relationships. In *Pathology in marine science,* 1st ed. (T. C. Cheng and F. O. Perkins, eds.), vol. 1, pp. 291–298. San Diego: Academic Press.

Coustau, C., Robbins, I., Delay, B., Renaud, F. and Mathieu, M. (1993). The parasitic castration of the mussel *Mytilus edulis* by the trematode parasite *Prosorhynchus squamatus.* Specificity and partial characterization of endogenous and parasite-induced antimitotic activities. *Comp. Biochem. Physiol.* **104A:**229–233.

Coustau, C. and Yoshino, T. P. (1994). Surface membrane polypeptides associated with hemocytes from *Schistosoma mansonis*–susceptible and -resistant strains of *Biomphalaria glabrata* (Gastropoda). *J. Invertebr. Pathol.* **63:**82–89.

Crews, A. E. and Yoshino, T. P. (1991). *Schistosoma mansoni:* Influence of infection on levels of translatable mRNA and on polypeptide synthesis in the ovotestis and albumen gland of *Biomphalaria glabrata. Exp. Parasitol.* **72:**368–380.

De Gaffe, G. (1994). Mechanisms of trematode-induced interference: Correlating *in vitro* assays with *in vivo* realities. M.S. Thesis, University of New Mexico.

De Jong-Brink, M. (1992). Neuroendocrine mechanisms underlying the effects of schistosome parasites on their intermediate snail host. *Invertebr. Reprod. Dev.* **22:**127–138.

Dirosa, I., Contenti, S., Fagotti, A., Simoncelli, F., Principato, B., Panara, F. and Pascolini, R. (1994). Fibronectin from the hemolymph of the marine bivalve *Mytilus galloprovincialis.* Purification, immunological characterization, and immunocytochemical localization. *Comp. Biochem. Physiol.* **107B:**625–632.

Dissous, C., Torpier, G., Duvaux-Miret, O. and Capron, A. (1990). Structural homology of tropomyosins from the human trematode *Schistosoma mansoni* and its intermediate host *Biomphalaria glabrata. Mol. Biochem. Parasitol.* **43:**245–256.

Drickamer, K. (1995). Increasing diversity of animal lectin structures. *Curr. Opin. Struct. Biol.* **5:**612–616.

Fryer, S. E. and Bayne, C. J. (1989). Opsonization of yeast by the plasma of *Biomphalaria glabrata* (Gastropoda): A strain-specific, time-dependent process. *Parasite Immunol.* **11:**269–278.

———. (1995). Cell-mediated killing of metazoan parasites by molluscan hemolymph. In *Techniques in fish immunology,* 4th ed. (J. S. Stolen, T. C. Fletcher, S. A. Smith, J. T. Zelikoff, S. L. Kaattari, R. S. Anderson, K. Soderhall and B. A. Weeks-Perkins, eds.), pp. 123–131. Fair Haven, NJ: SOS Publications.

Fryer, S. E., Hull, C. J. and Bayne, C. J. (1989). Phagocytosis of yeast by *Biomphalaria glabrata:* carbohydrate specificity of hemocyte receptors and a plasma opsonin. *Dev. Comp. Immunol.* **13:**9–16.

Granath, W. O. and Aspevig, J. E. (1993). Comparison of hemocyte components from *Schistosoma mansoni* (Trematoda)-susceptible and -resistant *Biomphalaria glabrata* (Gastropoda) that cross-react with larval schistosome surface proteins. *Comp. Biochem. Physiol.* **104B:**675–680.

Granath, W. O., Connors, V. A. and Tarleton, R. L. (1994). Interleukin-1 activity in hemolymph from strains of the snail *Biomphalaria glabrata* varying in susceptibility to the human blood fluke, *Schistosoma mansoni*–presence, differential expression, and biological function. *Cytokine.* **6:**21–27.

Granath, W. O. and Yoshino, T. P. (1983). Lysosomal enzyme activites in susceptible and refractory strains of *Biomphalaria glabrata* during the course of infection with *Schistosoma mansoni. J. Parasitol.* **69:**1018–1026.

———. (1984). *Schistosoma mansoni:* Passive transfer of resistance by serum in the vector snail, *Biomphalaria glabrata. Exp. Parasitol.* **58:**188–193.

Ham, P. J., Hagen, H. E., Baxter, A. J. And Grunwald, J. (1995). Mechanisms of resistance to *Onchocerca* infection in blackflies. *Parasitology Today* **11:**63–67.

Hertel, L. A., Stricker, S. A., Monroy, F. P., Wilson, W. D. and Loker, E. S. (1994). *Biomphalaria glabrata* hemolymph lectins: Binding to bacteria, mammalian erythrocytes, and to sporocysts and rediae of *Echinostoma paraensei. J. Invertebr. Pathol.* **64:**52–61.

Hultmark, D. (1993). Immune reactions in *Drosophila* and other insects–A model for innate immunity. *TIG.* **9:**178–183.

Iwanaga, S. (1993). The *Limulus* clotting reaction. *Curr. Opin. Immunol.* **5:**74–82.

Jourdane, J. and Theron, A. (1987). Larval development: Eggs to cercariae. In *The Biology of schistosomes: from genes to latrines,* 1st ed. (D. Rollinson and A. J. G. Simpson, eds.) pp. 83–113. London: Academic Press.

Knibbs, R. N., Osborne, S. E., Glick, G. D. and Goldstein, I. J. (1993). Binding determinants of the sialic acid–specific lectin from the slug *Limax flavus. J. Biol. Chem.* **268:**18524–18531.

Leippe, M. and Renwrantz, L. (1988). Release of cytotoxic and agglutinating molecules by *Mytilus* hemocytes. *Dev. Comp. Immunol.* **12:**297–308.

Lie, K. J. (1982). Survival of *Schistosoma mansoni* and other trematode larvae in the snail *Biomphalaria glabrata.* A discussion of the interference theory. *Trop. Geogr. Med.* **34:**111–122.

Lie, K. J. and Basch, P. F. (1967). The life history of *Echinostoma paraensei* sp. n. (Trematoda: Echinostomatidae). *J. Parasitol.* **53:**1192–1199.

Lie, K. J., Jeong, K. H. and Heyneman, D. (1987). Molluscan host reactions to helminthic infections. In *Immune responses in parasitic infections.* (E. J. L. Soulsby, ed.), pp.211–270. Boca Raton, FLA: CRC Press.

Lim, H. K. and Heyneman, D. (1972). Intramolluscan inter-trematode anatagonism: A review of factors influencing the host–parasite system and its possible role in biological control. *Advan. Parasitol.* **10:**191–268.

Lodes, M. J. and Yoshino, T. P. (1993). Polypeptides synthesized *in vitro* by *Biomphalaria glabrata* hemocytes bind to *Schistosoma mansoni* primary sporocysts. *J. Invertebr. Pathol.* **61:**117–122.

Loker, E. S. (1994). On being a parasite in an invertebrate host: A short survival course. *J. Parasitol.* **80:**728–747.

Loker, E. S. and Adema, C. M. (1995). Schistosomes, echinostomes, and snails: Comparative immunobiology. *Parasitology Today* **11:**120–124.

Loker, E. S. and Bayne, C. J. (1982). *In vitro* encounters between *Schistosoma mansoni* primary sporocysts and hemolymph components of susceptible and resistant strains of *Biomphalaria glabrata. Am. J. Trop. Med. Hyg.* **31:**999–1005.

Loker, E. S., Cimino, D. F. and Hertel, L. A. (1992). Excretory-secretory products of *Echinostoma paraensei* sporocysts mediate interference with *Biomphalaria glabrata* hemocyte functions. *J. Parasitol.* **78:**104–115.

Loker, E. S., Cimino, D. F., Stryker, G. A. and Hertel, L. A. (1987). The effect of size of M line *Biomphalaria glabrata* on the course of development of *Echinostoma paraensei. J. Parasitol.* **73:**1090–1098.

Loker, E. S., Couch, L. and Hertel, L. A. (1994). Elevated agglutination titres in plasma of *Biomphalaria glabrata* exposed to *Echinostoma paraensei:* Characterization and functional relevance of a trematode-induced response. *Parasitology* **108:**17–26.

Loker, E. S. and Hertel, L. A. (1987). Alterations in *Biomphalaria glabrata* plasma induced by infection with the digenetic trematode *Echinostoma paraensei. J. Parasitol.* **73:**503–513.

Loker, E. S., Yui, M. A. and Bayne, C. J. (1984). *Schistosoma mansoni:* Agglutination of sporocysts, and formation of gels on miracidia transforming in plasma of *Biomphalaria glabrata. Exp. Parasitol.* **58:**56–62.

Marchalonis, J. J. and Schluter, S. F. (1990). Origins of immunoglobulins and immune recognition molecules. *Bioscience.* **40:**758–768.

Millar, D. A. and Ratcliffe, N. A. (1994). Invertebrates. In *Immunology: A comparative approach.* (R. J. Turner, ed.), pp. 29–68. Chichester: John Wiley and Sons, Ltd.

Monroy, F. P., Hertel, L. A. and Loker, E. S. (1992). Carbohydrate-binding plasma proteins from the gastropod *Biomphalaria glabrata:* Strain specificity and the effects of trematode infection. *Dev. Comp. Immunol.* **16:**355–366.

Moore, J. (1995). The behavior of parasitized animals. *Bioscience.* **45:**89–96.

Mouritsen, K. N. and Jensen, K. T. (1994). The enigma of gigantism: Effect of larval trematodes on growth, fecundity, egestion, and locomotion in *Hydrobia ulvae* (Pennant) (Gastropoda: Prosobranchia). *J. Exp. Mar. Biol. Ecol.* **181:**53–66.

Noda, S. and Loker, E. S. (1988). Effects of infection with *Echinostoma paraensei* on the circulating hemocyte population of the host snail *Biomphalaria glabrata. Parasitol.* **98:**35–41.

Olafsen, J. A. (1986). Invertebrate lectins: Biochemical heterogeneity as a possible key to their biological function. In *Immunity in invertebrates: Cells, molecules, and defense reactions.* (M. Brehelin, ed.), pp. 94–111. Berlin: Springer-Verlag.

Otsukafuchino, H., Watanabe, Y., Hirakawa, C., Tamiya, T., Matsumoto, J. J. and Tsuchiya, T. (1992). Bactericidal action of a glycoprotein from the body surface mucus of giant African snail. *Comp. Biochem. Physiol.* **101C:**607–613.

Ouwemissioukemboyer, O., Porchet, E., Capron, A. and Dissous, C. (1994). Characterization of immunoreactive TNF alpha molecules in the gastropod *Biomphalaria glabrata. Dev. Comp. Immunol.* **18:**211–218.

Preston, T. M., and Southgate, V. R. (1994). The species specificity of *Bulinus-Schistosoma* interactions. *Parasitol. Today.* **10:**69–73.

Renwrantz, L. (1986). Lectins in molluscs and arthropods: their occurrence, origin, and roles in immunity. In *Immune mechanisms in invertebrate vectors.* (A. Lackie, ed.), Symposium of the Zoological Society of London, pp. 81–93. Oxford: Clarendon Press.

Renwrantz, L. and Stahmer, A. (1983). Opsonizing properties of an isolated hemolymph agglutinin and demonstration of lectin-like recognition molecules at the surface of hemocytes from *Mytilus edulis. J. Comp. Physiol.* **149:**535–546.

Richards, C. S., Knight, M. and Lewis, F. A. (1992). Genetics of *Biomphalaria glabrata* and its effect on the outcome of *Schistosoma mansoni* infection. *Parasitol. Today.* **8:**171–174.

Richards, E. H. and Renwrantz, L. R. (1991). Two lectins on the surface of *Helix pomatia* haemocytes: A Ca^{2+}-dependent, GalNac-specific lectin and a Ca^{2+}-independent, mannose 6-phosphate–specific lectin which recognizes activated homologous opsonins. *J. Comp. Physiol.* **161B:**43–54.

Roberts, L. S. and Janovy, J., Jr. (1996). *Foundations of parasitology,* 5th ed. Dubuque: Wm C. Brown.

Smith, V. J. and Soderhall, K. (1991). A comparison of phenoloxidase activity in the blood of marine invertebrates. *Dev. Comp. Immunol.* **15:**251–261.

Sousa, W. P. (1993). Interspecific antagonism and species coexistence in a diverse guild of larval trematode parasites. *Ecol. Monogr.* **63:**103–128.

Spray, F. J. and Granath, W. O., Jr. (1990). Differential binding of hemolymph proteins from schistosome-resistant and -susceptible *Biomphalaria glabrata* to *Schistosoma mansoni* sporocysts. *J. Parasitol.* **76:**225–229.

Stibbs, H. H., Owczarzak, A., Bayne, C. J. and Dewan, P. C. (1979). Schistosome sporocyst-killing amoebae isolated from *Biomphalaria glabrata. J. Invertebr. Pathol.* **33:**159–170.

Terwilliger, N. B., Terwilliger, R. C. and Schabtach, E. (1976). The quaternary structure of a molluscan *(Helisoma trivolvis)* extracellular hemoglobin. *Biochim. Biophys. Acta* **453:**101–110.

Uchikawa, R. and Loker, E. S. (1992). *Echinostoma paraensei* and *Schistosoma mansoni:* Adherence of unaltered and modified latex beads to hemocytes of the host snail *Biomphalaria glabrata. Exp. Parasitol.* **75:**223–232.

Van der Knaap, W. P. W. and Loker E. S. (1990). Immune mechanisms in trematode–snail interactions. *Parasitol. Today* **6:**175–182.

Weston, D. S. and Kemp, W. M. (1993). *Schistosoma mansoni:* Comparison of cloned tropomyosin antigens shared between adult parasites and *Biomphalaria glabrata. Exp. Parasitol.* **76:**358–370.

Weston, D., Allen, B., Thakur, A., Loverde, P. T. and Kemp, W. M. (1994). Invertebrate host-parasite relationships: Convergent evolution of a tropomyosin epitope between

Schistosoma sp., *Fasciola hepatica,* and certain pulmonate snails. *Exp. Parasitol.* **78:**269–278.

Woodruff, D. S. (1978). Biological control of schistosomiasis by genetic manipulation of intermediate-host snail populations. *Proc. Int. Conf. Schistosomiasis,* vol. 1, pp. 13–22. Cairo, Egypt: October 18–25, 1975.

Yang, R. and Yoshino, T. P. (1990a). Immunorecognition in the freshwater bivalve, *Corbicula fluminea.* I. Electrophoretic and immunologic analyses of opsonic plasma components. *Dev. Comp. Immunol.* **14:**385–395.

———. (1990b). Immunorecognition in the freshwater bivalve, *Corbicula fluminea.* II. Isolation and characterization of a plasma opsonin with hemagglutinating activity. *Dev. Comp. Immunol.* **14:**397–404.

Yoshino, T. P. and Boswell, C. A. (1986). Antigen sharing between larval trematodes and their snail hosts: How real a phenomenon in immune evasion? In *Immune mechanisms in invertebrate vectors.* (A. Lackie, ed.), Symposium of the Zoological Society of London, pp. 221–238. Oxford: Clarendon Press.

Yoshino, T. P. and Lodes, M. J. (1988). Secretory protein biosynthesis in snail hemocytes: *In vitro* modulation by larval schistosome excretory-secretory products. *J. Parasitol.* **74:**538–547.

5

The Growth Hormone-Like Factor from Plerocercoids of the Tapeworm *Spirometra mansonoides* is a Multifunctional Protein

C. Kirk Phares

I. Introduction

Parasitologists realize that any parasite is more interesting to study than almost any free-living animal. Beyond studying parasites as independent organisms, possibly the most interesting topic in parasitology is the complex interaction between parasites and their hosts. The examples of parasitic perturbations of the host's endocrine system and behavior presented in this volume are excellent examples of truly fascinating phenomena. Considering the importance of the endocrine system in regulation of both the immune system and behavior of animals, it is not difficult to imagine survival advantages for parasites that can selectively manipulate the endocrine system of their hosts.

The pseudophyllidian tapeworm *Spirometra mansonoides* expresses several unusual characteristics in its relationships with its hosts. The best known characteristic of *S. mansonoides* is the ability of the plerocercoid larvae ("spargana") to stimulate growth of their hosts. Plerocercoids produce and release a substance that binds growth hormone (GH) receptors from a number of species, increases insulin-like growth factor (IGF-1), and stimulates growth. In addition to growth enhancement resulting from binding GH receptors and increasing IGF–1 is a significant reduction in endogenous levels of GH in plerocercoid-infected animals. Therefore, as plerocercoid growth factor (PGF) can duplicate some, but not all, of the functions of GH, animals treated with PGF grow at an accelerated rate, but are deficient with respect to other GH functions. This unusual phenomenon has been observed in both the reproductive system (Ramaley and Phares, 1980) and immune system (Sharp et al., 1982) of PGF-treated animals.

The growth-promoting phenomenon was discovered by Justus Mueller, whose contributions to knowledge of *S. mansonoides*, in particular, and the field of host-parasite relationships, in general (Mueller, 1966a), are significant. Mueller clarified the taxonomic relationships between diphyllobothrid and spirometrid tapeworms when he established *S. erinacei* as the type species for a separate generic designation for *Spirometra* (Mueller, 1937). Mueller's interests were in *S. mansonoides*, which he found to regularly infect water snakes (plerocercoid stage) and domestic cats (adult worm) around Syracuse, New York. A remarkable accomplishment was to establish and sustain the entire complex life cycle of *S. mansonoides* in the laboratory (Mueller, 1966b).

Life Cycle of S. mansonoides

Under natural conditions, adults of this species are intestinal dwellers in cats, both domestic and wild species. Individual immature eggs are released from the worm into the intestine and become part of the fecal stream. If the feces is deposited in fresh water, the tapeworm embryo will develop into an oncosphere in two to three weeks, under ideal conditions. When fully developed, the oncosphere becomes very active inside the "egg shell" and the operculum is forced open, allowing the coracidium to emerge and swim freely. The coracidium does not feed and must find the first intermediate host, a cyclopoid copepod (usually *Cyclops vernalis*), within 24 hours or it dies. Once in the gut of the copepod, the coracidium quickly penetrates out of the gut and develops into a procercoid in the hemocoel. Under ideal growth conditions, procercoids are fully developed and infective for the next host in about two weeks. Tadpoles, frogs, snakes, and mammals may become infected by drinking water containing procercoid-infected copepods. In Louisiana, where *S. mansonoides* is common, various species of water snakes are a common host for the plerocercoid stage. When ingested, procercoids exit the intestine into the peritoneal cavity and develop into plerocercoids, which become infective for the final host within one week. Plerocercoids grow continuously and migrate extensively throughout the hosts' tissues and organs and, if ingested by an appropriate final host such as a cat, they mature into the adult form and begin to shed eggs within two weeks.

Prior to the discovery of the unique effect on growth stimulation, plerocercoids were best known for being the cause of human sparganosis. Plerocercoids cannot be definitively identified based on their morphology. In order to prove that the plerocercoids (spargana) of the tapeworm he found in snakes in Syracuse were the same as the plerocercoids that cause human sparganosis, Mueller was surgically infected and persuaded a colleague to do the same (Mueller and Coulston, 1941). The plerocercoids thrived and grew between 4–6 cm in just over 2 months. The plerocercoids were surgically removed and one was fed to a cat and a normal adult *S. mansonoides* developed.

The natural life cycle of *S. mansonoides* is not out of the ordinary for pseudophyllidian tapeworms. Morphologically, the stages are unspectacular and their similarity in appearance to diphyllobothrid tapeworms easily explains their inclusion in that group for so long. A unique characteristic of the life cycle is the remarkable diversity in host selection for both adult and plerocercoid stages.

Like other tapeworms, adults live in the small intestine of their hosts, but are not limited to a single species or genus. Rather, they are able to live and reproduce in several carnivores and omnivores. A wide variety of species of felines and canines support infections with adult *S. mansonoides*. Raccoons and opossums not only are suitable hosts for the adult tapeworms in their intestines, but may also harbor infections with the tissue-invading plerocercoid stage as well. An explanation for this phenomenon is that if procercoids are ingested, a plerocercoid infection will result, but if plerocercoids are ingested by an appropriate host, an adult tapeworm will develop (Mueller, 1971). This was shown to be the case when domestic kittens were fed procercoids and they developed plerocercoid infections, but when fed plerocercoids an adult infection resulted (Mueller, 1966a).

Whereas the adult is fairly inflexible in its selection of hosts, the plerocercoid stage is able to establish infections in species of all classes of vertebrates, except fish. Helminths, such as the trematode *Fasciola hepatica* and adults of the cestode *Diphillobothrium latum*, can infect a variety of mammalian species, but there are very few, if any, other species of helminth that, after being ingested, have the ability to penetrate out of the gut and successfully evade the defense mechanisms of amphibians, reptiles, birds, and mammals including humans. These animals become infected with plerocercoids from ingesting procercoid-infected copepods or by ingesting plerocercoid-infected vertebrates.

Not only is the plerocercoid stage extremely adaptable with respect to host preference, it is very patient in its wait for an appropriate host to enable it to complete the life cycle. It is reasonable to expect that plerocercoids could be passed from a tadpole or a frog to a snake, to a mammal such as a rat, back to a snake, before being ingested by an acceptable final host. In fact, plerocercoids from a single group were transferred annually through many generations of mice between 1957 and 1974 before several were fed to a cat and a normal adult infection occurred (Mueller, 1974).

The ability of plerocercoids to pass from one host to another across entire classes of vertebrates suggests that *S. mansonoides* evolved a mechanism(s) that allows plerocercoids to effectively escape the ability of the defense systems of all classes of vertebrates except fish to protect against infection with this organism. As fish are the most likely of all vertebrates to eat procercoid-infected copepods, one of the most interesting questions one might consider is what is different about fish that enable them to resist infection with plerocercoids of *S. mansonoides*.

II. Discovery of Plerocercoid Growth Factor (PGF)

Almost 30 years after his initial report on *Spirometra*, Mueller discovered that plerocercoids stimulate growth in their hosts (Mueller, 1963). His efforts to find a more suitable animal model than water snakes to serve as intermediate hosts for the plerocercoid stage resulted in finding that common laboratory mice served as very acceptable maintenance hosts. During these early studies, Mueller observed that mice seemed to suffer few ill effects from infections with moderate numbers of plerocercoids. In fact, plerocercoid-infected mice grew at a much faster rate than uninfected age-matched controls. The increase in body weight was not due to an inflammatory response or to increases in body fluids, nor could it be attributed to increases in parasite mass. Rather the increase was due to accelerated body growth, including increases of bone and muscle masses. We and others subsequently showed that the growth-promoting effect was due to some substance released into the environment of the worms (Phares and Ruegamer, 1972; Steelman et al., 1971).

The similarities between the effects of plerocercoid infection and the effects of GH are remarkable. Even more remarkable is the fact that several actions of PGF are more similar to those reported for human growth hormone (hGH) than to any other GH (Table 5-1). GH expresses multiple biological properties that have been difficult to reconcile with one common mechanism of action of the hormone. In addition to growth and other anabolic effects, all GHs are diabetogenic and in GH-deficient subjects, GH produces transient insulin-like effects (Kostyo, 1986).

Whereas the similarities between the actions of GH and PGF are remarkable, it is important to recognize that there are also distinct differences in their activities.

Table 5-1 Comparison of Some Effects of hGH[1] and PGF[2] on Growth and Metabolism

	hGH	PGF
Weight gain	increase	increase
Body length	increase	increase
Skeletal growth	increase	increase
IGF-1[3]	increase	increase
Serum GH	decrease	decrease
Pituitary GH	decrease	decrease
Lipolytic	yes	no
Lipogenic	no	yes
Diabetogenic	yes	no
Lactogenic[4]	yes	yes

[1]The effects are those observed following injections of hGH in animals or addition of hGH to tissues from animals *in vitro.*

[2]The effects of PGF were observed following injections of PGF or plerocercoid infections in animals or PGF incubated with tissues from animals.

[3]Increases in IGF-1 due to injections of hGH or PGF were observed in hypophysectomized rats and hypophysectomized rhesus monkeys.

[4]Lactogenic effects were observed in the pigeon crop sac bioassay for prolactin.

One of the most significant differences is in the duration and potency of PGF in stimulation of growth. Daily injections of GH are required to stimulate a sustained growth response in GH-deficient animals. Serum prepared from plerocercoid-infected hypophysectomized rat blood was shown to stimulate growth of other hypophysectomized rats. More remarkably, one injection of small amounts (0.1–0.5 ml) of PGF-containing serum stimulated growth responses in hypophysectomized rats that persisted for up to 2 weeks after the single injection (Garland and Daughday, 1972; Steelman, et al., 1970). Serum from GH-treated hypophysectomized rats had neither short-term nor extended growth-promoting activity. It is well-established that GH-stimulated growth is associated with reduction in stored fat. This contrasts distinctly to the growth response due to PGF which is associated with a significant increase in stored fat (Steelman, et al., 1971; Phares and Carroll, 1977). Furthermore, all vertebrate GHs can induce hyperglycemia and insulin resistance. Salem and Phares (1989) compared the effects of purified PGF and hGH on blood glucose level and glucose tolerance. As expected, hGH caused a significant increase in fasting blood glucose levels and a dramatic impairment of glucose tolerance, but PGF had no effect on either fasting glucose levels or glucose tolerance.

It is clear that PGF can duplicate some, but not all, of the actions of GH. As PGF causes a significant suppression of GH levels in both the pituitary and serum (Garland and Daughday, 1972; Phares, 1982) but does not replace all of the functions of GH, some of the alterations in the endocrine, defense, and metabolic activities of PGF-treated animals may be due to a combination of direct PGF effects and GH deficiency.

Growth hormone was discovered in the 1920s and bioassays in animal models were established in the 1940s. However, since GH from animals did not stimulate growth in humans, it was widely believed that humans did not need a GH for growth. Nevertheless, humans, like all vertebrates, do produce GH. Whereas GH is not essential for life, it is essential for growth. The reason that non-primate GH has no effect in humans was revealed when the species specificity of human GH (hGH) was discovered (Li and Papkoff, 1956). Simply stated, hGH is active in all vertebrates, but only hGH (or other primate GH) is active in humans. Unlike diabetes, which can be treated with non-human insulins (pork, beef, etc.), human dwarfism must be treated with human GH. In light of these facts, the ability of a larval tapeworm to produce a substance that mimics several actions thought to be unique to hGH is even more remarkable.

The possibility that a factor from any source, including a tapeworm, might substitute for hGH, or be a growth-promoting agent in domestic animals, stimulated interest in its isolation and characterization. We showed that active PGF could be detected in simple media in which plerocercoids were incubated (Phares and Ruegamer, 1972). However, early studies to characterize PGF were hindered by the lack of a sensitive assay. Growth-promoting bioassays in intact or hypophysectomized rodents are very expensive, time-consuming, and are not nearly sensitive enough to detect the low concentrations of GH-like activity in plerocercoid incubation media.

Tsushima and Friesen (1973) reported the development of the first radioreceptor assay (RRA) specific for hGH. This assay was able to detect nanogram (ng) amounts of hGH compared to the milligram (mg) amounts required for growth promoting bioassays. Shortly after the development of the RRA for hGH, Tsushima and coworkers (1974) reported that PGF competitively inhibited binding of ^{125}I hGH to rabbit liver receptors and produced a competitive inhibition curve parallel to that of hGH, making the RRA for hGH available for studies to isolate and characterize PGF.

III. Purification of PGF

Although PGF can be detected by RRA in media conditioned by incubated plerocercoids, the total binding activity of these preparations is low and significant activity is lost during various methods of concentration. We found that compared to using plerocercoid-conditioned media, the total amount of PGF activity could be greatly increased with a significant reduction of total volume of the starting material for purification after solubilization of plerocercoid membranes in a nonionic detergent (Phares, 1984). Homogenization of plerocercoid membranes in the absence of detergent is ineffective in extraction of active PGF.

The availability of a RRA with a reliable sensitivity to about 30 ng/ml and extracts of plerocercoid membranes containing microgram (μg) amounts of PGF activity finally made purification and characterization a reasonable possibility. In the RRA, hGH is used as the standard of comparison and PGF activity is expressed as equivalent units to the hGH standard in competitive inhibition of ^{125}I hGH binding. Any purification scheme involving multiple chromatographic procedures usually results in significant loss of total activity with each step. Therefore, when the mass of starting material is a limiting factor, as it always is with plerocercoids, it is important to reduce the number of steps and employ procedures that cause minimal dilution of the sample. With these limitations, a scheme was designed to increase the purity of PGF by using a single chromatographic procedure consisting of a hGH receptor affinity column (Phares, 1988). Although this procedure resulted in greater than 1,000-fold increase in specific activity, the product was not pure; SDS-PAGE of the product showed a major band at 27.5 kD and minor bands at 22 kD and 17 kD. However, the partially purified PGF was very active in the RRA, and daily injections of affinity-purified PGF for 10 days stimulated a highly significant dose-dependent growth response in hypophysectomized rats (Phares, 1988).

It is interesting to note that compared to the GH standard the growth-promoting potency of injections of PGF was much greater than its binding activity as detected in the RRA would predict. Daily injections of purified PGF with binding activity equivalent to as little as 100 ng of hGH stimulated a significant growth response and 400 ng eq/day of PGF stimulated a growth response of the same magnitude as daily injections of 250 μg of bovine GH. Differences in RRA activity, growth-promoting potency, and the duration of the growth-promoting effects between PGF and GH were previously observed. Tsushima and coworkers (1974) reported that a single injection of 1 ml of incubation medium containing 200 ng eq of PGF binding activity by RRA stimulated growth in hypophysectomized rats to the same degree as daily injections of 10 μg (50 times more) of bovine GH. An earlier report showed that infections of hypophysectomized rats with only one plerocercoid stimulated significant growth responses (Mueller, 1970). These data suggest that the mechanisms involved in binding and activation of the GH receptor by PGF may be different from that of GH.

Although receptor affinity chromatography was a powerful method to purify PGF, it was a time- and labor-intensive project: six rabbit livers produced only 12 μg of purified hGH receptor and the binding capacity of the GH receptor column degraded after only a few purification procedures. A more conventional scheme was developed involving solubilization of plerocercoid membranes followed by a series of steps, including chromatofocusing, ion exchange, and molecular sieving chromatography. This scheme enabled isolation of PGF activity detected by RRA consisting of a single stained band at 27.5 kD by SDS-PAGE. The presence of the 27.5 kD protein in all preparations containing PGF activity and the resolution of binding activity to one protein band strongly suggested that the 27.5 kD protein is PGF.

IV. Molecular Characterization of PGF

Considering the remarkable ability of PGF to competitively inhibit binding of hGH to its receptors and to mimic biological actions believed to be unique to hGH, it was assumed that PGF and hGH must share significant structural homology. It seemed obvious that any putative PGF would be confirmed as the hGH-like factor by simply comparing the primary amino acid sequences of these two substances and significant homology between the two would be obvious. Our approach to obtain the sequence of PGF was to isolate messenger RNA (mRNA) from plerocercoids, prepare a complementary DNA (cDNA) library, design a probe to specifically identify the cDNA encoded for the GH-like factor, amplify the cDNA for PGF, sequence the cDNA, and deduce the full-length amino acid sequence of the precursor and mature forms of the PGF protein.

Limited trypsin digestion of the purified 27.5 kD protein was necessary as the amino-terminal of the 27.5 kD protein was blocked and unavailable for direct sequencing. Several peptide fragments were sequenced, none of which showed any homology to GH. However, sequences were selected and used to design oligonucleotide probes to identify the cDNA for the 27.5 kD protein. A plerocercoid cDNA library was constructed with bacteriophage lambda gt11 DNA in *E. coli*. The polymerase chain reaction using synthetic oligonucleotides was performed to amplify cDNA for the 27.5 kD protein. A sequence of 1040 bases of the cDNA was obtained. The predicted amino acid sequence of the mature 27.5 kD plerocercoid protein is shown in Table 5-2. A computer search of both nucleotide and peptide sequence data bases revealed no homology to any known GH. However, there is significant (≈ 50 percent) homology to several cysteine proteinases. Highest homologies are to the mammalian lysosomal enzymes, cathepsin L, and cathepsin S. Comparison of consensus sequences of two highly conserved regions of cysteine proteinases (North et al., 1990) to the sequence of the 27.5 kD plerocercoid protein shows total identity to one of the highly conserved regions and only one amino acid difference in the other region. The sequences of these two conserved regions in the 27.5 kD protein and in human cathepsin L are 100% identical.

Subsequent studies confirm that the 27.5 kD plerocercoid protein is a neutral cysteine proteinase (Phares and Kubik, 1996). Confirmation was based on substrate and inhibitor specificity, activation by thiol-containing compounds, and zymogram analysis. The plerocercoid proteinase was very active against a collagen substrate and the 27.5 kD proteinase is glycosylated. As the proteinase is membrane-associated and released into the environment of plerocercoids, and is very active against collagen, an important function must be in tissue penetration to allow rapid migration of plerocercoids.

As stated above, the biological characteristics of PGF strongly predict that there should be a significant molecular homology to hGH. Continuing efforts have failed to reveal any plerocercoid protein with structural homology to any GH. Careful retrospective analysis and additional research strongly suggest that our initial approach was not flawed and did not mistakenly identify the 27.5 kD proteinase as the hGH-like factor. Rather, the data strongly suggest that the 27.5 kD proteinase expresses multiple activities, including binding and activation of GH receptors and is, in fact, the GH-like factor.

V. Rationale for the Enzyme and Growth Factor Being the Same Molecule

Identification of PGF during purification was based on competitive inhibition of hGH in a RRA *in vitro*. When sufficient amounts of material could be purified, 10-day growth-promoting bioassays in hypophysectomized rats or normal mice were conducted for confirmation. In RRAs, specific binding (and, therefore, competitive inhibition) is determined by comparing the amount of receptor-bound ^{125}I hormone to the amount of unbound (free) ^{125}I hormone. A proteolytic enzyme such as a neutral cysteine proteinase might digest the ^{125}I hormone or its receptor to cause an apparent reduction in specifically bound ^{125}I hormone and could give the false impression that the enzyme was a competitive inhibitor of hormone-receptor interaction. If the apparent binding of PGF was due only to its proteolytic activity, then it should also express "binding" activity in RRAs for other peptide

Table 5-2 Amino Acid Sequence[1] of the 27.5 kD Cysteine Proteinase/PGF from Plerocercoids of S. mansonoides

1	LEU	PRO	ASP	SER	VAL	ASN	TRP	HIS	GLU	LYS	GLY	ALA	VAL	THR	SER	VAL	LYS	ASN	GLN	GLY
21	GLN	CYS	GLY	SER	CYS	TRP	SER	PHE	SER	ALA	ASN	GLY	ALA	ILE	GLU	GLY	ALA	ILE	GLN	ILE
41	LYS	MET	GLY	ILE	LEU	PRO	THR	LEU	SER	GLU	GLN	GLN	LEU	VAL	ASP	CYS	SER	TRP	GLU	TYR
61	GLY	GLN	GLY	CYS	ASN	GLY	GLY	PHE	MET	SER	LEU	ALA	PHE	GLN	TYR	ALA	GLN	ARG	TYR	GLY
81	VAL	GLU	ALA	GLU	VAL	ASP	TYR	ARG	TYR	THR	ALA	LYS	ASP	GLY	PHE	CYS	ARG	TYR	GLN	GLN
101	ASP	MET	VAL	VAL	ALA	ASN	VAL	THR	GLY	TYR	ALA	GLU	LEU	PRO	GLN	GLY	ASP	GLU	ALA	SER
121	LEU	GLN	ARG	ALA	VAL	ALA	VAL	ILE	GLY	PRO	ILE	SER	VAL	GLY	ILE	ASP	ALA	ASN	ASP	PRO
141	GLY	PHE	MET	SER	TYR	SER	HIS	GLY	VAL	PHE	VAL	SER	LYS	THR	CYS	SER	PRO	ASP	ASP	ILE
161	ASN	HIS	GLY	VAL	LEU	VAL	ILE	GLY	TYR	GLY	THR	GLU	ASN	ASP	GLU	PRO	TYR	TRP	LEU	VAL
181	LYS	ASN	SER	TRP	GLY	ARG	SER	TRP	GLY	GLU	GLN	GLY	TYR	VAL	LYS	MET	ALA	ARG	ASN	LYS
201	ASN	ASN	MET	CYS	GLY	ILE	ALA	SER	VAL	ALA	SER	TYR	PRO	THR	VAL					

[1]The amino acid sequence was deduced from the nucleotide sequence of the cDNA for PGF.

hormones. If the enzyme were digesting the receptors, the binding capacity of enzyme-degraded receptors should be reduced after exposure to the proteinase. Furthermore, if the binding activity is due to a distinct protein and not associated with the proteinase, then total inhibition of the enzyme should not influence binding activity in the RRA. Review of the original report (Tsushima et al., 1974) of PGF activity in the hGH RRA reveals that PGF displaced ^{125}I hGH in parallel to the dose response curve of the hGH standard, but samples from the same preparation of PGF had no effect on ^{125}I prolactin binding to its receptors in rabbit mammary gland. Furthermore, Salem and Phares (1989) reported that highly purified PGF (27.5 kD protein) stimulated an insulin-like metabolic effect *in vitro* in rat adipose tissue via the GH receptor. The purified 27.5 kD PGF inhibited ^{125}I hGH binding to adipocytes, but did not compete with ^{125}I insulin for binding to its receptors in the same cells.

Binding studies with ^{125}I hGH and PGF have been conducted using liver membrane receptors for GH from rats (Phares and Booth, 1986a) and hamsters (Phares and Booth, 1986b). Competitive inhibition curves for PGF and hGH were indistinguishable in these studies. To determine if PGF was degrading the ability of receptors to bind hGH, liver membrane receptors were exposed to PGF in the absence of added proteinase inhibitor and incubated for six hours at room temperature. The chaotropic agent, $MgCl_2$, was added to the membranes to desaturate the receptors of any bound ligand (Kelly et al. 1979). Total, nonspecific, and specific binding of hGH was determined before and after exposure to PGF and 4M $MgCl_2$. Binding of ^{125}I hGH to its receptors was unaffected by six hours of exposure to PGF in both studies.

Solubilized plerocercoid membrane proteins (crude PGF) contains RRA, growth-promoting, and proteinase activity. Addition of the cysteine proteinase inhibitor E64 to crude PGF totally inhibited enzymatic activity, reduced RRA activity by as much as 80 percent, and reduced the growth-promoting activity in intact mice (unpublished data).

These data considered together suggest that neither degradation of ^{125}I hGH or its receptors can account for competitive inhibition of hGH binding by PGF. Inhibition of proteinase activity resulted in significant reductions in RRA and growth-promoting activity, suggesting that GH receptor-binding and growth-promoting activity expressed by plerocercoid infections and injections of PGF are characteristics of the 27.5 kD cysteine proteinase/PGF and not another plerocercoid protein.

VI. Discussion

Is this possible? Can an enzyme from a tapeworm stimulate GH-like responses in mammals? Normal functioning of the endocrine system is dependent on highly specific interactions between hormones and their receptors. Hormones like GH that act at the cell surface must not only bind the receptor in a very precise way, but must also activate the receptor to initiate a transmembrane signalling mechanism. However, hormonal agonists and antagonists exist that bind hormone receptors. Although these substances act via specific hormone receptors, they may not share any structural similarity to the natural hormone. Well-known examples of this phenomenon are the result of autoimmune reactions to hormone receptors. Antireceptor antibodies may bind the receptor and stimulate or inhibit hormone responses. It is possible that PGF is a GH agonist that is able to bind and activate GH receptors while sharing no sequence homology to any GH.

To confuse, or perhaps to clarify, the hypothesis that the proteinase and GH-like activity reside in the same molecule, the interaction of GH with its receptors has recently been shown to be much more complex than the concept that one molecule of hormone binds to one receptor molecule to initiate a hormonal signal. Contrary to early studies that reported a single binding site in hGH, recent work clearly indicates that each hGH molecule has two distinct non-overlapping binding sites that enable one molecule of hGH to dimerize two identical receptors by a sequential binding mechanism (Cunningham et al., 1989). Furthermore, mutants of hGH that are able to bind one receptor, but are unable to induce receptor dimerization, are biologically inactive (Cunningham and Wells, 1989). It is possible that a highly specific "lock and key" interaction of GH with its receptor is less important than dimerization of two receptors for induction of GH-like responses.

The possibility exists that PGF stimulates GH-like responses not through precise molecular mimicry of hGH, but by inducing dimerization of GH receptors and acting as a potent hormonal agonist. The distinct differences between some biological actions of PGF and hGH described above support the concept of nonidentical mechanisms for receptor activation.

The author of the current work is unaware of any studies reporting stimulation of body growth by injections of cysteine proteinases in animals. It is possible that no such studies have been performed. However, there are reports of mitogenic effects of proteinases on cells in culture (Vetvicka et al., 1994) and the relationship between high levels of cysteine proteinases to growth and metastasis of several cancers is well-established (Sloan and Honn, 1984).

Although much research remains to be done before the hypothesis that the proteinase and hGH-like activities associated with plerocercoids of *S. mansonoides* are characteristics of a single molecule is firmly established, there is sufficient data to continue to pursue this intriguing possibility. Final proof will come when recombinant PGF is produced and is shown to bind hGH receptors, stimulate growth, and express other hGH-like activities. Confirmation that the proteinase and PGF are one and the same will open an entirely new area of research with *S. mansonoides* to discover the molecular mechanisms involved in the stimulation of GH-like responses.

Whereas the ability of plerocercoids to stimulate growth is an interesting phenomenon, other more important adaptive roles for PGF are possible as well. In addition to facilitating rapid tissue penetration by plerocercoids and stimulating host growth, it is likely that this molecule is at least partially responsible for the ability of plerocercoids to evade the immune responses of its hosts. We have found that the 27.5 kD proteinase cleaves IgG (unpublished data). Since collagen is an excellent substrate for the proteinase, PGF may also be involved in the inactivation of the host complement system, as suggested for cysteine proteinases in other helminths (Leid and Mitchell, 1982).

Plerocercoids of *S. mansonoides* express several unusual characteristics. These characteristics include the ability to infect hosts from all classes of vertebrates except fish and to stimulate growth while manipulating the endocrine system in their hosts. The hypothesis presented in this chapter suggests that the ability of plerocercoids to stimulate growth and to evade the defenses of hosts by rapid, constant migration through tissues while disabling the host's immune effector mechanisms reside in a single substance, PGF.

VII. Acknowledgments

The author's research, which contributed to this chapter, was supported over a period of time by the National Institutes of Health (AM–17226 and AM–37030). The author wishes to acknowledge Dr. Richard Davis of Case Western Reserve University for his contributions to this work by establishing the cDNA library for plerocercoids and sequencing the cDNA for the 27.5 kD proteinase/PGF. Dr. Fritz Rottman of Case Western Reserve University was also instrumental in the cloning and sequencing of PGF. The technical expertise of Ms. Jacy Kubik in conducting much of the research and the secretarial assistance of Ms. Michelle Erickson and Ms. Marilyn Thomason are greatly appreciated.

VIII. References

Chapman, C. B., and Mitchell, G. F. (1982). Proteolytic cleavage of immunoglobulin by enzymes released by *Fasciola hepatica. Vet. Parasitol.* **11:**165–178.

Garland, J. T., and Daughaday, W. H. (1972). Feedback inhibition of pituitary growth hormone in rats infected with *Spirometra mansonoides. Proc. Soc. Exp. Biol. Med.* **139:**497–499.

Garland, J. T., Ruegamer, W. R., and Daughaday, W. H. (1971). Induction of sulfation factor activity by infection of hypophysectomized rats with *Spirometra mansonoides. Endocrinology* **88:**924–927.

Kelly, P., Leblanc, G., and Djiane, I. (1979). Estimation of total prolactin binding sites after *in vitro* desaturation. *Endocrinology* **104:**1631–1637.

Kostyo, J. L. (1986). The multivalent nature of growth hormone. In *Human growth hormone* (S. Raiti, and R. A. Toman, eds.), pp. 449–453. New York: Plenum Co.

Leid, R. W., Suquet, C. M., and Tanigoshi, L. (1987). Parasite defense mechanisms for evasion of host attack: A review. *Vet. Parasitol.* **25:**147–162.

Li, C. H., and Papkoff, H. (1956). Preparation and properties of growth hormone from human and monkey pituitary glands. *Science* **124:**1293–1294.

Mueller, J. F. (1937). A reparition of the genus *Diphyllobothrium. J. Parasitol.* **23:**308–310.

———. (1963). Parasite-induced weight gain in mice. *Ann. NY Acad. Sci.* **113:**217–233.

———. (1966a) Host–parasite relationships as illustrated by the cestode *Spirometra mansonoides* as an experimental tool. In *Host–Parasite relationships* (J. E. McCauley, ed.), pp. 15–58. Oregon State University Press.

———. (1966b). The laboratory propagation of *S. mansonoides* (Mueller, 1935) as an experimental tool. VII. Improved techniques and additional notes on the biology of the cestode. *J. Parasitol.* **52:**437–443.

———. (1970). Quantitative relationship between stimulus and response in the growth-promoting effect of *Spirometra mansonoides* spargana on the hypophysectomized rat. *J. Parasitol.* **56:**840–842.

———. (1974). The biology of *Spirometra. J. Parasitol.* **60:**3–14.

Mueller, J. F., and Coulston, F. (1941). Experimental human infection with the sparganum larvae of *Spirometra mansonoides* (Mueller, 1935). *Am J. of Trop. Med.* **21:**399–425.

North, M. J., Mottram, J. C., and Coombs, G. H. (1990). Cysteine proteinases of parasitic protozoa. *Parasitol. Today* **6:**270–275.

Phares, C. K. (1982). The lipogenic effect of the growth factor produced by plerocercoids of the tapeworm, *Spirometra mansonoides,* is not the result of hypothyroidism. *J. Parasitol.* **68:**999–1003.

———. (1984). A method for solubilization of a human growth hormone analogue from plerocercoids of *Spirometra mansonoides. J. Parasitol.* **70:**840–842.

———. (1988). Use of receptor affinity chromatography in purification of the growth hormone–like factor produced by plerocercoids of the tapeworm *Spirometra mansonoides. J. Recept. Res.* **8:**645–665.

Phares, C. K., and Booth, B. J. M. (1986a). Reduction of lactogenic receptors in female hamster liver due to the hGH analogue produced by plerocercoids of the tapeworm, *Spirometra mansonoides. Endocrinology* **118:**1102–1109.

Phares, C. K., and Booth. B. J. M. (1986b). Suppression of receptors for prolactin and estrogen in rat liver due to treatment with the growth hormone analogue produced by the tapeworm *Spirometra mansonoides. J. Recept. Res.* **6:**425–446.

Phares, C. K., and Carroll, R. M. (1977). A lipogenic effect in intact male hamsters infected with plerocercoids of the tapeworm, *Spirometra mansonoides. J. Parasitol.* **63:**690–693.

Phares, C. K., and Kubik, J. (1996). The growth factor from plerocercoids of *Spirometra mansonoides* is both a growth hormone agonist and a cysteine proteinase. *J. Parasitol.* **82;2:**210–215.

Phares, C. K., and Ruegamer, W. R. (1972). *In vitro* preparation of a growth factor from plerocercoids of the tapeworm, *Spirometra mansonoides. Proc. Soc. Exp. Biol. Med.* **142:**374–377.

Ramaley, J. A., and Phares, C. K. (1980). Delay of puberty onset due to suppression of growth hormone. *Endocrinology* **106(6):**1989–1993.

Sloan, B. F., and Honn, K. V. (1984). Cysteine proteinases and metastasis. *Cancer Metastasis Rev.* **3:**249–263.

Salem, M. A. M., and Phares, C. K. (1989a). The growth factor from plerocercoid larvae of the tapeworm, *Spirometra mansonoides,* stimulates growth but is not diabetogenic. *Proc. Soc. Exp. Med. Biol.* **191:**187–192.

———. (1989b). *In vitro* insulin-like actions of the growth factor from the tapeworm, *Spirometra mansonoides. Proc. Soc. Exp. Med. Biol.* **190:**203–310.

Sharp, S. E., Phares, C. K., and Heidrick, M. L. (1982). Immunological aspects associated with suppression of hormone levels in rats infected with plerocercoids of *Spirometra mansonoides* (Cestoda). *J. Parasitol.* **68(6):**993–998.

Steelman, S. L., Glitzer, M. S., Ostlind, D. A., and Mueller, J. F. (1971). Biological properties of the growth hormone–like factor from the plerocercoid of *Spirometra mansonoides. Recent Prog. Horm. Res.* **27:**97–120.

Steelman, S. L., Morgan, E. R., Cuccaro, A. H., and Glitzer, M. S. (1970). Growth hormone–like activity in hypophysectomized rats implanted with *Spirometra mansonoides* spargana. *Proc. Soc. Exp. Biol. Med.* **133:**269–273.

———. (1971). Growth hormone–like activity in hypophysectomized rats implanted with *Spirometra mansonoides. Endocrinology* **88:**924–927.

Tsushima, T., and Friesen, H. G. (1973). Radioreceptor assay for growth hormone. *J. Cli. Endocrinol. Metabol.* **27:**334–348.

Tsushima, T., Friesen, G., Chang, T. W., and Raben, M. S. (1974). Identification of sparganum growth factor by a radioreceptor assay for growth hormone. *Biochem. Biophys. Res. Commun.* **59:**1062–1068.

Vetvicka, V., Vektvickova, J., and Fusek, M. (1994). Effect of human procathepsin D on proliferation of human cell lines. *Cancer Lett.* **79:**131–135.

6

Peptides: An Emerging Force in Host Responses to Parasitism

Ian Fairweather

I. Introduction and Framework of Review

This chapter provides an overview of the impact of parasites on the endocrine system of their host. The review is concerned primarily with helminth parasites of vertebrates and specifically parasites that inhabit the gastrointestinal tract, because that is the area in which the overwhelming bulk of the experimental work has been carried out. In preparing the review, however, it soon became apparent that to discuss endocrine responses in isolation would be a very limiting exercise. In addition to causing changes in the hormonal status of the host, parasites elicit immune responses from the host, induce pathological changes in host tissues, and alter the normal functioning of the gut by disrupting neuroendocrine control mechanisms, epithelial cell functioning, and differentiation of mucosal cells, for example. Moreover, it is becoming increasingly apparent that the various facets of the host response are interlinked with each other and—of particular significance to this review—are involved with peptides, because peptidergic molecules are common to the endocrine, nervous, and immune systems. Consequently, the chapter will follow four main themes. Initially, it will briefly review pathophysiological changes in the gut resulting from parasitism. It will then examine the roles of gut neuroendocrine and immune systems in these responses, highlighting the disruption of peptidergic mechanisms in both the gut endocrine and nervous systems. The review will then consider interactions between the neuroendocrine and immune systems, first establishing general principles, then discussing data relating specifically to peptides and cytokines. Finally, an attempt will be made to draw together the different themes of the review and explore what progress has been made in unravelling connections among the main tissues discussed (namely, endocrine system, nervous system, immune system, and epithelium plus muscle). A number of schemes have been proposed to integrate such interactions into the overall picture of the host response. The models will be examined to determine how comprehensive they are and any weaknesses exposed will serve to highlight potential lines of future research.

II. Changes in Gut Physiology and Morphology Resulting from Parasitism

The presence of a parasite within the digestive tract is likely to disrupt normal gut homeostasis, either by inflicting direct mechanical injury to gut tissues or by disturbing physiological mechanisms, or both. A variety of changes has been observed: the principal ones are highlighted below.

A. General Changes in Gut Morphology, Motility, Food Intake, and Secretory Activity

Parasites can cause severe morphological damage to the gut, including shortening of intestinal villi, loss of microvilli, villous atrophy, crypt cell hyperplasia, and hypertrophic responses (Symons, 1969; Castro, 1980). These changes will affect the absorptive area of the gut and brush border enzyme activity, for example.

Alterations in intestinal motility have been observed in a number of nematode infections: *Haemonchus contortus*, *Trichostrongylus colubriformis*, *Trichinella spiralis*, *Nippostrongylus brasiliensis,* and *Strongylus vulgaris* (Symons, 1969; Schanbacher et al., 1978; Bueno et al., 1979, 1982; Gregory et al., 1985). Hypertrophy of gut musculature occurs in *N. brasiliensis* and *T. spiralis* infections in the rat (Bowers et al., 1986; Castro, 1989), which may be a manifestation of the inflammatory response to infection that is characteristic of enteric parasitism (Castro, 1989). Inappetence and reduced food intake are characteristic of helminth infections, as in sheep infected with *T. colubriformis* (Symons and Hennessy, 1981; Gregory et al., 1985) and with *O. circumcincta* (Anderson et al., 1976, 1985). Decreased food intake, in turn, may affect gut motility and digesta flow and so interpretation of data relating to the latter must be approached with caution.

The secretory activity of the gut can be altered in a number of ways. For example, reduced gastric secretion and elevation of abomasal pH occur in sheep infected with the stomach roundworms *Ostertagia circumcincta* and *H. contortus* (Anderson et al., 1976, 1981, 1985; Dakkak et al., 1982; Nicholls et al., 1985). Gut hyperacidity and reduced enzyme secretion have also been observed in infections with the intestinal nematodes *N. brasiliensis*, *T. colubriformis,* and *Ancylostoma duodenale* (Symons, 1965, 1969; Shayo and Benz, 1979; Barker and Titchen, 1982; Pimparker et al. 1982). Altered gastric acid secretion and elevation of pH has been demonstrated in rats infected with the larval stage of the tapeworm *T. taeniaeformis*, although the larval parasite itself resides in the liver (Cook and Williams, 1981). Hypersecretion of hydrochloric acid by the gastric mucosa in humans has been associated with the fish tapeworm *Diphyllobothrium latum* (Siurala, 1954). Finally, decreased pancreatic secretion of enzymes, bicarbonate, and fluid has been demonstrated in dogs infected with *T. spiralis* (Dembinski et al., 1979a) and reduced bicarbonate output observed in patients infected with the human blood fluke *S. mansoni* (Mott et al., 1972).

B. Epithelial Cell Functioning

Epithelial ion and fluid transport may be affected by parasitic infection. Thus active secretion of chloride ions, which causes net fluid movement into the lumen, has been observed in a number of enteric parasite infections: the nematodes *Strongyloides stercoralis, T. spiralis,* and *N. brasiliensis* (Kane et al., 1984; Baird et al., 1985; Russell and Castro, 1985, 1989; Castro et al., 1987) and the liver fluke *F. hepatica* (O'Malley et al., 1993). The phenomenon may have an immunological basis, since it can be triggered by antigenic challenge and by mast cell mediators (5-HT and histamine). Fluid secretion may also be responsible for the secretory diarrhea that occurs in a number of parasitic infections; it has even been suggested that it may represent a protective mechanism for the host, aiding worm expulsion (Baird and O'Malley, 1993).

Fluid movement into the intestinal lumen may be associated with permeability changes in the intestinal mucosa. Increased permeability has been demonstrated in *N. brasiliensis* and *H. contortus* (Murray, 1969; Bloch et al., 1979; Cobden et al., 1979; Dakkak et al., 1982). Reduced brush border enzyme activity, leading to maldigestion and decreased absorption, may also stimulate fluid movement into the gut lumen. Lowered enzyme levels have been observed in infections with *N. brasiliensis* (Symons and Fairbairn, 1962), *T. spiralis* (Castro et al., 1967) and *T. colubriformis* (Symons and Jones, 1970). Malabsorption has been noted in patients with hookworms (Sheehy et al., 1962) and in pigs infected with *Ascaris suum* (Forsum et al., 1981; Stephenson et al.,1980). It has also been observed in other experimental infections, including *N. brasiliensis* in rats (Symons, 1965; Symons et al., 1971; Scofield, 1977, 1980; Nolla et al., 1985).

Increased epithelial cell turnover, as evidenced by enhanced mitotic activity, has been observed in *N. brasiliensis* infections in rats, for example (Symons, 1978). Altered differentiation of mucosal cells also occurs: much of the hyperplasia of the parasitic gastritis in *Ostertagia* infections is due to the production of mucus-secreting cells (Anderson et al., 1965; McLeay et al., 1973).

C. Goblet Cell Activity

Increased goblet cell activity has been observed in response to a variety of helminth infections. Changes in four parameters have been described: a physical increase in cell numbers (that is, a hyperplasia) and increases in rate of mucin synthesis, mucus secretion and qualitative changes in goblet cell mucins (Table 6-1). In *N. brasiliensis* infections, for example, there is a switch from neutral to acidic mucins, with sulphomucins predominating amongst acidic mucins, rather than sialomucins (Koninkx et al. 1988).

Table 6-1. Effects of Helminth Parasites on Goblet Cell Activity

Parasite	Host	Effects on goblet cells	References
1. Nematodes			
Nippostrongylus brasiliensis	Rat	Hyperplasia, increased mucin synthesis, qualitative changes in mucin	Koninkx et al., 1988; Miller, 1987; Miller et al., 1981; Miller and Nawa, 1979
Trichinella spiralis	Rat & mouse	Hyperplasia, increased mucin synthesis	Alizadeh and Wakelin, 1982b; Bell et al., 1984; McCoy, 1940
Trichostrongylus colubriformis	Sheep	Increased mucus secretion	Douch et al., 1983 Jones et al., 1990
Trichostrongylus vitrinus	Sheep	Hyperplasia	Jackson et al., 1983
Strongyloides ratti	Mouse	Hyperplasia	Carroll et al., 1984
Oesophagostomum columbianum	Sheep	Hyperplasia	Dobson, 1967
2. Cestodes			
Hymenolepis diminuta	Mouse	Hyperplasia, increased mucus secretion	McKay et al., 1990
3. Trematodes			
Echinostoma trivolvis	Mouse	Hyperplasia	Weinstein and Fried, 1991
Echinostoma hortense	Mouse	Hyperplasia	Tani and Yoshimura, 1988
Fasciola hepatica	Rat	Hyperplasia	Pfister and Meierhofer, 1986

Mucus secretion from goblet cells is stimulated by a number of secretagogues, including acetylcholine from enteric nerves, leukotrienes from inflammatory cells, and prostaglandins (particularly prostaglandins E and F). The role of mast cell mediators (e.g., 5-HT, histamine) remains controversial. Although mast cell and goblet cell responses may be linked, 5-HT and histamine may not act directly on mucus secretion, but indirectly via goblet cell proliferation (Lamont, 1992; Neutra and Forstner, 1987). The proportion of goblet cells within the overall population of epithelial cells may rise in infected animals (e.g., *N. brasiliensis* infection in rats: Miller and Nawa, 1979). The mechanism underlying this phenomenon is not known, although it has been suggested that circulating lymphocytes may "home" to the jejunum and release mediators or growth factors that in turn influence goblet cell differentiation (Lamont, 1992).

A temporal link between goblet cell hyperplasia and expulsion of a primary infection has been established for a number of gut parasites (e.g., Tani and Yoshimura, 1988; Rothwell, 1989; McKay et al., 1990; Weinstein and Fried, 1991). The hyperplasia occurs more quickly in response to a secondary infection and the resulting secretion of copious amounts of mucus prevents establishment of worms (Bell et al., 1984; Miller, 1987; McKay et al., 1990; Lamont, 1992). Goblet cell proliferation parallels mast cell hyperplasia in a number of nematode infections, although the two phenomena and their postulated association with rejection are not always linked (Rothwell, 1989). Nevertheless, where the two events are related, the mucus response is perhaps best viewed as part of a wider immunologically mediated host defence mechanism.

Mucus secreted onto the surface of the intestine forms a physical barrier to invading parasites, engulfing and trapping them. In this way, their attachment to the surface is prevented and their expulsion by peristalsis facilitated. Goblet cell responses to helminth infections have been divided into two categories, those resulting from tissue injury and those triggered by immunological stimulation (Tse and Chadee, 1991). The invasion of the gut mucosa by parasites such as *N. brasiliensis* causes damage to the epithelial cells, which can elicit an inflammatory response. The latter involves the recruitment of neutrophils, macrophages, and mast cells to the site of injury, their products stimulating goblet cells to proliferate and secrete mucus, thus limiting tissue injury and aiding in the expulsion of the parasite.

The goblet cell response to immunological stimulation during a primary infection forms a layer of mucus over the surface of the intestine that deters worm establishment or contains a substance that disrupts the normal motility and feeding behaviour of the parasite. The entrapment of the parasite in mucus allows the complexing of parasite antigens with IgG and IgE antibodies and sets in motion a number of local immune events in the lamina propria, leading to the release of goblet cell–stimulating factors from macrophages and T-cells.

There is an accelerated goblet cell response to secondary infections: in nematode infections it can occur within a matter of hours (Lamont, 1992). The challenge of hypersensitive mucosae by worm antigen leads to increased mucosal permeability and secretory changes that involve inflammatory mediators. Parasite antigens can cause mast cell degranulation and mucus release; the mast cell mediators may stimulate goblet cell release or, via permeability changes in the mucosa, leak into the lumen and gain more direct contact with the parasite. Once there, histamine and 5-HT may induce paralysis of the worm, aid in immobilizing it within a coating of mucus, or cause structural damage to the parasite. Leukotrienes and 5-HT can mediate direct damage to *T. colubriformis*, for example (Tse and Chadee, 1991). A systemic immune response may also be triggered, which allows for immunological stimulation by movement of IgG into the lumen. Antigen–antibody (Ag-Ab) complexes are known mucus secretagogues that can participate in immune expulsion of intestinal helminths (Tse and Chadee, 1991). As well as containing IgG, IgE, and complement, mucus contains secretory IgA (sIgA) and this, too, may be involved in host defense mechanisms against infection (Neutra and Forstner, 1987; Tse and Chadee, 1991).

III. The Role of the Neuroendocrine and Immune Systems in Gut Responses to Parasitism

The intake, movement, digestion, and absorption of food require sophisticated and well-coordinated control mechanisms involving complex interactions between the nervous and endocrine systems. Regulatory peptides are present in both systems and can act as hemocrine, neurocrine, and paracrine messengers. Moreover, their coexistence in nerve cells with classical neurotransmitters provides an extra dimension to regulatory processes, allowing them to operate in a neuromodulatory fashion. A useful review of the dominant role of peptides in digestive processes is given by Nicholl and colleagues (1985). Parasites disrupt peptidergic mechanisms in both the gut endocrine and nervous systems and each system will be dealt with separately.

The gut contains a particularly high concentration of immune cells, with approximately the same number of cells as in the rest of the body put together (O'Dorisio, 1986; Wood, 1991). The enteric immune system is responsible for mounting an effective immune response to invasion by parasites, a response that may ultimately result in their expulsion from the body. The role of immune effector cells such as eosinophils and mast cells in this response will be considered separately.

A. Effects of Parasites on the Gut Neuroendocrine System

1. Effects on Gut Hormones

A summary of the data pertaining to helminth parasites is presented in Table 6-2. As is evident from the table, hypergastrinemia is a common feature of gut nematode infections; it has also been observed in hepatic infections with the

Table 6-2. Effect of Gut Parasites on Host Peptide Levels

Hormone	Change	Parasite	Host	References
Gastrin	Increase	*Ostertagia* spp.	Sheep	Anderson et al., 1985
		Haemonchus contortus	Lamb	Nicholls et al., 1985
		Trichinella spiralis	Rat	Castro et al., 1976
		Strongyloides ransomi	Pig	Enigk and Dey-Hazra, 1978
		Taenia taeniaeformis	Rat	Cook et al., 1981
Gastrin	Decrease	*Trichostrongylus colubriformis*	Sheep	Titchen, 1982
Gastrin releasing peptide	Increase	*Echinostoma liei*	Mouse	Thorndyke et al., 1988
Cholecystokinin	Decrease	*Nippostrongylus brasiliensis*	Rat	Ovington et al., 1985
Insulin	Decrease	*N. brasiliensis*	Rat	Ovington et al., 1985
Secretin	Increase	*N. brasiliensis*	Rat	Ovington et al., 1985
Glucagon	Increase	*N. brasiliensis*	Rat	Ovington et al., 1985
Enteroglucagon	Increase	*N. brasiliensis*	Rat	Ovington et al., 1985
		H. contortus	Lamb	Nicholls et al., 1985
		Hymenolepis diminuta	Mouse	McKay et al., 1991
Pancreatic polypeptide	Decrease	*Ostertagia* spp.	Sheep	Anderson et al., 1985
Somatostatin	Decrease	*Ostertagia* spp.	Sheep	Titchen and Reid, 1988
Substance P	Increase	*H. diminuta*	Mouse	McKay et al., 1991
Neurotensin	Decrease	*H. diminuta*	Mouse	McKay et al., 1991
Vasoactive intestinal peptide	Decrease	*H. diminuta*	Mouse	McKay et al., 1991
		E. liei	Mouse	Thorndyke et al., 1988
Peptide histidine isoleucine	Decrease	*H. diminuta*	Mouse	McKay et al., 1991
β-endorphin	Increase	*Schistosoma mansoni*	Mouse	Isseroff et al., 1989

metacestode of *T. taeniaeformis* (Cook et al., 1981). However, it is not a universal phenomenon because gastrin levels are reduced in infections with *T. colubriformis* (Titchen, 1982), whereas no changes accompany infections with *N. brasiliensis* (Ovington et al., 1985). In the latter, levels of glucagon and enteroglucagon rise, while levels of CCK and insulin fall (Ovington et al., 1985). Elevated gastrin levels are not observed in *Hymenolepis diminuta* infections of the rat (normal host) (Castro et al., 1976) or the mouse (abnormal host) (McKay et al., 1991). Despite this, levels of other peptides are altered in the tapeworm–mouse interaction: enteroglucagon and SP levels are increased, whereas levels of neurotensin, VIP, and peptide histidine isoleucine (PHI) are reduced (McKay et al., 1991).

Although the gut is the largest endocrine gland in the body, the diffuse organization of the endocrine tissue has hampered understanding of hormonal control mechanisms. Consequently, it is difficult to assess the impact of changes in one or at best a few hormones out of the total complement of gut peptides. Changes in the activity of one hormone will undoubtedly alter the functioning of other hormones. The situation is further complicated by the role of peptides in the enteric nervous system and in cell-mediated immune responses. Bearing these points in mind, some of the changes in peptide hormone levels have been linked to specific pathophysiological effects of the kind discussed in Section II. Unfortunately, the information available to date is limited, being concerned with only four hormones: gastrin, enteroglucagon, CCK, and secretin (Table 6-3).

Gastrin is known to stimulate gastric acid secretion (Jönsson, 1989); consequently, elevated levels are likely to cause a rise in abomasal pH. In turn, the latter condition will lead to changes in intestinal motility and digesta flow, possibly resulting in the diarrhea that is typical of some nematode infections (for references, see Table 6-3). Gastrin is also a trophic factor, stimulating gastrointestinal growth (Goodlad and Wright, 1987). In consequence, increased gastrin levels may be linked to the gut hypertrophy observed in *Ostertagia* infections (and proliferation of goblet cells in particular) (Titchen and Reid, 1988) and the gut hyperplasia that occurs in *T. taeniaeformis* infections (Cook et al., 1981). Gastrin is known to have a trophic action on both parietal and mucus-secreting cells (Willems et al., 1977).

Enteroglucagon is another trophic gut hormone (Goodlad and Wright, 1987). *N. brasiliensis* infections trigger epithelial hyperplasia in the small intestine and it has been suggested that this response, triggered by elevated enteroglucagon levels, may compensate for the villous atrophy caused by the parasite (Ovington et al., 1985). Enteroglucagon has also been linked with the crypt cell hyperplasia that occurs in *H. diminuta* infections in mice (McKay et al. 1991).

Reduced levels of CCK and secretin have been linked with decreased pancreatic secretion in *T. spiralis* infections, although the association has yet to be confirmed (Dembinski et al., 1979a, b). Regulation of gastric acid and pepsin secretion is also impaired in this infection; this may be due to a failure of secretin-mediated mechanisms to inhibit gastrin-induced secretions (Dembinski et al., 1979b). CCK is regarded as a satiety factor in controlling food intake (Baile et al., 1986); this fact has been used as an argument to support the contention that the anorexia observed in sheep infected with *T. colubriformis* is mediated by increased circulatory levels of CCK (Symons and Hennessy, 1981).

2. *Effects on the Enteric Nervous System*

The enteric nervous system is involved in coordinating the various digestive functions of the different gut regions and is extensive, containing as many neurons as the spinal cord; it effectively functions like an independent minibrain (Furness and Costa, 1986; Wood, 1991). There has been only one record of disruption of the peptidergic component of the enteric nervous system by a helminth parasite, namely, the human blood fluke, *S. mansoni*. During the course of schistosome infections in mice, granulomas form around entrapped eggs in the ileum and colon and cause focal destruction of the nerves in these organs. The damage to the nerves could alter bowel motility in the diarrhea that is commonly associated

Table 6-3. Pathophysiological Consequences of Elevated Hormone Levels in Intestinal Parasitism

Hormone	Pathophysiological change	Parasite	Host	References
Gastrin	Stimulation of gastric acid secretion, leaing to increased gut acidity and changes in intestinal motility and digesta flow. May be responsible for diarrhea.	*Ostertagia* spp.	Sheep	Anderson et al., 1985 Titchen, 1982; Titchen and Reid, 1988
		Trichinella spiralis	Rat	Castro et al., 1976
		Haemonchus contortus	Lamb	Nicholls et al., 1985
Gastrin	Hypertrophy of abomasal and small intestinal mucosa. Hyperplasia of stomach and small intestine.	*Ostertagia circumcincta*	Cow, sheep	Titchen and Reid, 1988
		Taenia taeniaeformis	Rat	Cook et al., 1981
Enteroglucagon	Epithelial hyperplasia in small intestine.	*Nippostrongylus brasiliensis*	Rat	Ovington et al., 1985
	Crypt cell hyperplasia in small intestine.	*Hymenolepis diminuta*	Mouse	McKay et al., 1991
Cholecystokinin	Anorexia.	*Trichostrongylus colubriformis*	Sheep	Symons and Hennessy, 1981
Cholecystokinin and secretin (reduced levels)	Decreased pancreatic secretion (bicarbonate and protein output).	*Trichinella spiralis*	Dog	Dembinski et al., 1979a, b

with schistosomiasis. Immunostaining for VIP has yielded an increased reaction in nerve cell bodies belonging to nerve ganglia adjacent to the granuloma (Varilek et al., 1991). Such an enhanced reaction may be linked to the pathology associated with the disease, because VIPomas are associated with watery diarrhea (Bloom and Polak, 1982).

More substantive information on gut neuropeptide changes has been obtained from studies involving the protozoan parasite *Trypanosoma cruzi*, the causative agent of Chagas' disease. Thus, reductions in the number and intensity of immunostaining of peptide-containing nerves (VIP, SP, enteroglucagon, and somatostatin) have been observed in rectal biopsy specimens from human patients (Long et al., 1980). In a parallel radioimmunoassay study, the levels of these four peptides were seen to drop to less than half their values in control samples (Long et al., 1980). A decrease in the SP content of the colon in infected mice has also been demonstrated and was correlated with reduction in the number of dense-cored vesicles in Auerbach's plexus of the colon (Almeida et al., 1977). The results of these two studies indicate a profound disruption of the neural and endocrine regulation of normal gut functioning that may be linked to the gastrointestinal abnormalities associated with the disease (Long et al., 1980).

B. *Effects of Parasites on the Gut Immune System*

1. *Eosinophils*

Along with mastocytosis and increased IgE synthesis, eosinophilia is one of the immunological hallmarks of parasitic helminth infections (examples are given in Table 6-4). Recruitment of eosinophils to the site of infection is stimulated by the release of eosinophil chemotactic factor of anaphylaxis (ECF-A) released from mast cells (Rothwell, 1989). Upon degranulation, eosinophils release a product known as major basic protein which is potentially toxic to parasites; it has been shown that eosinophils can attach to the surface of parasites *in vitro* and extrude their granule contents, causing surface damage. Parasites susceptible to such damage include the schistosomula of *S. mansoni*, the microfilariae and larvae of some filarial parasites, and juvenile liver flukes (Butterworth, 1984; Capron et al. 1978; Davies and Goose, 1981; McLaren et al., 1977). Despite this, a direct protective role *in vivo* for eosinophils against helminth parasites remains controversial; some workers consider that eosinophilia may simply be an immunological manifestation of the dominant Th2 response induced by the parasites (Sher and Coffman, 1992).

2. *Mast Cells*

Mast cell hyperplasia ("mastocytosis") occurs in a variety of helminth infections (Table 6-5). Activation of mast cells and release of their mediators can occur in response to a number of stimuli, including lymphokines, eosinophil

Table 6-4. Induction of Eosinophilia by Helminth Parasites

Parasite	Host	References
1. Nematodes		
Nippostrongylus brasiliensis	Rat	Taliaferro and Sarles, 1939
Strongyloides ratti	Rat	Moqbel, 1980
Trichostrongylus colubriformis	Guinea pig	Rothwell and Dineen, 1972
Oesophagostomum columbianum	Sheep	Shelton and Griffiths, 1967
Bunostomum phlebotomum	Calves	Sprent, 1946
Ostertagia ostertagi	Calves	Baker et al., 1993
2. Cestodes		
Hymenolepis nana	Mouse	Bortoletti et al., 1985; Bortoletti et al., 1989
Taenia taeniaeformis	Rat	Cook and Williams, 1981
Taenia crassiceps	Mouse	Siebert et al., 1979
3. Trematodes		
Metagonimus yokogawai	Mouse	Ohnishi, 1987
Fasciola hepatica	Rat	Burden et al., 1983; Charbon et al., 1991; Doy et al., 1981b; Pfister and Meierhofer, 1986

products, and nervous stimuli (e.g., acetylcholine released from autonomic nerve fibres innervating mast cells) (Rothwell, 1989). Mast cells release a battery of mediators: histamine, 5-HT, prostaglandins, leukotrienes, enzymes, cytokines, and chemotactic factors for eosinophils and neutrophils (Lee et al., 1986; Rothwell, 1989; Gordon et al., 1990; Galli, 1993). Histamine plays a role in inflammatory responses by recruiting eosinophils and neutrophils to the site of infection and influencing their activity; it also acts on T-cells via T-cell H_2 receptors (Lee et al., 1986). 5-HT affects vascular permeability and lymphocyte functions (Lee et al., 1986). Aside from its immunological functions, 5-HT exerts a variety of effects *in vitro* on the locomotory responses and metabolism of helminth parasites (see reviews by Rothwell, 1989; Webb, 1988, 1991). Such effects may influence parasite survival *in vivo*. Elevated levels of histamine and 5-HT in small intestine mucosal extracts have been observed in a number of nematode infections. However, the increases were not significant and reached their peak after expulsion (Rothwell, 1989). Leukotrienes have been shown to mediate direct damage to trichostrongyle larvae *in vitro* (Douch et al., 1983); the role of prostaglandins has yet to be established (Lee et al., 1986; Rothwell, 1989).

Although the possession of hydrolytic enzymes by mast cells implies that they have the capacity to damage and destroy parasites, there is little evidence to support this. Rather, the adherence of mast cells to parasites may draw eosinophils, neutrophils, and platelets to the parasites, where they become activated; it is these cells, not the mast cells, that are responsible for killing the parasite (Capron et al., 1978; Mackenzie et al., 1981).

Table 6-5. Induction of Mast Cell Hyperplasia (Mastocytosis) by Helminth Parasites

Parasite	Host	References
1. Nematodes		
Nippostrongylus brasiliensis	Rat	Arizono and Nakao, 1988p; Cheema and Scofield, 1985; Nawa and Miller, 1979
Trichinella spiralis	Mouse	Alizadeh and Wakelin, 1982a; Tuohy et al., 1990
Strongyloides ratti	Rat	Olson and Schiller, 1978; Mimori et al., 1982
Strongyloides ratti	Mouse	Abe and Nawa, 1988
Strongyloides venezuelensis	Mouse	Khan et al., 1993
Trichuris muris	Mouse	Lee and Wakelin, 1982
Trichostrongylus colubriformis	Guinea pig	Rothwell and Dineen, 1972
Ostertagia astertagi	Calves	Baker et al., 1993
2. Cestodes		
Hymenolepis diminuta	Rat	Featherston and Copeman, 1990; Featherston et al., 1992; Hindsbo et al., 1982; Ishih, 1992
Hymenolepis diminuta	Mouse	Andreassen et al., 1978
Hymenolepis microstoma	Mouse	Novak and Nombrado, 1988
Hymenolepis nana	Mouse	Bortoletti et al., 1989
Taenia taeniaeformis	Rat	Cook and Williams, 1981
3. Trematodes		
Metagonimus yokogawai	Mouse	Ohnishi and Taufan, 1984
Fasciola hepatica	Rat	Charbon et al., 1991; Doy et al., 1981a, b; Pfister and Meierhofer, 1986
	Sheep	Murray et al., 1968

Peak mast cell numbers have been correlated with worm expulsion in a number of nematode, cestode, and trematode infections (e.g., Andreassen et al., 1978; Ohnishi and Taufan, 1984; Rothwell, 1989). However, this is not a universal phenomenon, because mastocytosis can occur without expulsion and vice versa. The experimental evidence surrounding this relationship and the results of studies involving drugs that modify the activity of mast cell mediators are discussed in the reviews by Lee and colleagues (1986) and Rothwell (1989). Mast cell hyperplasia has also been linked with goblet cell hyperplasia in a number of infections but this is not an invariable association. Mast cell mediators may enter mucus and affect trapped worms in a number of ways, as described previously.

A number of hypotheses have been advanced to try to account for the role of effector cells such as mast cells in the expulsion process. A favoured idea is that expulsion occurs due to deterioration of the gut microenvironment as a result of immune-mediated inflammation in the gut mucosa. Mediators may trigger mucus hypersecretion, hypermotility, epithelial cell changes, increased fluid secretion, increased mucosal permeability, and other effects that, together with more direct metabolic damage to the parasite, make life intolerable for the parasite.

IV. Interplay between the Neuroendocrine and Immune Systems

Just a few years ago, endocrine influences over the immune system were considered to be simple, being limited to glucocorticoids and certain neurotransmitter molecules (histamine, serotonin, and acetylcholine). Now it is recognized that the two systems share a much wider range of molecules—immune cells produce peptides normally associated with the neuroendocrine system and neuroendocrine cells produce peptides classically linked to immune cells. Moreover, they possess the same array of receptors for shared ligands. The potential of this framework for bidirectional communication between the two systems is realized through the hormonal influence of immunologic peptides and the immunoregulatory functions of neuroendocrine peptides. This concept has led to the creation of a new interdisciplinary research area, termed neuroimmunoendocrinology (Blalock, 1989). Such interplay between the two systems is not surprising when one considers the often intimate association between immune cells and neuroendocrine factors. Immune cells are exposed *in vivo* to hormones, neurotransmitters, and neuropeptides, while immunologic organs are innervated. For more background information on neuroimmunoendocrinology, the reader is referred to reviews by O'Dorisio, 1986; Shanahan and Anton, 1988; Blalock, 1989; and O'Dorisio and Panerai, 1990.

The two-way communication between the immune and nervous systems has led to the hypothesis that one of the principal functions of the immune system may be to serve as a sensory organ (Blalock, 1989). According to this idea, the immune system senses noncognitive stimuli (e.g., bacteria, viruses, and antigens) that are not recognized by the nervous system. The recognition of such stimuli by immune cells is converted into information in the form of peptide hormones, lymphokines, and monokines that is conveyed to the neuroendocrine system and a physiological change occurs. Alternatively, neural recognition of cognitive stimuli results in similar hormonal information being conveyed to and recognized by hormone receptors on immune cells and an immunologic change results. It seems probable, then, that the sensory function of the immune system may mimic the neuroendocrine system in terms of a given stimulus evoking a particular set of hormones and hence physiological responses. If this is so, then the pathophysiology that is associated with a particular infectious agent (such as a parasite), antigen, or tumour could be related to the particular hormone or set of hormones that are produced by the immune system. For example, leukocyte ACTH may be responsible for the well-known increase in circulatory corticosteroid levels observed during viral and bacterial infections. The next two sections will consider what evidence is available for the involvement of peptides and cytokines in the responses to helminth infections.

A. The Role of Peptides

Substance P (SP) has been implicated in the response to a number of helminth infections (Table 6-6). For example, it is produced by eosinophil cells in hepatic egg granulomas in murine *S. mansoni* infections (Weinstock and Blum, 1989). Moreover, SP is the predominant tachykinin produced (rather than neurokinin A and neuropeptide K) even though all three peptides are coded for in a single preprotachykinin gene (Nawa et al., 1983). The result indicates that the eosinophils express mainly α-preprotachykinin mRNA. Since SP is known to play a modulatory role in a variety of immune responses, it is possible that it may be involved in the regulation of the granulomatous response in schistosomiasis (Weinstock and Blum, 1989). Elevated gut tissue levels of SP correspond to peak rejection times in primary and secondary infections of mice with *H. diminuta* (McKay et al., 1991), which correlates with increased 5-HT–positive enterochromaffin cells (McKay et al., 1990) known to contain SP (Coupland and Fujita, 1976). Thus, SP may contribute to the immune response that leads to worm rejection in this abnormal host, although it is probably not solely responsible for rejection since increased levels of SP can also be recorded in infected rats in which the rejection phenomenon is not evident (McKay et al., 1991).

As described previously, mast cell hyperplasia is a typical feature of gut nematode infections, since the mast cells orchestrate the immune response and are involved in worm repulsion (Lee et al., 1986; Rothwell, 1989). In *N. brasiliensis*-infected rats, a direct innervation of intestinal mucosal mast cells by nerve fibers containing SP (and calcitonin gene-related peptide (CGRP)) has been demonstrated (Stead et al., 1987), providing evidence for possible communication between the nervous and immune systems. That communication may be one-way, two-way, or both, but neural influence over the immune system is certainly a possibility, because SP has been shown to stimulate histamine release from mast cells isolated from *N. brasiliensis* - infected intestinal tissue (Shanahan et al., 1985). Other neuropeptides, including neurotensin, VIP, and somatostatin, can trigger the release of mast cell mediators (Lee et al., 1986).

VIP is known to exert an inhibitory influence over immune effector cells; consequently, the reduced tissue levels of VIP recorded in infections with *H. diminuta* and *E. liei* may facilitate an immune response to these parasites (McKay et al., 1991; Thorndyke et al., 1988). Long-term infections of *S. mansoni* in hamsters leads to activation of the endogenous opiate system (Kavaliers et al., 1984). Several life cycle stages (miracidium, cercaria, schistosomule, and adult) of *S. mansoni* synthesise and release proopiomelanocortin (POMC)-derived peptides (namely, ACTH, MSH, and β-endorphin) and met-enkephalin (Duvaux-Miret et al., 1992a, 1993). It has been suggested that release of these peptides may serve to interfere with the immune response in both the snail and mammalian hosts by stimulating Th2-cell responses, while at the same time suppressing the potentially more dangerous Th1-cell responses (Duvaux-Miret et al., 1992b). This is

Table 6-6 Role of Peptides in the Immune Response to Helminth Parasites

Peptide	Parasite	Host	Observation	References
Substance P	*Schistosoma mansoni*	Mouse	Produced by eosinophils in hepatic egg granulomas.	Weinstock and Blum, 1989
Substance P	*Hymenolepis diminuta*	Mouse	Peak tissue levels correspond with rejection times in primary and secondary infections.	McKay et al., 1991
Substance P Calcitonin gene-related peptide (CGRP)	*Nippostrongylus brasiliensis*	Rat	Innervation of mucosal mast cells and triggering of histamine release (by SP) from mast cells.	Shanahan et al., 1985 Stead et al., 1987
Vasoactive intestinal peptide	*Hymenolepis diminuta* *Echinostoma liei*	Mouse Mouse	Facilitation of immune response.	McKay et al., 1991 Thorndyke et al., 1988
Opioid peptides	*Schistosoma mansoni*	Hamster	Activation of opioid system in long-term infection. Possible immunomodulatory role for opioids.	Kavaliers and Colwell, 1992; Kavaliers et al., 1984
Angiotensins I and II	*Schistosoma mansoni*	Mouse	Produced by macrophages surrounding hepatic egg granulomas. Possible role in regulating granolomatous inflammation.	Hirayama et al., 1990; Khayyal et al., 1991
ACTH, α-MSH, β-endorphin, metenkephalin	*Schistosoma mansoni*	Mouse Snail	Released by parasite and may serve to modulate the immune response in both mammalian and snail hosts.	Duvaux-Miret et al., 1992a, b, 1993
Plerocercoid growth factor (PGF)	*Spirometra mansonoides*	Mouse	Isolated human lymphocytes possess growth hormone receptors that bind PGF.	Phares and Watts, 1988

an interesting example of molecular mimicry in which the parasite and host communicate with each other via common signal molecules and the parasite making use of the molecules to evade the host's immune response. Hepatic granulomas in murine schistosmiasis produce angiotensin I and II (AI and AII, respectively) and possess high angiotensin converting enzyme (ACE) activity, which may contribute to the elevated serum ACE levels observed in infected animals (Hirayama et al., 1990; Khayyal et al., 1991). The peptides and converting enzymes are produced by the macrophages that surround the granulomas (Hirayama et al., 1990). Since AII is believed to be involved in inflammatory responses, it is possible that ACE activity plays a role in the regulation of granulomatous inflammation in schistosomiasis (Hirayama et al., 1990). Finally, it has been shown that human lymphocytes possess GH receptors that bind PGF produced by *S. mansonoides* (Phares and Watts, 1988; Phares, 1992, Chapter 5).

Gut peptides appear to have little effect on mucus secretion. This may be due to their operation via cAMP pathways, whereas mucus secretion occurs by a cAMP-independent pathway (Neutra and Forstner, 1987). In intestinal infections with the protozoan *Entamoeba histolytica* it has been suggested that peptides such as neurotensin and VIP exert an indirect effect on mucus secretion via stimulation of prostaglandin and leukotriene synthesis (Tse and Chadee, 1991).

B. The Role of Cytokines

Cytokines produced by the Th2 subset of $CD4^+$ T-cells are responsible for the eosinophilia and mast cell hyperplasia that are typical of helminth infections: the former is stimulated by interleukin–5 (IL-5) and the latter by interleukins 3, 4, and 10 (IL-3, IL-4, and IL-10, respectively) (Finkelman et al., 1991; Cox and Liew, 1992; Scott and Sher, 1992; Sher and Coffman, 1992). As mentioned previously, lymphokines can promote activation of mast cells and release of their mediators. Goblet cell hyperplasia is also regarded as a T-cell–dependent phenomenon, lymphokines activating macrophages whose products (prostaglandins and proteinases) are known mucus secretagogues.

The influence of mast cell mediators over enhanced fluid secretion in some enteric parasite infections may operate directly or indirectly via enteric nerves (Russell and Castro, 1987; Castro, 1989; Baird and O'Malley, 1993). Again, the increased permeability associated with fluid movement has been linked with enhanced mast cell activity (Russell and Castro, 1987). However, the specific involvement of cytokines in these neuroimmune relationships has yet to be established.

V. Connections and Conclusions

The review has examined, albeit briefly, some of the major components of the gut wall—endocrine cells, immune cells (eosinophils and mast cells), epithelial cells (e.g., goblet cells), nerve cells, and muscle—and demonstrated how the functioning of these tissues can change during parasitic infection. Moreover, it has attempted to show how responses from different tissues are interlinked with each other. For example, immune cells can influence epithelial cell physiology, nerve cells, endocrine cells, and muscle, and the various immune cells can communicate with each other; nerve cells can affect gut enterocytes, immune cells, endocrine cells, and muscle; and the endocrine system can exert an influence over epithelial cells, muscle, immune cells, and nerve cells. The particularly high concentration of different cell types within the gut allows for complex interactions between the cells, and the interactions can be one-way, two-way, direct, or indirect. The overall regulation of gut function is achieved by the integrated and coordinated activity of these cells—it is not compartmentalized into separate and independent endocrine, nervous, and immunological control systems as was believed just a few years ago. The connections between different elements of the control systems become disrupted or perhaps exaggerated during parasitic infection. These intrinsic mechanisms are themselves controlled by extrinsic factors from the central and autonomic nervous systems and from the endocrine system. The coordinated activity of the intrinsic and extrinsic systems comprises the host's overall response to invasion by the parasite, perhaps leading to its expulsion.

The gut tissues examined in this review produce a bewildering variety of chemical mediators, and there is a growing awareness that peptides play important roles in the intercellular communications that operate between them. Peptides are produced by endocrine, nerve, and immune cells and the cells possess receptors for them. Gross changes in peptide levels occur during parasitic infection and peptides have been linked with immune responses by, and physiological changes in, the host.

A number of models have attempted to account for the integration of gut responses to infection. That advanced by Perdue and McKay (1993, Fig. 2.8. See also Fig. 1, McKay and Perdue, 1993) emphasized interactions among nerves, immune cells, and epithelial cell functioning (including fibroblasts) but did not incorporate endocrine cells and goblet cells, for example, within the overall scheme. The model proposed by Castro (1992, Fig. 2. See also Fig. 1, Russell and Castro, 1987) essentially focused on inflammatory effects on muscle and epithelial cell functioning and on hormonal mechanisms that, in turn, affect gut secretions. The two-way nerve and endocrine interactions were not included. The scope of the models reflects both the emphasis of studies to date, that is, on immune influences over gut physiology, and the current state of our overall understanding of host responses to infection. The impact of parasites on the gut endocrine system has been examined for only a small number of peptides out of the total complement of hormones. There is a need to examine a greater range of hormones from both tissue and plasma samples and to examine tissues from other sites in the body invaded by parasites. The consequences of the changes to gut physiology and the role of peptides in immune responses are only beginning to be appreciated. Our perception of physiological responses is perhaps hampered by lack of knowledge of the precise actions of gut hormones themselves. It is these areas, then, that need to be addressed. They present an enormous challenge to physiologists, immunologists, and parasitologists alike, but will have a profound impact not just on our understanding of the host–parasite relationship, but of other disease states, too.

VI. References

Abe, T., and Nawa, Y. (1988). Worm expulsion and mucosal mast cell response induced by repetitive IL-3 administration in *Strongyloides ratti*–infected nude mice. *Immunology* **63:**181–185.

Alizadeh, H., and Wakelin, D. (1982a). Genetic factors controlling the intestinal mast cell response in mice infected with *Trichinella spiralis. Clin. Exp. Immunol.* **49:**331–337.

———. (1982b). Comparison of rapid expulsion of *Trichinella spiralis* in mice and rats. *Int. J. Parasitol.* **12:**65–73.

Almeida, H. O., Tafuri, W. L., Cunha-Melo, J. R., Freire-Maia, L., Raso, P., and Brener, Z. (1977). Studies on the vesicular component of the Auerbach's plexus and the substance P content of the mouse colon in the acute phase of the experimental *Trypanosoma cruzi* infection. *Virchows Arch. A Path. Anat. and Histol.* **376:**353–360.

Anderson, N., Armour, J., Jarrett, W. R. H., Jennings, F. W., Ritchie, J. D. S., and Urquart, G. M. (1965). A field study of parasitic gastritis in cattle. *Vet. Rec.* **77:**1196–1204.

Anderson, N., Blake, R., and Titchen, D. A. (1976). Effects of a series of infections of *Ostertagia circumcincta* on gastric secretion of sheep. *Parasitol.* **72:**1–12.

Anderson, N., Hansky, J., and Titchen, D. A. (1981). Effects of *Ostertagia circumcincta* infections on plasma gastrin in sheep. *Parasitol.* **82:**401–410.

———. (1985). Effects on plasma pepsinogen gastrin and pancreatic polypeptide of *Ostertagia spp.* transferred directly into the abomasum of sheep. *Int. J. Parasitol.* **15:**159–165.

Andreassen, J., Hindsbo, O., and Ruitenberg, E. J. (1978). *Hymenolepis diminuta* infections in congenitally athymic (nude) mice: Worm kinetics and intestinal histopathology. *Immunol.* **34:**105–113.

Arizono, N., and Nakao, S. (1988). Kinetics and staining properties of mast cells proliferating in rat small intestine tunica muscularis and subserosa following infection with *Nippostrongylus brasiliensis. APMIS* **96:**964–970.

Baile, C. A., McLaughlin, C. L., and Della-Fera, M. A. (1986). Role of cholecystokinin and opioid peptides in control of food intake. *Physiol. Rev.* **66:**172–234.

Baird, A. W., Cuthbert, A. W., and Pearce, F. L. (1985). Immediate hypersensitivity reactions in epithelia from rats infected with *Nippostrongylus brasiliensis. Br. J. Pharmacol.* **85:**787–795.

Baird, A. W., and O'Malley, K. E. (1993). Epithelial ion transport – possible contribution to parasite expulsion. *Parasitol. Today* **9:**141–143.

Baker, D. G., Gershwin, L. J., and Hyde, D. M. (1993). Cellular and chemical mediators of type 1 hypersensitivity in calves infected with *Ostertagia ostertagi:* Mast cells and eosinophils. *Int. J. Parasitol.* **23:**327–332.

Barker, I. K., and Titchen, D. A. (1982). Gastric dysfunction in sheep infected with *Trichostrongylus colubriformis,* a nematode inhabiting the small intestine. *Int. J. Parasitol.* **12:**345–356.

Bell, R. G., Adams, L. S., and Ogden, R. W. (1984). Intestinal mucus trapping in the rapid expulsion of *Trichinella spiralis* by rats: Induction and expression analyzed by quantitative worm recovery. *Infect. Immun.* **45:**267–272.

Blalock, J. E. (1989). A molecular basis for bidirectional communication between the immune and neuroendocrine systems. *Physiol. Rev.* **69:**1–32.

Bloch, K. J., Bloch, D. B., Stearns, M., and Walker, W. A. (1979). Intestinal uptake of macromolecules, VI. Uptake of protein antigen *in vivo* in normal rats and in rats infected with *Nippostrongylus brasiliensis* or subjected to mild systematic anaphylaxis. *Gastroenterology* **77:**1039–1044.

Bloom, S. R., and Polak, J. M. (1982). Clinical aspects of gut hormones and neuropeptides. *Br. Med. Bull.* **38:**233–238.

Bortoletti, G., Conchedda, M., and Ferretti, G. (1985). Damage and early destruction of *Taenia taeniaeformis* larvae in resistant hosts, and anomalous development in susceptible hosts: A light microscopic and ultrastructural study. *Int. J. Parasitol.* **15:**377–384.

Bortoletti, G., Gabriele, F., and Palmas, C. (1989). Kinetics of mast cells, eosinophils, and phosopholipase B activity in the spontaneous-cure response of two strains of mice (rapid and slow responder) to the cestode *Hymenolepis nana. Parasitol. Res.* **75:**465–469.

Bowers, R. L., Castro, G. A., and Weisbrodt, N. W. (1986). Alterations in intestinal smooth muscle of the rat induced by *Trichinella spiralis. Gastroenterology* **90:**1353.

Bueno, L., Dakkak, A., and Fioramonti, J. (1982). Gastroduodenal motor and transit disturbances associated with *Haemonchus contortus* infection in sheep. *Parasitol.* **84:**367–374.

Bueno, L., Ruckebusch, Y., and Dorchies, P. (1979). Disturbances of digestive motility in horses associated with strongyle infection. *Vet. Parasitol.* **5:**253–260.

Burden, D. J., Bland, A. P., Hammet, N. C., and Hughes, D. L. (1983). *Fasciola hepatica:* Migration of newly excysted juveniles in resistant rats. *Exp. Parasitol.* **56:**277–288.

Butterworth, A. E. (1984). Cell-mediated damage to helminths. *Adv. Parasitol.* **23:**143–235.

Capron, M., Capron, A., Torpier, G., Bazin, H., Bout, D., and Joseph, M. (1978). Eosinophil-dependent cytotoxicity in rat schistosomiasis. Involvement of IgG2a antibody and role of mast cells. *Eur. J. Immunol.* **8:**127–133.

Carroll, S. M. Mayrhofer, G., Dawkins, H. J. S., and Grove, D. I. (1984). Kinetics of intestinal lamina propria mast cells, globule leucocytes, intraepithelial lymphocytes, goblet cells, and eosinophils in murine strongyloidiasis. *Int. Arch. Allergy Appl. Immunol.* **74:**311–317.

Castro, G. A. (1980). Regulation of pathogenesis in disease caused by gastrointestinal parasites. In *The host invader interplay* (H. van den Bossche, ed.), pp. 457–467. Amsterdam: Elsevier/North-Holland Biomedical Press.

Castro, G. A. (1989). Immunophysiology of enteric parasitism. *Parasitol. Today* **5:**11–19.

Castro, G. A. (1992). Intestinal physiology in the parasitized host: Integration, disintegration, and reconstruction of systems. *Ann. N.Y. Acad. Sci.* **664:**369–379.

Castro, G. A., Copeland, E. M., Dudrick, S. J., and Johnson, L. R. (1976). Serum and antral gastrin levels in rats infected with intestinal parasites. *Am. J. Trop. Med. Hyg.* **25:**848–853.

Castro, G. A., Harari, Y., and Russell, D. (1987). Mediators of anaphylaxis-induced ion transport changes in small intestine. *Am. J. Physiol.* **253:**B540–B548.

Castro, G. A., Olson, L. J., and Baker, R. D. (1967). Glucose malabsorption and intestinal histopathology in *Trichinella spiralis*-infected guinea pigs. *J. Parasitol.* **53:**595–612.

Charbon, J. L., Spähni, M., Wicki, P., and Pfister, K. (1991). Cellular reactions in the small intestine of rats after infection with *Fasciola hepatica. Parasitol. Res.* **77:**425–429.

Cheema, K. J., and Scofield, A. M. (1985). The influence of level of infection of rats with *Nippostrongylus brasiliensis* on the hematology and phospholipase activity and mast cell numbers in the small intestine and colon. *Int. J. Parasitol.* **15:**55–60.

Cobden, I., Rothwell, J., and Axon, A. T. R. (1979). Intestinal permeability in rats infected with *Nippostrongylus brasiliensis. Gut* **20:**716–721.

Cook, R. W., and Williams, J. F. (1981). Pathology of *Taenia taeniaeformis* infection in the rat: Gastrointestinal changes. *J. Comp. Path.* **91:**205–217.

Cook, R. W., Williams, J. F., and Lichtenberger, L. M. (1981). Hyperplastic gastropathy in the rat due to *Taenia taeniaeformis* infection: Parabiotic transfer and hypergastrinemia. *Gastroenterology* **80:**728–734.

Cox, F. E. G., and Liew, E. Y. (1992). T-cell subsets and cytokines in parasitic infections. *Parasitol. Today* **8:**371–374.

Dakkak, A., Fioramonti, J., and Bueno, L. (1982). *Haemonchus contortus:* Abomasal transmural potential difference and permeability changes associated with experimental infection in sheep. *Exp. Parasitol.* **53:**209–216.

Davies, C., and Goose, J. (1981). Killing of newly excysted juveniles of *Fasciola hepatica* in sensitized rats. *Parasite Immunol.* **3:**81–96.

Dembinski, A. B., Johnson, L. R., and Castro, G. A. (1979a). Influence of enteric parasitism on hormone-regulated pancreatic secretion in dogs. *Am. J. Physiol.* **237:**R232–R238.

———. (1979b). Influence of parasitism on secretin-inhibited gastric secretion. *Am. J. Trop. Med. Hyg.* **28:**854–859.

Dobson, C. (1967). Changes in the protein content of the serum and intestinal mucus of sheep with reference to the histology of the gut and immunological response to *Oesophagostomum columbianum* infections. *Parasitol.* **57:**201–219.

Douch, P. G. C., Harrison, G. B. L., Buchanan, L. L., and Greer, K. S. (1983). *In vitro* bioassay of sheep gastrointestinal mucus for nematode-paralysing activity mediated by a substance with some properties characteristic of SRS-A. *Int. J. Parasitol.* **13:**207–212.

Doy, T. G., Hughes, D. L., and Harness, E. (1981a). Hypersensitivity in rats infected with *Fasciola hepatica:* Lack of correlation between serum reaginic antibody levels and rejection of flukes. *Res. Vet. Sci.* **30:**357–359.

———. (1981b). Hypersensitivity in rats infected with *Fasciola hepatica:* Possible role in protection against a challenge infection. *Res. Vet. Sci.* **30:**360–363.

Duvaux-Miret, O., Leung, M. K., Capron, A., and Stefano, G. B. (1993). *Schistosoma mansoni:* An enkephalinergic system that may participate in internal and host–parasite signaling. *Exp. Parasitol.* **76:**76–84.

Duvaux-Miret, O., Stefano, G. B., Smith, E. M., and Capron, A. (1992a). Neuroimmunology of host–peptide interactions: Proopiomelanocortin-derived peptides in the infection by *Schistosoma mansoni. Adv. Neuroimmunol.* **2:**297–311.

Duvaux-Miret, O., Stefano, G. B., Smith, E. M., Dissous, C., and Capron, A. (1992b). Immunosuppression in the definitive and intermediate hosts of the human parasite *Schistosoma mansoni* by release of immunoactive neuropeptides. *Proc. Natl. Acad. Sci. USA* **89:**778–781.

Enigk, K., and Dey-Hazra, A. (1978). Influence of a *Strongyloides* infection on the formation of gastrin and other hormones in the pig. *Zbl. Vet. Med. B* **25:**697–706.

Featherston, D. W., and Copeman, C. N. (1990). Mucosal mast cells in Sprague-Dawley rats infected with *Hymenolepis diminuta* tapeworms. *Int. J. Parasitol.* **20:**401–403.

Featherston, D. W., Wakelin, D., and Lammas, D. A. (1992). Inflammatory responses in the intestine during tapeworm infections. Mucosal mast cells and mucosal mast cell proteases in Sprague-Dawley rats infected with *Hymenolepis diminuta. Int. J. Parasitol.* **22:**961–966.

Finkelman, F. D., Pearce, E. J., Urban, J. F., Jr., and Sher, A. (1991). Regulation and biological function of helminth-induced cytokine responses. In *Immunoparasitology today* (C. Ash and R. B. Gallagher, eds.), Vol. 7, pp. A62–A66. Cambridge: Elsevier Trends Journals.

Forsum, E., Nesheim, M. C., and Crompton, D. W. T. (1981). Nutritional aspects of *Ascaris* infection in young protein-deficient pigs. *Parasitol.* **83:**497–512.

Furness, J. B., and Costa, M. (1980). Types of nerves in the enteric nervous system. *Neuroscience* **5:**1–20.

Galli, S. J. (1993). New concepts about the mast cell. *N. Engl. J. Med.* **328:**257–265.

Goodlad, R. A., and Wright, N. A. (1987). Peptides and epithelial growth regulation. *Experientia* **43:**780–784.

Gordon, J. R., Burd, P. R., and Galli, S. J. (1990). Mast cells as a source of multifunctional cytokines. *Immunol. Today* **11:**458–464.

Gregory, P. C., Wenham, G., Poppi, D., Coop, R. L., MacRae, J. C., and Miller, S. T. (1985). The influence of a chronic subclinical infection of *Trichostrongylus colubriformis* on gastrointestinal motility and digesta flow in sheep. *Parasitol.* **91:**381–396.

Hindsbo, O., Andreassen, J., and Ruitenberg, J. (1982). Immunological and histopathological reactions of the rat against the tapeworm *Hymenolepis diminuta* and the effects of antithymocyte serum. *Parasite Immunol.* **4:**59–76.

Hirayama, K., Tukwyama, K., and Epstein, W. L. (1990). Angiotensin II–producing proteases from granulomatous tissue reaction in mice infected with *Schistosoma mansoni. Comp. Biochem. Physiol.* **96B:**553–557.

Ishih, A. (1992). Mucosal mast cell response to *Hymenolepis diminuta* infection in different rat strains. *Int. J. Parasitol.* **22:**1033–1035.

Isseroff, H., Sylvester, P. W., Bessette, C. L., Jones, P. L., Fisher, W. G., Rynkowski, T. A., and Gregor, K. R. (1989). Schistosomiasis: Role of endogenous opioids in suppression of gonadal steroid secretion. *Comp. Biochem. Physiol.* **94A:**41–45.

Jackson, F., Angus, K. W., and Coop, R. L. (1983). The development of morphological changes in the small intestine of lambs continuously infected with *Trichostrongylus vitrinus. Res. Vet. Sci.* **34:**301–304.

Jones, W. O., Windon, R. G., Steel, J. W., and Outteridge, P. M. (1990). Histamine and leukotriene concentrations in duodenal tissue and mucus of lambs selected for high and low responsiveness to vaccination and challenge with *Trichostrongylus colubriformis. Int. J. Parasitol.* **20:**1075–1079.

Jönsson, A.-C. (1989). Gastrin/cholecystokinin-related peptides—comparative aspects. In *The comparative physiology of regulatory peptides* (S. Holmgren, ed.), pp. 61–86. London: Chapman and Hall Ltd.

Kane, M. G., Luby, J. P., and Krejs, G. J. (1984). Intestinal secretion as a cause of hypokalemia and cardiac arrest in a patient with strongyloidiasis. *Dig. Dis. Sci.* **29:**768–772.

Kavaliers, M., and Colwell, D. D. (1992). Parasitism, opioid systems, and host behavior. *Adv. Neuroimmunol.* **2:**287–295.

Kavaliers, M., Podesta, R. B., Hirst, M., and Young, B. (1984). Evidence for the activation of the endogenous opiate system in hamsters infected with human blood flukes, *Schistosoma mansoni. Life Sci.* **35:**2365–2373.

Khan, A. I., Horii, Y., Tiuria, R., Sato, Y., and Nawa, Y. (1993). Mucosal mast cells and the expulsive mechanisms of mice against *Strongyloides venezuelensis. Int. J. Parasitol.* **23:**551–555.

Khayyal, M. T., Saleh, S., Metwally, A. A., Botros, S. S., and Mahmoud, M. R. (1991). *Schistosoma mansoni:* Angiotensin converting enzyme activity in mice under the influence of praziquantel and/or captopril. *Exp. Parasitol.* **73:**117–126.

Koninkx, J. F. J. G., Mirck, M. H., Hendriks, H. G. C. J. M., Mouwen, J. M. V. M., and Van Dijk, J. E. (1988). *Nippostrongylus brasiliensis:* Histochemical changes in the composition of mucus in goblet calls during infection in rats. *Exp. Parasitol.* **65:**84–90.

Lamont, J. T. (1992). Mucus: The front line of intestinal mucosal defence. *Ann. N.Y. Acad. Sci.* **664:**190–201.

Lee, T. D. G., Swieter, M., and Befus, A. D. (1986). Mast cell responses to helminth infection. *Parasitol. Today* **2:**186–191.

Lee, T. D. G., and Wakelin, D. (1982). The use of host strain variation to assess the significance of mucosal mast cells in the spontaneous-cure response of mice to the nematode *Trichuris muris*. *Int. Arch. Allergy Appl. Immunol.* **67:**302–305.

Long, R. G., Bishop, A. E., Barnes, A. J., Albuquerque, R. H., O'Shaughnessy, D. J., McGregor, G. P., Bannister, R., Polak, J. M., and Bloom, S. R. (1980). Neural and hormonal peptides in rectal biopsy specimens from patients with Chagas' disease and chronic autonomic failure. *The Lancet* **1:**559–562.

Mackenzie, C. D., Jungery, M., Taylor, P. M., and Ogilvie, B. M. (1981). The *in vitro* interaction of eosinophils, neutropohils, macrophages, and mast cells with nematode surfaces in the presence of complement or antibodies. *J. Pathol.* **133:**161–175.

McCoy, O. R. (1940). Rapid loss of trichinella larvae fed to immune rats and its bearing on the mechanism of immunity. *Am. J. Hyg.* **32D:**105–116.

McKay, D. M., Halton, D. W., Johnston, C. F., Shaw, C., Fairweather, I., and Buchanan, K. D. (1991). *Hymenolepis diminuta:* Changes in the levels of certain intestinal regulatory peptides in infected C57 mice. *Exp. Parasitol.* **73:**15–26.

McKay, D. M., Halton, D. W., McCaigue, M. D., Johnston, C. F., Fairweather, I., and Shaw, C. (1990). *Hymenolepis diminuta:* Intestinal goblet cell response to infection in male C57 mice. *Exp. Parasitol.* **71:**9–20.

McKay, D. M., and Perdue, M. H. (1993). Intestinal epithelial functions: The case for immunophysiological regulation. Implications for disease (second of two parts). *Dig. Dis. Sci.* **38:**1735–1745.

McLaren, D. J., Mackenzie, C. D., and Ramalho-Pinto, F. J. (1977). Ultrastructural observations on the *in vitro* interaction between rat eosinophils and some parasitic helminths (*Schistosoma mansoni, Trichinella spiralis, Nippostrongylus brasiliensis*). *Clin. Exp. Immunol.* **30:**105–118.

McLeay, L. M., Anderson, N., Bingley, J. B., and Titchen, D. A. (1973). Effects on abomasal function of *Ostertagia circumcincta* infections in sheep. *Parasitol.* **66:**241–257.

Miller, H. R. P. (1987). Gastrointestinal mucus, a medium for survival and for elimination of parasitic nematodes and protozoa. *Parasitol.* **94:**S77–S100.

Miller, H. R. P., Huntley, J. F., and Wallace, G. R. (1981). Immune exclusion and mucus trapping during the rapid expulsion of *Nippostrongylus brasiliensis* from primed rats. *Immunol.* **44:**419–429.

Miller, H. R. P., and Nawa, Y. (1979). *Nippostrongylus brasiliensis:* Intestinal goblet cell response in adoptively immunised rats. *Exp. Parasitol.* **47:**81–90.

Mimori, T., Nawa, Y., Korenaga, M., and Tada, I. (1982). *Strongyloides ratti:* Mast cell and goblet cell responses in the small intestine of infected rats. *Exp. Parasitol.* **54:**366–370.

Moqbel, R. (1980). Histopathological changes following primary, secondary, and repeated infection of rats with *Strongyloides ratti,* with special reference to tissue eosinophils. *Parasite Immunol.* **2:**11–27.

Mott, C. B., Neves, D. P., Okumura, M., De Brito, T., and Bettarello, A. (1972). Histologic and functional alterations of human exocrine pancreas in Manson's schistosmiasis. *Dig. Dis.* **17:**583–590.

Murray, M. (1969). Structural changes in bovine ostertagiasis associated with increased permeability of the bowel wall to macromolecules. *Gastroenterology* **56:**763–772.

Murray, M., Miller, H. R. P., and Jarrett, W. F. H. (1968). The globule leukocyte and its derivation from the subepithelial mast cell. *Lab. Invest.* **19:**222–234.

Nawa, Y., and Miller, H. R. P. (1979). Adoptive transfer of intestinal mast cell response in rats infected with *Nippostrongylus brasiliensis. Cell. Immunol.* **42:**225–239.

Neutra, M. R., and Forstner, J. F. (1987). Gastrointestinal mucus: Synthesis, secretion, and function. In *Physiology of the gastrointestinal tract,* 2nd ed. (L. R. Johnson, ed.), Vol. 2, pp. 975–1009. New York: Raven Press.

Nicholl, C. G., Polak, J. M., and Bloom, S. R. (1985). The hormonal regulation of food intake, digestion, and absorption. *Ann. Rev. Nutr.* **5:**213–239.

Nicholls, C. D., Lee, D. L., Adrian, T. E., Bloom, S. R., and Carr, A. D. (1985). Endocrine effects of *Haemonchus contortus* infection in lambs. *Regul. Pept.* **13:**85.

Nolla, H., Bristol, J. R., and Mayberry, L. F. (1985). *Nippostrongylus brasiliensis:* Malabsorption in experimentally infected rats. *Exp. Parasitol.* **59:**180–184.

Novak, M., and Nombrado, S. (1988). Mast cell responses to *Hymenolepis microstoma* infection in mice. *J. Parasitol.* **74:**81–88.

O'Dorisio, M. S. (1986). Neuropeptides and gastrointestinal immunity. *Am. J. Med.* **81**(Suppl. 6B):74–81.

O'Dorisio, M. S., and Panerai, A. (1990). Neuropeptides and immunopeptides: Messengers in a neuroimmune axis. *Ann. N.Y. Acad. Sci.* **594:**1–503.

O'Malley, K. E., Sloan, T., Joyce, P., and Baird, A. W. (1993). Type I hypersensitivity reactions in intestinal mucosae from rats infected with *Fasciola hepatica. Parasite Immunol.* **15:**449–453.

Ohnishi, Y. (1987). Eosinophil responses in mice infected with *Metagonimus yokogawai. Jpn. J. Parasitol.* **36:**271–275.

Ohnishi, Y., and Taufan, M. (1984). Increase of permeability in the intestinal mucosa of mice infected with *Metagonimus yokogawai. Jpn. J. Vet. Sci.* **46:**885–887.

Olson, C. E., and Schiller, E. L. (1978). *Strongyloides ratti* infections in rats. I. Immunopathology. *Am. J. Trop. Med. Hyg.* **27:**521–526.

Ovington, K. S., Bacarese-Hamilton, A. J., and Bloom, S. R. (1985). *Nippostrongylus brasiliensis:* Changes in plasma levels of gastrointestinal hormones in the infected rat. *Exp. Parasitol.* **60:**276–284.

Perdue, M. H., and McKay, D. M. (1993). Immunomodulation of the gastrointestinal epithelium. In *The handbook of immunopharmacology—Immunopharmacology of the gastrointestinal system,* (J. L. Wallace, ed.), pp. 15–39. New York: Academic Press Ltd.

Pfister, K., and Meierhofer, B. (1986). Cellular responses in the small intestine and liver of *Fasciola hepatica*-infected rats. *Mitt. Osterr. Ges. Tropenmed. Parasitol.* **8:**73–82.

Phares, C. K. (1992). Biological characteristics of the growth hormone–like factor from plerocercoids of the tapeworm *Spirometra mansonoides. Adv. Neuroimmunol.* **2:**235–247.

Phares, C. K., and Watts, D. J. (1988). The growth hormone–like factor produced by the tapeworm *Spirometra mansonoides* specifically binds receptors on cultured human lymphocytes. *J. Parasitol.* **74:**896–898.

Pimparker, B. D., Sharma, P., Satoskar, R. S., Raghavan, P., and Kinare, S. G. (1982). Anemia and gastrointestinal function in ancylostomiasis. *J. Postgrad. Med.* **28:**51–63.

Rothwell, T. L. W. (1989). Immune expulsion of parasitic nematodes from the alimentary tract. *Int. J. Parasitol.* **19:**139–168.

Rothwell, T. L. W., and Dineen, J. K. (1972). Cellular reactions in guinea pigs following primary and challenge infection with *Trichostrongylus colubriformis* with special reference to the roles played by eosinphils and basophils in rejection of the parasite. *Immunol.* **22:**733–745.

Russell, D. A., and Castro, G. A. (1985). Anaphylactic-like reaction of small intestinal epithelium in parasitized guinea pigs. *Immunol.* **54:**573–579.

———. (1987). Physiology of the gastrointestinal tract in the parasitized host. In *Physiology of the gastrointestinal tract,* 2nd ed. (L. R. Johnson, ed.), Vol. 2, pp. 1749–1780. New York: Raven Press.

———. (1989). Immunological regulation of colonic ion transport. *Am. J. Physiol.* **256:**G396–G403.

Schanbacher, L. M., Nations, J. K., Weisbrodt, N. W., and Castro, G. A. (1978). Intestinal myoelectric activity in parasitized dogs. *Am. J. Physiol.* **234:**R188–R195.

Scofield, A. M. (1977). Intestinal absorption of hexose in rats infected with *Nippostrongylus brasiliensis. Int. J. Parasitol.* **7:**159–165.

———. (1980). Effect of level of infection with *Nippostrongylus brasiliensis* on intestinal absorption of hexoses in rats. *Int. J. Parasitol.* **10:**375–380.

Scott, P. A., and Sher, A. (1993). Immunoparasitology. In *Fundamental immunology,* 3rd ed. (W. E. Paul, ed.), pp. 1179–1210. New York: Raven Press Ltd.

Shanahan, F., Denburg, J. A., Fox, J., Bienenstock, J., and Befus, D. (1985). Mast cell heterogeneity: Effects of neuroenteric peptides on histamine release. *J. Immunol.* **135:**1331–1337.

Shayo, M. E., and Benz, B. W. (1979). Histopathologic and histochemic changes in the small intestine of calves infected with *Trichostrongylus colubriformis. Vet. Parasitol.* **5:**353–364.

Sheehy, T. W., Meroney, W. H., Cox, R. S., Jr., and Soler, J. E. (1962). Hookworm disease and malabsorption. *Gastoenterology* **42:**148–156.

Shelton, G. C., and Griffiths, H. J. (1967). *Oesophagostomum columbianum:* Experimental infections in lambs. Effects of different types of exposure on the intestinal lesions. *Pathol. Vet.* **4:**413–434.

Sher, A., and Coffman, R. L. (1992). Regulation of immunity to parasites by T cells and T cell–derived cytokines. *Ann. Rev. Immunol.* **10:**385–409.

Siebert, A. E., Jr., Good, A. H. R., and Simmons, J. E. (1979). Ultrastructural aspects of the host cellular immune response to *Taenia crassiceps* metacestodes. *Int. J. Parasitol.* **9:**323–331.

Siurala, M. (1954). Gastric lesion in some megaloblastic anemias, with special reference to the mucosal lesion in pernicious tapeworm anemia. *Acta Med. Scand.* **151**(Suppl. 299):1–47.

Sprent, J. F. A. (1946). Immunological phenomena in the calf, following experimental infection with *Bunostomum phlebotomum. J. Comp. Pathol. Ther.* **56:**286–297.

Stead, R. H., Tomioka, M. Quinonez, G., Simon, G. T., Felten, S. Y., and Bienenstock, J. (1987). Intestinal mucosal mast cells in normal and nematode-infected rat intestines are in intimate contact with peptidergic nerves. *Proc. Nat. Acad. Sci. USA* **84:**2975–2979.

Stephenson, L. S., Pond, W. G., Nesheim, M. C., Krook, L. P., and Crompton, D. W. T. (1980). *Ascaris suum:* Nutrient absorption, growth, and intestinal pathology in young pigs experimentally infected with 15-day-old larvae. *Exp. Parasitol.* **49:**15–25.

Symons, L. E. A. (1965). Kinetics of the epithelial cells, and morphology of villi and crypts in the jejunum of the rat infected by the nematode *Nippostrongylus brasiliensis. Gastroenterology* **49:**158–168.

Symons, L. E. A. (1969). Pathology of gastrointestinal helminthiases. *Int. Rev. Trop. Med.* **3:**49–100.

Symons, L. E. A., and Fairbairn, D. (1962). Pathology, absorption, transport, and activity of digestive enzymes in rat jejunum parasitized by the nematode *Nippostrongylus brasiliensis. Fed. Proc.* **21:**913–918.

Symons, L. E. A., Gibbins, J. K., and Jones, W. O. (1971). Jejunal malabsorption in the rat infected by the nematode *Nippostrongylus brasiliensis. Int. J. Parasitol.* **1:**179–187.

Symons, L. E. A., and Hennessy, D. R. (1981). Cholecystokinin and anorexia in sheep infected by the intestinal nematode *Trichostrongylus colubriformis. Int. J. Parasitol.* **11:**55–58.

Symons, L. E. A., and Jones, W. O. (1970). *Nematospiroides dubius, Nippostrongylus brasiliensis,* and *Trichostrongylus colubriformis:* Protein digestion in infected mammals. *Exp. Parasitol.* **27:**496–506.

Taliaferro, W. H., and Sarles, M. P. (1939). The cellular reactions in the skin, lungs, and intestine of normal and immune rats after infection with *Nippostrongylus muris. J. Infect. Dis.* **64:**157–192.

Tani, S., and Yoshimura, K. (1988). Spontaneous expulsion of *Echinostoma hortense* Asada, 1926 (Trematoda: Echinostomatidae) in mice. *Parasitol. Res.* **74:**495–497.

Thorndyke, M. C., Riddell, J. H., Dimaline, R., Balogun, D., and Whitfield, P. J. (1988). Changes in ileal vasoactive intestinal polypeptide and gastrin releasing peptide/bombesin levels associated with chronic infections of the digenean helminth *Echinostoma liei. Regul. Pept.* **22:**435.

Titchen, D. A. (1982). Hormonal and physiological changes in helminth infestations. In *Biology and control of endoparasites* (L. E. A. Symons, A. D. Donald and J. K. Dineen, eds.), pp. 257–275. Sydney: Academic Press Inc.

Titchen, D. A., and Reid, A. M. (1988). Putative roles of peptides in the genesis and control of parasitic diseases. In *Aspects of digestive physiology in ruminants* (A. Dobson and M. J. Dobson, eds.), pp. 217–237. Ithaca: Comstock Publishing Associates.

Tse, S.-K., and Chadee, K. (1991). The interaction between intestinal mucus glycoproteins and enteric infections. *Parasitol. Today* **7:**163–172.

Tuohy, M., Lammas, D. A., Wakelin, D., Huntley, J. F., Newlands, G. F. J., and Miller,

H. R. P. (1990). Functional correlations between mucosal mast cell activity and immunity to *Trichinella spiralis* in high and low responder mice. *Parasite Immunol.* **12:**675–685.

Varilek, G. W., Weinstock, J. V., Williams, T. H., and Jew, J. (1991). Alterations of the intestinal innervation in mice infected with *Schistosoma mansoni. J. Parasitol.* **77:**472–478.

Webb, R. A. (1988). Endocrinology of acoelomates. In *Invertebrate endocrinology* (R. G. H. Downer and H. Laufer, eds.), Vol. 2, pp. 31–62. New York: Alan R. Liss.

Webb, R. A. (1991). Serotonin—A ubiquitous neuroactive agent in platyhelminths. In *Neurobiology and endocrinology of selected invertebrates* (B. G. Loughton and A. S. M. Saleuddin, eds.), pp. 145–162. Captus University Press.

Weinstein, M. S., and Fried, B. (1991). The expulsion of *Echinostoma trivolvis* and retention of *Echinostoma caproni* in the ICR mouse: Pathological effects. *Int. J. Parasitol.* **21:**255–257.

Weinstock, J. V., and Blum, A. M. (1989). Tachykinin production in granulomas of murine *Schistosomiasis mansoni. J. Immunol.* **142:**3256–3261.

Willems, G., Gepts, W., and Bremner, A. (1977). Endogenous hypergastrinemia and cell proliferation in the fundic mucosa in dogs. *Am. J. Dig. Dis.* **22:**419–423.

Wood, J. D. (1991). Communication between minibrain in gut and enteric immune system. *News Physiol. Sci.* **6:**64–69.

PART 2

Parasitism and Reproduction

7

Testosterone and Immunosuppression in Vertebrates: Implications for Parasite-Mediated Sexual Selection

Nigella Hillgarth and John. C. Wingfield

I. Introduction

Females of many species prefer to mate with males that have the fewest parasites (Møller, 1990; Read,1990; Kirkpatrick and Ryan, 1991; Clayton, 1991; McLennen and Brooks, 1991; Clayton et al., 1992; Sullivan, 1991; Zuk, 1992). Females appear to discriminate among males on the basis of the expression of secondary sex characters, such as comb color and feather lenth in junglefowl (*Gallus gallus*) (Zuk et al., 1990), red plumage in house finches (*Cardodacus mexicanus)* (Hill, 1991); eye spot number on the tail-like train of peafowl (*Pavo cristatus*) (Petrie et al., 1991), or tail length in barn swallows (*Hirundo rustica*) (Møller, 1991). Experimental manipulation of parasite load can alter and hinder the expression of secondary sex characters (Zuk et al., 1990, Møller, 1991, Hillgarth, 1990). Therefore, females choosing males with well-developed secondary sex characters are likely to choose relatively parasite-free mates.

Parasite infection disproportionately affects the development of secondary sex traits. Parasitized jungle fowl (Zuk et al., 1990) were similar in adult size and weight. However, combs were consistently smaller in parasitized males, compared to controls. Zuk et al., (1990) showed that female jungle fowl preferred to mate with the uninfected control males and that choice was significantly associated with traits mostly affected by parasites. Males with high mating success and well developed secondary sex characters often have lower parasite intensity. However, a majority of these studies have been conducted on birds; no published work on mammals exists that we are aware of, although female mice (*Mus musculus)* do prefer males with a major histocompatability complex (MHC) different from their own (Potts and Wakeland, 1993). This may indicate a preference for maximizing MHC polymorphism, thereby increasing the possibility of disease resistance in the offspring.

Parasite-mediated sexual selection studies have been carried out in other vertebrates including guppies (*Poecilia reticulata*) (McMinn, 1990), and sticklebacks (*Gasterosteus aculeatus*) (Milinski and Bakker, 1990). In these two studies, parasites altered the showiness of the males and influenced female choice. Female sticklebacks (Milinski and Bakker, 1990) discriminate against males formerly parasitized by *Ichthyophthirus* ciliates and recognize such males by the lower intensity of their breeding coloration. Results from work on amphibians and reptiles are inconclusive. In fence lizards (*Sceloporus occidentalis*) parasites do affect fitness and alter male appearance but females do not prefer unparasitized males (Ressel and Schall, 1989). Parasites also do not appear to influence female choice in amphibians such as tree frogs (*Hyla versilcolor*) (Hausfauter et al., 1990) or spadefoot toads (*Scaphiopus couchii*) (Tinsley, 1990).

The mechanisms behind parasite-mediated sexual selection are now being explored. The immunocompetence hypothesis proposes that parasites affect secondary sex traits through interaction of endocrine and immune systems. Folstad and Karter (1992) suggest that males may have to produce high levels of the sex-steroid testosterone to express fully developed secondary sex characters and appropriate courtship behavior. However, testosterone also has immunosuppressive properties, resulting in elevated vulnerablity to parasite infection. Only individuals that have good genetic resistance to infection will be able to express a well-developed secondary sex character and cope simultaneously with high testosterone levels. Note that parasite infestation can lower circulating testosterone levels in the host (Isserhoff, et al., 1986), and thus infected animals may not be able to develop adequate secondary sex characters if these traits are regulated by testosterone. Many secondary sex characters are not dependent on testosterone (see other chapters in this volume; Owens and Short, 1995; Hillgarth and Wingfield, 1997). Even in cases where testosterone is known to regulate expression of secondary sex characters, such as chicken combs and wattles, there is no good evidence that expression is dose dependent, or indeed that much steroid is needed at all (Hillgarth and Wingfield, 1997; see also Chapter 14).

The immunosuppressive properties of testosterone, and other hormones, involved in the production of secondary sex characters are complex. Any animal that evolves a mechanism in which the expression of a secondary sex character does not require an immunosuppressive hormone will not suffer the possibly costly consequences of compromising immune responses. The reallocation hypothesis (Wedekind, 1992, 1994; Wedekind and Folstad, 1994) proposes that androgens such as testosterone suppress areas of the immune system that are not needed to combat infection so that costly resources such as rare metabolites can be reallocated for the development of good secondary sex characters. This is an intriguing hypothesis; however, the immunosuppressive properties of testosterone appear to be so widespread in vertebrates that it appears unlikely that reallocation of resources from the immune system for the expression of secondary sex characters can be the main function of androgen immunosuppression.

Here we examine some of the immunosuppressive properties of testosterone, and explore the relationship between hormone levels and parasitism. We first present a brief overview of the immune system, then outline the evidence for immunosuppression of the vertebrate immune system by testosterone, and implications for mechanisms of parasite-mediated sexual selection. We also examine the effects of parasite infections on circulating steroid levels.

II. Interactions of Steroids with Immunity

A. The Immune System in Vertebrates

There is considerable difference in the immune responses mobilized against microparasites such as protozoans that reproduce inside the host, and macroparasites like helminths that mostly do not reproduce in the host and are usually a serious threat only following repeated exposure to infection (Cox, 1993). In both cases the host immune responses are often inadequate to completely conquer the parasite, leading to chronic infection, and often symptoms of immunopathology are induced. For example, damage to host tissue may be caused by an immune response to nematode worm larval stages embedded in mammalian gut tissue (Cox, 1993). The extent of chronic parasitic infections might suggest ineffective parasite immune responses in vertebrates.

Here we simply wish to discuss mechanisms of sex steroid regulation of the immune system and the implications for sexual selection. (For a general introduction to vertebrate immunology see Roitt, 1994; or Roitt et al., 1993). The host immune response example outlined below is designed to introduce several components of the immune system that are referred to in relation to sex-steroid-immune interactions: In mammals, when a microparasite such as a protozoan invades the host it may be taken up by a macrophage, where it is broken down and the resulting fragments are placed outside the macrophage. This foreign protein fragment is an antigen, and it is bound to the surface of the cell by an MHC (major histocompatability complex) molecule. T cell helper lymphocytes have T cell receptors that clasp on to the MHC molecule. One group of T cells interacts with B lymphocytes and helps them divide, differentiate, and make antibodies. Another group of cells called T helper cells interact with macrophages and other phagocytic cells. A third group kills host cells infected with intracellular pathogens, and are called cytotoxic T cells. There are also T suppressor cells that appear to be involved in the modulation of a variety of humoral and cellular responses. Via various messangers such as cytokines, other cell groups are activated and B lymphocytes produce antibodies specific to the invading organism. Even after the infection is terminated, memory cells remain that can be activated should the host become reinfected with the same parasite, and an even more rapid immune response arises. The above example of a classic response to an infective organism is a mixture of the so-called innate and acquired immune systems. The acquired response involves presenting part of the invading organism to cells that can manufacture specific antibody against the organism, and then having the memory to use this humoral response against recurring infections. So innate and acquired areas of the immune system work in tandem to fight infection.

Sex steroids have long been known to have immunosuppressive and immunocytotoxic effects in vertebrates at high doses (Grossman, 1985); however, it is only recently that the varying and regulatory role of gonadal steroids in the vertebrate immune system has been appreciated (Marsh and Scanes, 1994). Mammalian females are known to have a significantly more active immune system than males (Talal, 1992). Females have higher antibody responses to many antigens and have greater cellular immune responses. However, they are more susceptible to autoimmune conditions than males (Grossman, 1985; Talal, 1992). Castrated male animals become similar to females in their immune responses, suggesting that androgens suppress immune function. This is an area of active research because of clinical implications for the treatment of autoimmune diseases in humans. There is now evidence that estrogen modulates T cells, particularly repressing suppressor cells, and estrogen receptors have indeed been found on CD8+ T cells (Talal, 1992). Mice treated with estrogen have less active suppressor cells, and less of the cytokine IL–2 is produced. On the other hand, testosterone stimulates suppressor cell function (Talal, 1992; Ahmed et al., 1985).

Staples et al. (1983) added, *in vitro,* several different types of steroids to peripheral sheep or goat lymphocytes stimulated by mitogens, and found that response was inhibited by many steroids, particularly progestins, corticosteroids, and the androgens testosterone and androstenedione. However, *in vivo,* only progesterone inhibited lymphocyte response at the highest natural physiological levels occuring in sheep. The effect of steroids appeared to be dose dependent with cortiosteroids having more effect at lower doses than the other steroids. In mice Ahmed et al. (1985) found that short-term administration of testosterone enhanced suppressor cell activity and eostrogen decreased suppressor cell activity. There are numerous examples demonstrating the immunosuppressive effects of testosterone on mammalian parasite virulence. In many instances males have considerably more severe infections than females or castrated males. High physiological levels of testosterone prevent male mice healing from infections of *Plasmodium chabaudi,* or acquiring immunity from secondary infections. However, castrated mice self-heal, and develop immunity. Castrates implanted with testosterone respond in a similar way to noncastrated males (Benton et al., 1992). However, other androgens such DHT and androsterone given to castrated mice do not prevent self-healing (Benton et al., 1992). Harder (1992) found that testosterone had a considerable effect on a very different parasite from *Plasmodium. Heterakis spumosa,* a nematode worm, grows faster, produces more eggs, and lives longer in male mice that have naturally high physiological levels of testosterone. There is often variation in parasite load within testosterone-treated males. A recent study by Harder et al. (1994) shows that mice with different MHC haplotypes appear to have significantly different responses to nematode worm infections that only become apparent after treatment with testosterone. This suggests that genetic variation may account for some of the variation in expression of sex steroid-related immunosuppression. For example, males with a particular MHC haplotype may have high levels of circulating testosterone without corresponding immunosuppressive effects. The mechanism whereby testosterone and other steroids actually increase levels of parasite infection has not been explored. However, in at least one study, testosterone was shown to have a direct positive effect on parasite growth *in vitro* (Fleming, 1985).

Whether steroids have a direct effect on parasites *in vivo* as well as indirect effects via immune responses requires careful experimentation. Barnard et al. (1994) have shown that social behavior appears to influence susceptibility to infection with the protozoan *Babasia microti.* Dominant and subordinate males have the highest mean infection levels whereas midranking males have low infection rates. This may be explained by the immunosuppressive effects of the high physiological levels of circulating testosterone in the dominant males, and high levels of stress steroids in the subordinate males. Midranking males have relatively low levels of immunosuppressive hormones and so have lower infection rates.

B. Birds

The avian immune system has been well investigated especially in the chicken and other commercially important species such as turkey, quail, geese and ducks. On the other hand, remarkably little is known about immunity in passerine birds, or any wild bird species for that matter. However, we do know that birds have responses similar to that of mammals with some unique features including some differences in immunoglobulin structure and MHC arrangement, and a bursa of Fabricus which produces B cells (Glick, 1986; Jurd, 1994). The bursa may be a way of enabling birds to produce antibody responses soon after hatching and may reflect the fact that fewer maternal antibodies pass through into the egg than into the mammalian fetus (Wakelin and Appanus, 1997).

Testosterone has been shown to have a suppressive effect on autoimmune responses in young chickens (Gause and Marsh, 1986; Marsh and Scanes 1994). Chickens with spontaneous autoimmune thyroiditis (SAT), and implanted with low levels of testosterone for several weeks soon after hatching, showed significantly less severe symptoms of SAT than controls. Importantly, testosterone levels were within physiological levels (6 ng/ml of plasma in treated birds and 2–3 ng/ml in controls). In this case testosterone suppresses the development of several types of T cells such as T helper cells which affect the production of the immunoglobulin IgG. The switch from IgM to IgG immunoglobulin is mediated by T helper cells. It appears that testosterone stimulates suppressor T cells rather than having a direct inhibitory effect on the development of T cells (Lehmann et al., 1988). These results are similar to a study by Hirota et al. (1976) in which testosterone treated chicken embryos had significantly lower IgG levels but normal IgM production.

Redig et al. (1985) found that turkey lymphocytes with added pharmacological levels of testosterone did not respond as effectively to stimulation by mitogens as controls, and that this might explain why male turkeys have significantly more problems with the fungal disease aspergillosis than females. However, the responses of gonadectomized male turkeys and controls did not differ significantly in response to infection by the bacterium *Pasturella multocida* (Friedlander et al., 1992). Pharmological steroid levels may have very different effects on lymphocytes than physiological levels of the same steroid. Therefore, although it is possible the differences in aspergillus susceptibility between male and female turkeys is due to higher plasma levels of testosterone in males, other factors might be involved. For example, circulating corticosteroids increase in males of some avian species when testosterone levels rise (Astheimer et al., 1994; Hillgarth and Wingfield, 1997), or estrogens in females may have immuno-stimulatory effects (Grossman, 1990). The lack of difference in response to *Pasturella* in gonadectomized and normal male turkeys does not show that testosterone is not immunosuppressive. It may be that testosterone does not suppress the major immune responses to a bacterial infection like *Pasturella* whereas it does suppress immune responses to the fungal infection *Aspergillosis.* We have used these two examples to show that measuring the effects of testosterone and other steroids on the immune system is very complex and easily misleading as well as being poorly understood (see Hillgarth and Wingfield, 1997 for other examples). Studies should examine the immunosuppressive effects of testosterone within physiological levels, and controls and manipulated subjects should be treated in a similar fashion.

C. Other vertebrates

Reptiles have many different types of white blood cells that mediate immunity, however their homology with mammalian cells types is still not fully understood. In all species that have been examined, there are antibody responses to infection. The current, general picture of immune responses in reptiles suggests a system of considerable complexity and finesse as in other vertebrates but most of the specifics have not yet been worked out (Jurd, 1994).

The immune system in amphibians has been considerably better studied than that of reptiles because of the interest in the effects of metamorphosis on the immune response. Immunologically, metamorphosis is a difficult period for amphibians as the system has to learn to recognize the new stage of the organism as self and not produce damaging autoimmune responses (Horton, 1994). It appears that self tolerance is generated at metamorphosis because suppressor cells dampen immune responses at this time (Horton, 1994). Otherwise amphibian immune responses appear to be similar to those of other vertebrates.

In reptiles and amphibians white blood cells become less responsive to mitogen stimulus coinciding with ovarian development. This may be due to increased sex steroids in females at this time (Garrido et al., 1989). Also, in reptiles high levels of testosterone correlate with thymus involution suggesting immunosuppressive effects. Lizards injected during the breeding season with testosterone have reduced lymphocytes and delayed humoral responses (Saad et al., 1990). Slater and Schreck (1993) found that testosterone added *in vitro* suppressed development of salmon leukocytes. This provides good evidence that testosterone is immunosuppressive in fish. However, Slater and Schreck (1993) demonstrated that a combination of testosterone and corticosterone produced an even greater immunosuppressive effect on leukocytes than either steroid on its own.

Fish have a large range of innate responses to disease, many of them analogous to those of other vertebrates (Manning, 1994). Specific responses may be less varied than in mammals, for example the antibody repertoire is restricted compared to mammals although antibody production appears analogous to mammals and specific cell-mediated responses have many mammalian attributes. Fish have an MHC complex, however it appears to be considerably less diverse than that of birds or mammals (Manning, 1994). In summary, the immune responses in fish may be more restricted than those in other vertebrates. However, there is considerable variation among species in sex differences in immune function. For example, in rock fish, *Sebasticus marmoratus,* the female thymus atrophies before spawning while the males does not change. This is opposite in viviparous surf perch (Zapata et al., 1992). Interspecific differential immune responses may be due to variation in reproductive strategies with fish that give birth to live young differing from those that do not.

III. Parasite Effects on Steroid Levels in the Host

Once a vertebrate becomes infected with parasites it may have lower circulating levels of sex steroids (Spindler, 1988). The mechanism whereby parasite infection lowers these steroids is not fully understood. In mice infected with the trematode *Schistosoma mansoni,* testosterone levels fall dramatically. This is apparently due to increased activation of the endogenous opiate system, which decreases the gonadatrophin lutienizing hormone production, which stimulates production of testosterone (Spindler, 1988). Even ectoparasite infections may affect steroid levels. De Vaney et al. (1977) showed that male chickens infected with mites had decreased circulating testosterone levels as a function of the mite population size. On the other hand, corticosteroid levels usually rise with parasite infection. Rats (*Rattus rattus*) infected with the cestode *Mesocestoides cortii*, or the nematode *Strongyloides ratti* had increased corticosterone levels compared with controls (Chernin & Morinan, 1985). Little is known about the effects of parasites on hormones in other vertebrates, but fence lizards infected with *Plasmodium* have lower basal testosterone levels and higher stress response plasma levels of corticosterone than uninfected lizards (Dunlap and Schall, 1995) These studies suggest that at least some species of parasites, directly or indirectly, cause a change in the level of circulating steroid which is consistent with the immunocompetence hypothesis of parasite-mediated sexual selection.

IV. Discussion

The immunocompetence hypothesis (Folstad & Karter 1992) states that parasitic infection will influence the expression of host secondary sex characters by lowering the hormone, such as testosterone, that regulates the character. Although there is growing evidence that testosterone levels are decreased with parasitic infection (Dunlap & Schall, 1995; Spindler, 1988; De Vaney et al., 1977) this will influence only secondary sex characters that are primarily regulated by testosterone, and in a dose-dependent way. (Hillgarth and Wingfield 1997; Hews and Moore this volume).

In this review we have outlined the evidence for some sex steroids and immunosuppression. We concentrated on testosterone as this hormone is widely thought to be immunosuppressive to some degree in several vertebrate groups. However, other androgens that may play a part in the expression of secondary sex characters and male breeding behavior, such as DHT or androsterone, may not be as immunosuppressive (Benton et al., 1992). More research needs to be carried out on the immunosuppressive properties of any hormones that are involved in the production, or supression, of secondary sex characters such as estrogen, and the gonadotrophin luteinizing hormone (Hillgarth and Wingfield, 1997).

We found that there were many examples in the literature of the immunosuppressive properties of testosterone such as increased development of the nematode *Heterakis spumosa* in mice with high physiological levels of testosterone (Harder, 1992). However, most studies used high doses and pharmological levels of testosterone. Clearly, considerably more research needs to be done on the immunosuppressive effects of natural levels of circulating testosterone, and steroid interactions with different types of infective challenge. No difference was found between males with natural levels of testosterone and castrated turkeys, indicating that physiological levels of testosterone were possibly not immunosuppressive in this case (Friedlander *et al.*, 1992).

The complexity involved in endocrine immune interactions includes many environmental factors that alter circulating hormone levels. Alteration of sex steroids due to seasonality or social interaction may have an effect on the level of hormone-related immunosuppression. Research on immunosuppressive effects of testosterone and other androgens should include studies under natural conditions as well as in the laboratory.

Factors such as mating system and testosterone function in the regulation of aggression may be influential in the relative importance of hormones in immunosuppression. For example, testosterone is known to be important for the regulation of aggression in many birds as well as some mammals (Wingfield et al., 1994). Birds have much lower relative levels of testosterone than mammals. Are the peaks of testosterone seen in some birds after extended periods of aggressive encounters high enough to significantly depress the immune system? Some studies have found no correlation with aggression and elevated testosterone levels, suggesting that aggression may not be controlled by testosterone in all birds (Wingfield et al., 1994). This raises the possibility that animals in which testosterone does not regulate aggression will be less immunosuppressed during periods of conflict such as territorial defense or mate guarding than species in which testosterone does not regulate aggression.

Several studies in birds have shown that immune response decreases as the reproductive season progresses. Norris et al. (1994) found that prevalence of hematazoan parasites in great tits was higher in females at the beginning of the breeding period, but increased significantly in males later in the season, particularly when birds had large clutches. When clutch size was artificially increased hematazoan prevalence also increased. Gustaffasson et al. (1994) found that as reproductive effort increased, immune parameters decreased. In pied flycatchers, prebreeding nutritional status correlated positively with reproductive success and negatively with parasite infection and immune response. Birds with artificially increased broods also had increased parasitism. These studies suggest that reproductive effort increases the likelihood of immunosuppression in some form. This is distinct from immune parameters changing at different seasons, due to alterations in sex steroid levels during the reproductive season.

In summary, the data provide mixed evidence for the immunosuppressive effects of testosterone at physiological levels; however, there is evidence that testosterone is immunosuppressive, at least under laboratory conditions, which supports hypotheses of the mechanisms of parasite-mediated sexual selection. (Folstad and Karter, 1992; Wedekind and Folstad, 1994; Zuk, 1992). However, considerably more laboratory and field studies are needed before we can say that *in vivo* levels of testosterone are immunosuppressive under natural conditions.

V. References

Ahmed, S. A., Dauphinee, J., and Talal, N. (1985). Effects of short-term administration of sex hormones on normal and autoimmune mice. *J. Immunol.* **134:**204–210.

Astheimer, L. B., Buttemer, W. A., and Wingfield, J. C. (1994). Gender and seasonal differences in the adrenocortical response to ACTH challenge in an arctic passerine, *Zonotricia leucophrys gambelii. Gen. Comp. Endocrinol.* **94:**33–44.

Barnard, C. J., Behnke, J. M., and Sewell, J. (1994). Social behaviour and susceptibility to infection in house mice (*Mus musculus*): Effects of group size, aggressive behaviour and status-related hormonal responses prior to infection on resistance to *Babesia microti. Parasitology* **108:**487–496.

Benton, W. P. M., Wunderlick, F., and Mossmann, H. (1992). Testosterone-induced suppression of self-healing *Plasmodium chabaudi* malaria: An effect not mediated by androgen receptors? *J. Endocrinol.* **135:**407–413.

Chernin, J., and Morinan, A. (1985). Analysis of six serum components from rats infected with tetrathyridia of *Mesocestoides corti. Parasitol.* **90:**441–447.

Cox, F. E. G. (1993). Immunology. In: *Modern parasitology.* F. E. G. Cox, ed. Oxford: Blackwell Scientific Publications. 193–218.

Clayton, D. H. (1991). The influence of parasites on host sexual selection. *Parasitol. Today.* **7:**329–334.

Clayton, D. H., Pruett-Jones, S. G., and Lande, R. (1992). Reappraisal of the interspecific prediction of parasite-mediated sexual selection: Opportunity knocks. *J. Theor. Biol.* **157:**95–108.

De Vaney, J. A., Elissalde, M. H., Steel, E. G., Hogan, B. F., and Del var Petersen, H. (1977). Effect of the Northern fowl mite *Ornithonyssus sylviarum* on White Leghorn roosters. *Poult. Sci.* **56:**1585–1590.

Deviche, P., Balthazart, J., Malacare, G., and Hendrick, J. C. (1982). Effects of in vitro corticosterone treatment on the metabolism of testosterone in the comb and brain of the young male chicken. *Gen. Comp. Endocrinol.* **48:**398–402.

Dunlap, K. P., and Schall, J. J. (1995). Hormonal alterations and reproductive inhibition in male fence lizards (*Sceloporus occidentalis*) infected with the malarial parasite *Plasmodium mexicanum. Physiol. Zool.* **68:**608–621.

Fleming, M. W. (1985). Steroidal enhancement of growth in parasitic larvae of *Acaris suum:* Validation of a bioassey. *J. Exp. Zool.* **233:**229–233.

Folstad, I., and Karter, A. J. (1992). Parasites, bright males, and the immunocompetence handicap. *Am. Nat.* **139:**603–622.

Fowles, J. R., Fairbrother, A., Fix, M., Schiller, S., and Kerkvliet, N. I. (1993). Glucocorticoid effects on natural and humoral immunity in mallards. *Dev. Comp. Immunol.* **17:**165–177.

Friedlander, R. C., Olsen, L. D., and McCune, E. L. (1992). Comparative susceptability of caponized and uncaponized tom turkeys to *Pasteurella multicodia. Avian Dis.* **10:**97–100.

Garrido, E., Gomariz, R. P., Lecata. J, and Zapata, A. (1989) Different sensitivity to the dexamethasone treatment of the lymphoid organs of Ranaperezi in two different seasons. *Dev. Comp. Immunol.* **13:**57–64.

Gause, W. C., and Marsh, J. A. (1986). Effect of testosterone treatments for varying periods on autoimmune development and on specific infiltrating leukocyte populations in the thyroid gland of obese strain chickens. *Clin. Immunol. Immunopathol.* **39:**464–478.

Glick, B. (1986). Immunophysiology. In *Avian physiology* (P. D. Sturkie, ed.), pp. 87–101. New York: Springer-Verlag.

Greenburg, N., and Wingfield, J. C. (1987). Stress and reproduction: Reciprocal relationships. In *Hormones and reproduction in fishes, amphibians and reptiles* (D. O. Norris and R. E. Jones, eds.), pp. 389–426. New York: Plenum.

Grossman, C. J. (1990). Are there underlying immune-neuroendocrine interactions responsible for immunological sexual dimorphism? *Prog. NeuroEndocrinImmunol.* **3:**75–82.

Grossman, C. J. (1985). Interactions between the gonadal steroids and the immune system. *Science.* **227:**257–261.

Gustafsson, L., Nordling, D., Andersson, M. S., Sheldon, B. C., and Qvarnstrom, A. (1994). Infectious diseases, reproductive effort and the cost of reproduction in birds. *Phil. Trans. R. Soc. Lond. B. Biol. Sci.* **260:**323–331.

Hamilton, W. D., and Zuk, M. (1982). Heritable true fitness and bright birds: A role for parasites? *Science* **218:**384–387.

Harder, A. (1992). Effects of testosterone on *Heterakis spumosa* infections in mice. *Parasitology* **105:**335–342.

Harder, A., Danneschewski, A., and Wunderlick, F. (1994). Genes of the mouse H-2 complex control the efficacy of testosterone to suppress immunity against the intestinal nematode *Heterakis spumosa. Parasitol. Res.* **80:**446–8.

Hausfauter, G., Gerhardt, H. C., and Klump, G. M. (1990). Parasites and mate choice in green treefrogs, *Hyla versicolor. Am. Zool.* **30:**299–312.

Hill, G. E. (1991). Plumage coloration is a sexually selected indicator of male quality. *Nature* (Lond.) **350:**337–339.

Hillgarth, N. (1990). Parasites and female choice in the ring-necked pheasant. *Am. Zool.* **30:**227–233.

Hillgarth, N., and Wingfield, J. C. (1997). Parasite-mediated sexual selection: Endocrine aspects. In *Host-parasite evolution: General principles and avian models* (D. Clayton and J. E. Moore, eds.), pp. 78–104. Oxford: Oxford University Press.

Hirota, Y., Suzuki, Y., Chazono, Y., and Bito, Y. (1976). Humoral immune responses characteristic of testosterone-oropionate-treated chickens. *Immunology* **30:**341–348.

Horton, J. D., Amphibians. (1994). In *Immunology: A comparative approach* (R. J. Turner, ed.) pp. 101–136. Chichester: Wiley.

Isserhoff, H., Sylvester, P. W., and Held, W. A. (1986). Effects of *Schistosoma mansoni* on androgen regulated gene expression in the mouse. *Mol. Biochem. Parasitol.* **18:**401–412.

Jurd, R. D. (1994). Reptiles and birds. In *Immunology: a comparative approach* (R. J. Turner, ed.), pp. 137–172. Chichester: Wiley.

Kirkpatrick, M., and Ryan, M. J. (1991). The evolution of mating preferences and the paradox of the lek. *Nature* (Lond.) **350:**33–38.

Lehmann, D., Siebold, K., Emmons, L. R., and Muller, H. J. (1988). Androgens inhibit proliferation of human peripheral blood lymphocytes *in vitro. Clin. Immunol. Immunopathol.* **46:**122–128.

Manning, M. J. (1994). Fishes. In *Immunology: A comparative approach.* (R. J. Turner, ed.) pp. 69–100. Chichester: Wiley.

Marsh, J. A., and Scanes, C. G. (1994). Neuroendocrine-immune interactions. *Poult. Sci.* **73:**1049–1061.

McLennan, D. A., and Brooks, D. R. (1991). Parasites and sexual selection: A macroevolutionary perspective. *Quart. Rev. Biol.* **66:**255–286.

McMinn, H. (1990). Effects of the nematode parasite *Camallanus cotti* on the sexual and non-sexual behaviors in the guppy (*Poecilia reticulata*). *Am. Zool.* **30:**245–249.

Milinski, M., and Bakker, T. C. M. (1990). Female sticklebacks use male coloration in mate choice and hence avoid parasitised males. *Nature* (Lond.) **344:**330–333.

Møller, A. P. (1988). Female choice selects for male sexual tail ornaments in the monogamous swallow. *Nature* (Lond.) **332:**640–642.

Møller, A. P. (1990). Parasites and sexual selection: current status of the Hamilton and Zuk hypothesis. *J. Evol. Biol.* **3:**319–328.

Møller, A. P. (1991). Parasites, sexual ornaments and mate choice in the barn swallow. In *Bird-parasite interactions. ecology, evolution and behaviour* (J. E. Loye and M. Zuk, eds.) pp. 328–343. Oxford: Oxford University Press.

Norris, K., Anwar, M., and Read, A. F. (1994). Reproductive effort influences the prevalence of haematozoan parasites in great tits. *J. Anim. Ecol.* **63:**601–610.

Owens, I. P. E., and Short, R. V. (1995). Hormonal basis for sexual dimorphism in birds: implications for new theories of sexual selection. *Trends Ecol. Evol.* **10:**44–47.

Petrie, M., Halliday, T., and Sanders, C. (1991). Peahens prefer peacocks with elaborate trains. *Anim. Behav.* **41:**323–331.

Potts, W. K., and Wakeland, E. K. (1993). Evolution of MHC genetic diversity: A tale of incest, pestilence and sexual preference. *Trends Genet.* **9:**408–412.

Read, A. F. (1990). Parasites and the evolution of host sexual behavior. In *Parasitism and Host Behaviour* (C. J. Barnard and J. M. Behnke, eds.), pp. 78–99. London: Taylor and Francis.

Read, A. F., and Harvey, P. H. (1989). Reassessment of comparative evidence for Hamilton and Zuk theory on the evolution of secondary sexual characters. *Nature* (Lond.) **339:**618–620.

Redig, P. T., Dunnette, J. L., Mauro, L., Sivanandan, V., and Markham, F. (1985). The *in vitro* response of turkey lymphocytes to steroid hormones. *Avian Dis.* **29:**373–383.

Ressell, S., and Schall, J. J. (1989). Parasites and showy males. Malarial infection and color variation in fenze lizards. *Oecologica,* **78:**158–64.

Roitt, I. M. (1994). *Essential immunology.* Oxford: Blackwell Scientific.

Roitt, I. M., Brostoff, J., and Male, D. K. (1993). *Immunology.* London: Mosby.

Saad, A. H., Khalek, N. A., and El Ridi, R. (1990). Blood testosterone level: A season-dependent factor regulating immune reactivity in lizards. *Immunobiology* **180:**184–194.

Slater, C. H., and Schreck, C. B. (1993). Testosterone alters the immune response of chinook salmon *Oncorhynchus tshawytscha. Gen. Comp. Endocrinol.* **89:**291–298.

Spindler, K. D. (1988). Parasites and hormones. In *Parasitology in focus* (H. Mehlhorn and D. Bunnag, eds.) pp. 134–148.

Staples, L. D., Binnes, R. M., and Heap, R. B. (1983). Influences of certain steroids on lymphocyte transformation in sheep and goats studied *in vitro. J. Endocrinol.* **98:**55–69.

Sullivan, B. K., (1991). Parasites and sexual selection: Separating causes and effects. *Herpetologica.* **47:**250–264.

Talal, N. (1992). In *Encyclopedia of immunology* (I. M. Roitt and P. J. Delves, eds.), pp. 1112–1114. London: Academic Press.

Tinsley, R. C. (1990). The influence of parasite infection on mating success in spadefoot toads, *Scaphiopus couchii. Am. Zool.* **30:**313–325.

Turner, R. J. Mammals. (1994). In *Immunology: A comparative approach.* (R. J. Turner, ed.) pp. 173–214. Chichester: Wiley.

Wakelin, D., and Appanus, V. (1997). Genetic control of immune responses. In *Host-parasite evolution: General principles and avian models* (D. Clayton and J. E. Moore, eds.), pp. 30–58. Oxford: Oxford University Press.

Wedekind, C. (1992). Detailed information about parasites revealed by sexual ornamentation. *Proc. R. Soc. of Lond. Biol. Sci.* **247:**169–174.

Wedekind, C. (1994). Mate choice and maternal selection for specific parasite resistences before, during and after fertilization. *Proc. R. Soc. of Lond. Biol. Sci.* **249:**303–311.

Wedekind, C., and Folstad, I. (1994). Adaptive or non-adaptive immunosuppression by sex hormones? *Am. Nat.* **143:**936–938.

Wingfield, J. C. (1990). Interrelationships of androgens, aggression, and mating systems. In *Endocrinology of birds: Molecular to behavioral* (M. Wada, S. Ishii, and C. G. Scanes, eds.) pp. 187–205. Tokyo: Japan Scientific Societies Press/Berlin: Springer-Verlag.

Wingfield, J. C. (1994). Modulation of the adrenocortical response to stress in birds. In *Perspectives in comparative endocrinology* (K. G. Davey, R. E. Peter, and S. S. Tobe, eds.) pp. 1012–1022. Canada: National Research Council.

Wingfield, J. C., Whaling, C. S., and Marler, P. (1994). Communication in vertebrate aggression and reproduction: the role of hormones. In *The physiology of reproduction* (E. Knobil and J. D. Neill, eds.) pp. 303–342. New York: Raven Press.

Wingfield, J. C., Hegner, R. E., Dufty, A. M., and Ball, G. F. (1990). The "Challenge Hypothesis": Theoretical implications for patterns of testosterone secretion, mating systems, and breeding strategies. *Am. Nat.* **136:**829–846.

Zapata, A. G., Varas, A., and Torroba, M. (1992). Seasonal variations in the immune system of lower vertebrates. *Immunol. Today.* **113:**142–147.

Zuk, M. (1992). The role of parasites in sexual selection: current evidence and future directions. *Adv. Study Behav.* **21:**39–68.

Zuk, M., Thornhill, R., Ligon, J. D., and Johnson, K. (1990). Parasites and mate choice in red jungle fowl. *Am. Zool.* **30:**235–244.

8

Host Embryonic and Larval Castration as a Strategy for the Individual Castrator and the Species

John J. Brown and Darcy A. Reed

I. Natural Selection

Parasitic castration of hosts occurs throughout the animal kingdom, as well as being the result of insect infestations of plant tissues. Specific examples of disruption to the reproductive capability of invertebrate hosts by parasites have been the subject of several reviews (Hurd, 1993; Hurd, 1990a, b; Read, 1990). While these earlier monographs dealt with all classes of invertebrates, not just insects, this chapter will be restricted to insect parasitoid–insect host interactions. Our discussion will be limited to interactions between species of Hymenoptera (primarily bees and wasps), Lepidoptera (butterflies and moths), Coleoptera (beetles), and Strepsiptera (twisted-wing parasites) in the Class Hexapoda. Most of our information will be on hymenopteran parasitoids attacking lepidopteran hosts in their larval or egg stages. Generally, the effect of the parasitoid on the gonads of the host is maximized if the host is young when initially attacked. This is due, in part, to the continual and exponential development of host gonads.

Biological control of insect pest species has regained importance in entomology for several reasons. Applied entomologists and agriculturists are aware of the increasing reports of resistance in pest species to chemical controls, and they also are aware of the public's perception of agricultural chemicals as being harmful to the environment. The combined use of broad-spectrum insecticides and insect biological agents is often incompatible. However, new insect growth regulators, which selectively kill lepidopteran pests without killing ectoparasitic parasitoids (Brown, 1994), should find acceptance in integrated pest management (IPM) strategies of the future. An increased understanding of the basic biology of host regulating factors produced by parasitic Hymenoptera (Coudron, 1991) may explain how hosts are castrated, thus leading researchers toward the discovery of castration factor(s) which could be incorporated into recombinant microbial insecticides.

Every living organism has evolved towards mastering its niche; the alternative is extinction. Humans may represent the best example of animals that can dominate their environment. One flaw in our dominance, however, has been our inability to eradicate insect pest species with synthetic insecticides. It is not that these chemicals are inadequate to kill insects, but that insects have such a tremendous ability to develop resistance. Insect parasitoids have evolved various factors that aid in their ability to subdue hosts that in some cases may: 1) have a mass 1,000 times greater than the individual parasitoid; 2) have physical, chemical, and behavioral defenses; 3) be capable of rapid or long distance mobility; and most importantly, 4) have an ability to rapidly mutate via a short generation time and high rate of fecundity. Parasitoids and hosts share the latter three attributes in their evolutionary battle for survival. Parasitoids have evolved specific factors [e.g., venom (Jones and Coudron, 1993), polydnavirus (Stoltz, 1993; Fleming and Krell, 1993), teratocytes (Dahlman and Vinson, 1993); and various peptides (Jones et al., 1990)] to establish their dominance over their food source. One or a combination of these factors in chelonine parasitoids (a subfamily of egg–larval-attacking braconid wasps) is responsible for the elimination of their hosts' gonads. Irreversible destruction of host gonads or castration ensures that the host will remain susceptible to attack by parasitoid progeny. Even the "lucky" mutant host, capable of surviving the normally lethal effect of a parasitoid attack, would be incapable of transferring those genes to the next generation.

Castration has been described as the irreversible destruction of tissues involved in reproduction (Doutt, 1963; Hurd, 1990a). This can include: 1) physical destruction (direct ingestion, see Clausen, 1940; Baudoin, 1975) by the parasitoid's consumption of tissues; 2) biochemical destruction or interference by nutrient deprivation (indirect biochemical, see Doutt, 1963; Drea, 1968; Baudoin, 1975); 3) hormonal changes which halt tissue development or cause atrophy (hormonal castration, see Baudoin, 1975; Girardie and Girardie, 1977); as well as 4) alterations in morphology (Wülker, 1975) and behavior which result in the inability or reduced ability of the host to mate and reproduce (Read, 1990; Hurd, 1990a). Hurd's definition was inclusive of all examples of host castration, but host castration in insect-insect associations is rarely observed (Hurd, 1993). This is mainly because, in host-parasitoid relationships, the host is killed before reaching a reproductive age. Researchers have not looked closely at host gonads, because from an applied research perspective there was no need. The parasitoid larva egressed from an immature host that was killed and consumed or subsequently died, thereby, negating the question of whether gonadal development was affected. Kuris (1974) suggested that parasitoids and castrators can be differentiated. A parasitoid may castrate a host prior to killing the host, but a castrator simply destroys the reproductive tissue.

Milinski (1990) argues that parasitic castration, to prevent the host species from developing counteradaptations, is not possible without group selection. Simply preventing reproduction of the individual that has a mutation for a specific counteradaptation, such as behavioral avoidance, would benefit the next parasitoid generation, castrator and noncastrator unselectively. We would argue that a host species' perpetual susceptibility to a parasitoid species' attack is not the immediate "natural selection" reward for being able to castrate one's host. Our narrative explores a possible sequence of intermediate mutations that resulted in host castration, such as (1) diversion of energy reserves away from host reproductive tissue for the benefit of the parasitoid larva; (2) elimination of host testes which are involved with ecdysteroid metabolism, which could impair a parasitoid larva's ability to monitor the host's endocrine system in order to synchronize its own over-wintering survival and vernal development with that of its host; and (3) elimination of host testes which may interfere with the parasitoid's pre-egression control of apolysis of the host's integument, thereby impeding the ability of the newly molted third stadium parasitoid to egress from the host's hemocoel. These sequential mutations may have culminated in the ultimate reward for the species; i.e., the deterrence of host resistance to the parasitoid. The ability to castrate hosts is the end product of multiple steps. As a parasitoid species masters each of the three sequential mutations mentioned above, it would have more offspring than a competitor species which has not mastered its niche.

Our discussion will follow what we assume to be an evolutionary sequence, first addressing changes in host behavior and the possible diversion of host nutrients away from gonadal development for the benefit of the parasitoid larva(e). Then we report extensively on documented studies of larval (*cessation of testicular development*) and egg–larval (*embryonic castration*) parasitoids. Finally, in response to the question, Why castrate the host? we discuss endocrine communication between the host and parasitoid larva, and diapause-associated events that help explain the benefit to the parasitoid of being able to castrate its host.

II. Regulation of Host Physiology and Size

Earlier researchers assumed castration eliminated gonads which competed for nutrients (Sousa, 1983) or produced hormones that influenced host physiology and behavior. Essentially this is what humans do when they castrate domestic livestock. Castrated animals are less aggressive and gain mass faster than their sexually active siblings; however, this pattern is not always true for insect hosts.

Parasitized codling moth (*Cydia pomonella* L.) larvae are less aggressive toward their siblings. Healthy codling moth larvae are generally solitary feeders; seldom in the wild do you find more than one third, fourth, or fifth stadium larva in a single fruit. In the laboratory, when two or more neonates are placed in a plastic cup with adequate diet to support the growth of 10 codling moth larvae, generally only one, and occasionally two adults, will eclose. Healthy codling moth larvae become aggressive toward their siblings after they enter the third stadium (Ferro and Harwood, 1973). In the apple this is when larvae initiate feeding upon the oil- and protein-rich apple seeds. Artificial diet containing casein and wheat germ represents an excessive lipid and protein diet, but this still does not avert the aggressive behavior among healthy third stadium larvae. Codling moth larvae parasitized by *Ascogaster quadridentata* Wesmael tolerate crowded laboratory conditions and often five parasitoid adults can be collected from the same-sized cup and same amount of artificial diet used to rear one healthy codling moth larva. A similar situation was reported in shore crabs, in which dominance hierarchies common for the species are eliminated in parasitized animals (Caiger and Alexander, 1973). Parasitoid modification of host aggressiveness toward siblings may be evolutionarily sanctioned because parasitized larvae need less protein and many sibling hosts may have been parasitized by the same female wasp. It is normal for a parasitoid female to intensify her search for additional hosts in the immediate vicinity of the initial find. Therefore, host aggression toward siblings would be counterproductive to the survival of the parasitoid species. Castration may result in hosts tolerating crowded conditions.

Increased insect growth as a result of being parasitized is generally measured by increased larval body mass, since most parasitoids devour their host prior to adult eclosion. *Trichoplusia ni*, parasitized by *Copidosoma truncatellum*, consumed 35% more food than healthy larvae and converted some of that ingested food into increased body mass (Hunter and Stoner, 1975; Jones et al., 1982). An increased host body mass may be common in lepidopterans attacked by these polyembryonic encyrtid parasitoids; however, the increase may not be directly attributed to the massive parasitoid brood or castration of the host. Strand (1989) reported a 49% increase in larval weight in the parasitized population, compared to healthy *T. ni,* but only 13% of that weight gain was attributed to the mass of the *C. floridanum* brood. Testes are present in *T. ni* parasitized by *C. floridanum* (Strand, Personal Communication), so this increase in size is not related to host castration. The increased body mass of parasitized *T. ni* may be no more than the result of natural selection pressures favoring a parasitoid's ability to bring about an enlargement in the host body that would allow a greater number of parasitoid progeny to successfully develop within the larger hemocoel. In other words, the increased body mass of the host has no direct nutritional value to the host or developing parasitoid larvae, except upon final consumption of the host by the encyrtid larvae prior to pupation.

Much more common in insect host-parasitoid relationships is the situation where host growth is suppressed to prevent the host from exceeding some critical premetamorphic mass. The same host species (*T. ni*) parasitized by (polyembryonic) *C. floridanum* may gain a maximum weight of 432 mg and have a head capsule width of 1.93 mm (Strand, 1989). This same species parasitized by the egg-larval parasitoid *Chelonus* near *curvimaculatus* as an egg, seldom exceeds 17 mg before the fourth stadium (head capsule width 0.96 mm) host is consumed by a solitary parasitoid larva. Therefore, the development of *T. ni* parasitized by *Chelonus* falls far short of the critical mass (216 mg) and head capsule width (1.66 mm) needed for a healthy *T. ni* larvae to pupate (Jones et al., 1981a).

However, factors independent of the parasitoid larva are involved, because pseudoparasitized *T. ni*, where the parasitoid egg or larva died (see Jones, 1985 for definition of pseudoparasitism), weighed 55–60 mg and had a head capsule width of 1.34 mm (Jones et al., 1981b), still far short of the critical parameters needed for pupation. Both truly parasitized and pseudoparasitized *T. ni* are castrated.

In summary, there is no evident pattern that host castration in host-parasitoid situations in insects regulates host size; especially not along the lines we would think of concerning domesticated livestock. On the contrary, most documented cases of insect host castration have, thus far, reported smaller hosts, rather than larger parasitized individuals. This is not to negate the importance of a balance of host nutrition and parasitoid development; when compared to parasites of vertebrate hosts, parasitoids occupy a significantly greater proportion of their host hemocoel (Thompson, 1990). Alterations in host metabolic pathways for the benefit of parasitoid development (Thompson, 1993) are regulatory aspects of host-parasitoid interactions that are much more subtle than castration, and yet reproductive performance of the host still suffers.

In holometabolous insects, especially lepidopterans, the gonads tend to grow very slowly during the early larval stages, with the majority of growth occurring in the final larval stadium for males and maturing in the prepupal and pupal stages. The ovaries grow and mature principally during the pupal stage. Nutrient reserves accumulated in larvae are used for both adult tissues and for reproductive behavior. For insects attacked by larval parasitoids, there are no host reserves depleted by reproductive behavior, because the host does not survive to be an adult. Therefore, Baudoin's (1975) assertion that host castration can only benefit the castrator does not properly address larval or egg-larval parasitoids, because the timing of infection does not coincide with the host's utilization of energy reserves for reproductive activities. Even in adult-attacking parasitoids, such as strepsipterans, there is evidence that parasitoid development occurs long before host reserves normally used for reproduction would be available and therefore affected. Strepsipterans that attack adult hymenopteran hosts retard development of fat body, corpora allata, and ovaries, therefore effectively castrating female hosts (Strambi et al., 1982). Strepsipteran influences on all three of these tissues may be involved in reducing fecundity. Girardie and Girardie (1977) suggested that the lack of juvenile (or gonadotrophic) hormone in some parasitized insects may be due to the hypoactivity of the median neurosecretory cells, which, in turn, control the corpora allata in hosts. Thus, host castration by these strepsipterans does not necessarily mean a direct reduction or effect on the gonads, but rather an indirect effect through altering the host's endocrine system. This also may be the situation with *Perilitus rutilus* Nees, a braconid parasitoid of the pea weevil *Sitona lineata* L. that renders the host's ovaries functionless (Jackson, 1924), and essentially eliminated; however, testes of *Sitona* species were not noticeably affected by the braconid's attack (Loan and Holdaway, 1961).

III. Parasitoids That Attack Larval Hosts

Junnikkala (1985) reported that testes of *Pieris brassicae* (the cabbage butterfly larva) parasitized by *Apanteles glomeratus* (a small parasitic wasp now placed in the genus *Cotesia*) undergo steady growth and development, but at a much slower pace than those in nonparasitized males. Castration was not an "all or none block." The testicular envelope, which has a trophic function, was only 25 percent as thick as normal and pigmentation was lacking. Sperm bundles were formed but were highly disorganized leading Junnikkala to question their viability. Hosts which had been stung, but in which no parasitoid larva could be found (pseudoparasitized), possessed testes of normal volume, but the thickness of testicular wall tissue was reduced.

Parasitoids such as *Cotesia* spp which attack larvae, stop host testicular development at the time of attack, meaning hosts that are not attacked until the final larval stadium may have testes that have already reached roughly their maximum volume; however, secondary spermatogenesis and spermiogenesis are halted (Yagi and Tanaka, 1992). Tanaka et al. (1994) showed that control larvae of *Pseudaletia separata* (an army worm) do not develop testes of appreciable size until the fifth instar. Testes failed to increase in volume after the host was parasitized by *Cotesia kariyai* (Fig. 8-1).

Some parasitic wasps introduce not only their developing eggs when ovipositing into a host but also polydnaviruses in the surrounding calyx fluid (Stoltz, 1993). When the parasitoid egg hatches, it may also release teratocytes (Dahlman and Vinson, 1993). Some evidence suggests that the causative agent of host castration is a symbiotic polydnavirus introduced by larval-attacking parasitoid wasps upon oviposition. Yagi and Tanaka (1992) demonstrated that calyx fluid is adequate, but calyx fluid plus venom has a greater or more dramatic effect in inducing castration. Testes of animals injected with virus alone or venom alone were not significantly different from saline-injected controls, whereas the two components in combination reduced testicular growth dramatically (Fig. 8-2). Venom may be necessary to open channels in testicular septa and surrounding membranes in host larvae attacked by larval-attacking parasitoids. Once this membrane has formed, polydnavirus may not be able to penetrate without the help of venom. In larval parasitoids, part of the role of the venom may be to disrupt testicular wall tissue which would allow the penetration of the virus into testicular cells.

IV. Embryonic Castration

Salt (1968) proposed that parasitoids attacking very young animals have time to act as chronic subduer of host defenses. The ultimate in parasitoid attack, resulting

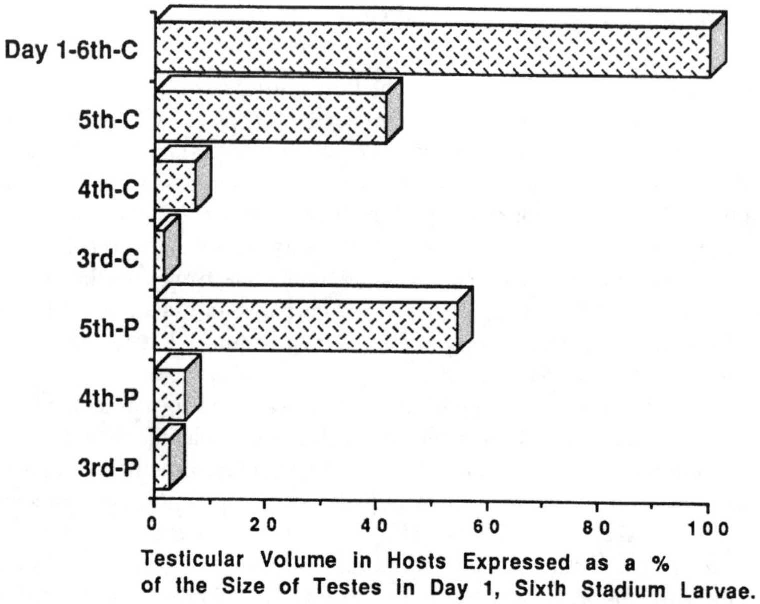

Figure 8-1. Suppression of testicular growth in *Pseudaletia separata* parasitized (P) by *Cotesia kariyai* in various instars compared to the size of testes in control larvae (C) for reference. Testicular volumes have been adjusted to reflect suppression of growth in parasitized individuals compared to the size of testes found in nonparasitized day 1 sixth stadium larva (Adapted with permission from Wiley-Liss from Tanaka, et al. 1994).

in host castration prior to any energy reserves being diverted to gonadal development, is the example of embryonic castration (Reed-Larsen and Brown, 1990) found in the chelonine group of braconid egg-larval endoparasitoids. Castration occurs before any substantial numbers of germ cells are formed that could deprive the parasitoid larva of any nutrients. A chelonine adult introduces its egg into the egg of a host. The parasitoid egg hatches within 30 h (25°C) of oviposition, and depending on the host age at the time of attack, this generally occurs before the host hatches. Therefore, the neonate host contains a first instar parasitoid within its hemocoel. The parasitoid remains in its first stadium, while the host larva molts three times. Lepidopteran hosts attacked during the egg stage, such as the codling moth by *Ascogaster quadridentata* develop into larvae which are castrated. *Adoxophyes* sp. attacked by *Ascogaster reticulatus* (Brown et al., 1993) and *T. ni* attacked by *Chelonus* near *curvimaculatus* (Jones et al., 1981b) have similar precocious development and both hosts are castrated. Host larvae undergo precocious metamorphosis, becoming behaviorally and physiologically mature in the penultimate stadium. The precocious parasitized fourth instar is similar in size to a nonparasitized fourth stadium larva and yet it wanders and spins a cocoon and begins accumulating storage proteins (Jones, 1989). During the host's

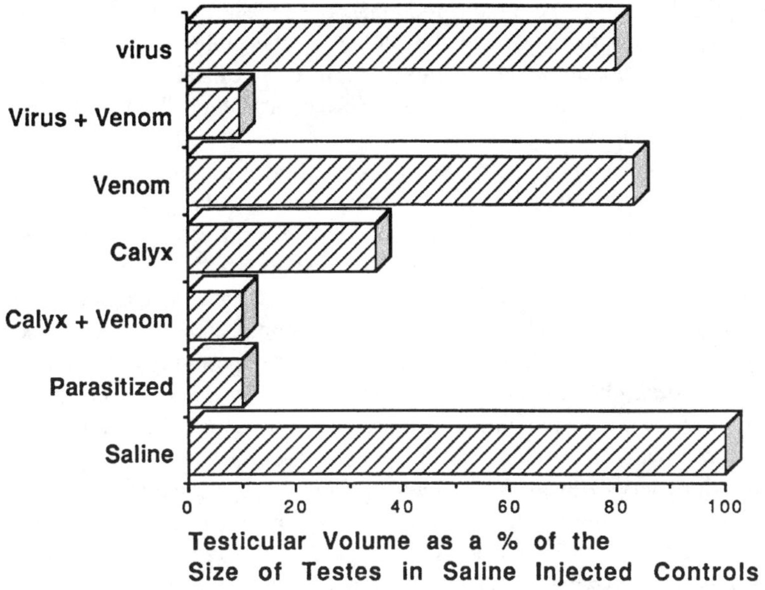

Figure 8-2. Effect of factors extracted from *Cotesia kariyai* and injected into *Pseudaletia separata* on host testicular development, compared to naturally parasitized hosts and hosts injected with saline only. Injections were made on day 3 of the fifth stadium of the host and dissection to evaluate the effects on testicular volume occurred six days later. (Adapted with permission from Wiley-Liss and Tanaka, et al. 1994).

fourth stadium, under nondiapausing conditions the parasitoid initiates its own molt to the second instar. Finally, concurrent to the parasitoid's second larval ecdysis, it punctures the host integument and completes its development externally, consuming the host's tissues.

Chelonine parasitoids do not discriminate between male and female hosts, both are castrated. If host eggs are attacked early in embryogenesis, the germ tissue, both testes and ovaries, is completely obliterated. Larvae which developed from eggs protected from parasitoid attack until 2–3 days after being oviposited have abnormal testes. There were often alterations in the normal morphology of four testicular lobes and bean-shaped appearance of the testes. All types of variations were witnessed: unilateral development, reduction in the number of follicular lobes unilaterally or bilaterally, with unequal numbers of lobes within each pair of testes. At the follicular, rather than the testicular, level embryonic castration is an "all or none" type of effect (Friedländer and Brown, 1995). Complete or partial gonadal destruction is dependent on the age and embryonic stage of development of the host embryo at the time of parasitization. The castrating factor(s) may act to destroy or atrophy groups of pole or primordial germ cells early in embryogenesis prior to dorsal closure. Dorsal closure is a developmental event where the ectodermal lobes meet, trapping yolk cells that eventually occupy the lumen of the midgut in the embryo. Following dorsal closure, the developing gonads exhibit tolerance or resistance to the castrating factor(s) and, having failed to affect cells of a specific follicle, host spermatogenesis proceeds regularly.

The accomplishment of complete host castration, meaning no gonadal tissue development, corresponds closely with the point in codling moth embryology when dorsal closure is observed. The uptake of yolk effected by dorsal closure may provide a passive mechanism for the entry of both the parasitoid larva (Hawlitzky 1972, 1979) and the castration factor(s) into developing embryos. Once dorsal closure is complete, nonparalytic venom with chitinase properties (Krishnan et al., 1994), might extend the window of opportunity for the castration factor(s) to penetrate the developing embryo. In fact, one could speculate that a failure of chelonine venom proteins to effectively function in larval hosts could have placed natural selection pressures on this group to parasitize younger and younger host larvae, even to the point of specializing on embryos. Almost all field-collected codling moth larvae parasitized by *Ascogaster* are completely castrated, suggesting that most hosts are attacked prior to dorsal closure. However, in the laboratory *Ascogaster* can successfully parasitize, but not castrate, hosts in the "black head" egg stage. In such eggs, the neonate is fully developed and is preparing to hatch, and the parasitoid wasp oviposits directly into the hemocoel, through the host's integument and the egg chorion.

V. Why Castrate the Host?

There is no way to conclusively answer this evolutionary question, but a theoretical narrative will be presented here. At some point in time, parasitic and predatory behavior were probably similar. Parasites were simply predators that oviposited eggs on or near a surplus of host carcasses after the female's own appetite had been satiated. Next, the parasitic group utilized venom to immobilize the host, which retained the host tissues longer, allowing for a longer developmental time for ectoparasitic larvae. The paralytic venom of these primitive entomophagous parasites may have simply halted metabolic activity, which, in turn, prevented gonadal use of energy reserves. However, the ectoparasitic larvae were still vulnerable to predatory and hyperparasitic attack. The next step rewarded by natural selection was the use of venom to defeat the host immune system, thereby allowing endoparasitic development. Larval endoparasitoids had an advantage over pupal parasitoids in that imaginal gonadal discs consumed little or no reserves. Likewise, larval parasitoids that attacked their hosts in early instars had an advantage over those attacking late larval stages, because gonads were less mature. If this analogy is continued to include egg-larval parasitoids, no host energies will be lost to gonadal development, plus the host embryo is relatively defenseless against attack (Strand, 1986). Host defenselessness negates the need for immobilizing venom and now venom can serve other needs of the parasitoid, perhaps aiding in the access of the castration factor(s) into target tissues. Age distribution favors parasitoids that can successfully attack earlier stages of development because, in addition to defenselessness, there are more hosts available. However, a parasitoid's earlier attack on the younger host requires a more intricate control over the host's physiology, behavior, and endocrine metabolism.

VI. Endocrine Communication

Testes of healthy larvae and *Ascogaster* larvae in hosts have similar growth patterns. Growth curves comparing the increase in parasitoid volume in castrated larvae and the point where testes fuse, indicate maximum testicular growth in healthy larvae of *Cydia* and *Adoxophyes,* two hosts for *Ascogaster* sibling species, are similar (Fig. 8-3). Testes of *Adoxophyes,* or *A. reticulatus* larvae in castrated hosts, initiate rapid growth and development almost immediately after the lepidopteran larva has ecdysed into its ultimate stadium. Whereas, testes in *Cydia* programmed to pupate, or an *A. quadridentata* larva in a castrated host, initiate

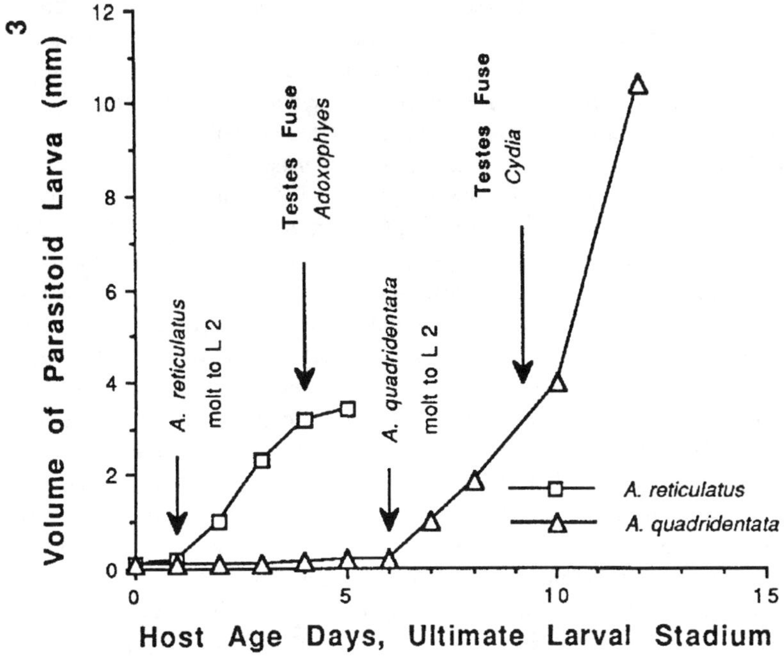

Figure 8-3. Growth of two species of *Ascogaster* during the ultimate larval stadium of two lepidopteran host species. Both species are castrated, but arrows denote the approximate time when testes fuse in healthy ultimate stadium larvae of both species (Modified from Brown and Kainoh, 1992).

development after the lepidopteran larva has initiated cocoon spinning activity, an event four to five days after ecdysis into the ultimate stadium (Brown and Kainoh, 1992). When these growth curves are compared, it appears that the parasitoids in castrated larvae mimic the growth of gonadal tissue in healthy host larvae. Parasitoid larvae initiate and complete their increase in body mass at the same time within their host's stadium as testes would be observed to initiate and complete development in healthy host larvae. The parasitoid larva may respond to the same endocrine cue(s) as testes do, lending support to the evolutionary theory that elimination of testes not only relieves competition for nutrients, but also controls endocrine messages.

In vitro studies suggest that the involvement of lepidopteran testes in ecdysteroid metabolism increases dramatically after the animal purges its gut, and fused testes even synthesize ecdysteroids (Gelman et al., 1989). Other tissues such as gut, fat body and malpighian tubules are also capable of ecdysteroid conversion, but not as much as testes (Delbeque et al., 1990). Jarvis et al., (1994) incubated various tissues, [e.g., prothoracic glands, testes, gut, carcass (integument), fat body and malpighian tubules] from last stadium *Spodoptera littoralis* with ecdysteroid precursors and found that the testes converted more of these precursors into ecdysone and 20-hydroxyecdysone than any of the other tissues, even more than the prothoracic glands. *In vitro* studies would suggest that testes take up ecdysteroids from the hemolymph, oxidize them (Gelman and Woods, 1985), and release active 20-hydroxyecdysone back into the hemolymph (Loeb et al., 1982).

Midway through the last larval stadium of lepidopterans (Fig. 8-4), a small pulse of ecdysteroids is released that causes epidermal cells to commit to pupation (Riddiford and Kiely, 1981; Benz and Ren, 1987). Testes of the codling moth also respond to this commitment peak of ecdysteroids or related factors (Jans et al., 1984) and initiate spermiogenesis. Spermiogenesis and metamorphosis require large amounts of nutrients and most larval and egg-larval parasitoids consume the host body before host maturation and transformation changes are initiated. Parasitoids with a life cycle similar to *A. quadridentata* remain in the first stadium while their host molts several times and only after the host wanders away from the diet and initiates cocoon spinning activity do the parasitoids resume their growth. These endoparasitoids float free in the host hemocoel and, in order for them to correctly monitor their host's physiology, they must be capable of distinguishing host releases of ecdysteroids in the presence or absence of juvenile hormone. Ecdysteroids in the absence of juvenile hormone would indicate a pending larval-pupal molt. Lawrence (1986) suggested that first stadium parasitoid larvae may be incapable of synthesizing their own ecdysteroids and may rely upon ecdysteroids absorbed from the host. Grossniklaus-Burgin et al., (1989) demonstrated that *Chelonus,* a sibling genus to *Ascogaster,* was capable of acquiring ^{14}C-ecdysteroids from the host hemolymph. Parasitoid larvae will respond to ecdysteroids *in vitro* by increasing their body length and occasionally molting (Brown et al. 1990). Therefore, we assume that the first larval–larval molt of *A.*

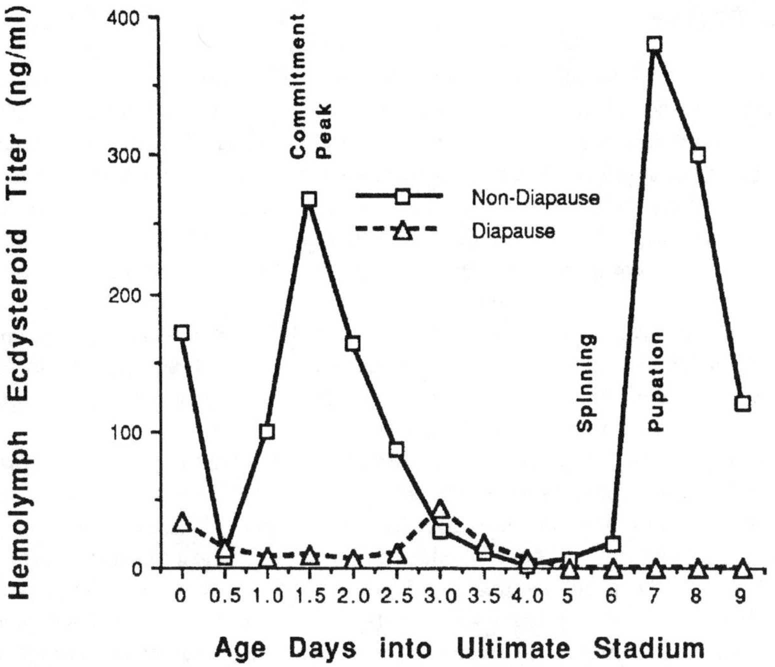

Figure 8-4. Hemolymph ecdysteroid titers in the ultimate larval stadia of codling moths programmed to pupate (nondiapause) and diapause. (Adapted from Benz and Ren (1987) with permission from Dr. G. Benz.)

quadridentata is in response to the host release of the commitment pulse of ecdysteroids in the absence of juvenile hormone, corresponding to host wandering behavior. If this is the case with *A. quadridentata,* which castrates its host, then host castration may be important to the evolutionary survival of this parasitoid species. Host castration would eliminate a tissue known to be active in ecdysteroid metabolism. Testes might compete for the uptake of the small pulse of ecdysteroids released midway through the last larval stadium or; conversely, testes could be the source of ecdysteroids, potentially confusing the parasitoid larva. In the absence of testes, the parasitoid larva can more accurately monitor slight changes in its host endocrine milieu.

Host castration is not essential to *A. quadridentata*'s survival. As mentioned earlier, if the host's egg is protected from parasitoid attack until after dorsal closure then testes do develop. Under photoregimes promoting continuous development and under optimum laboratory conditions, an *A. quadridentata* larva can survive in a host that has not been castrated. However, natural selection pressures that favored host castration were experienced under adverse, not optimum conditions.

VII. Diapause

Diapause is a physiological pause in development, favored by natural selection to synchronize an insect's development with its food source (Doutt et al., 1976). A parasitoid's diapause may be controlled by the diapause of the host (Strand, 1986). In two *Ascogaster* species development beyond the first stadium may be related to host programming to accept or avoid diapause. A sibling species, *A. reticulatus,* has a life cycle similar to *A. quadridentata,* but the first larval-larval molt of *A. reticulatus* is correlated with the molt of the host into its last stadium. This is unlike *A. quadridentata* whose molt is triggered by the host's midfinal stadium wandering and spinning activities (Fig. 8-3). What distinguishing aspect separates the life cycles of these two *Ascogaster* species? Obviously they develop in two different lepidopteran genera, and one physiological characteristic that separates the two hosts is how they overwinter.

Adoxophyes sp. (*orana* or *fasciata,* see Tamaki, 1991), the host of *A. reticulatus,* overwinters in various larval stages in Japan, whereas *A. orana* diapauses only in its early third stadium in Europe (Brown, 1991). Thus, in the case of *A. reticulatus,* once its host has molted to the fourth instar, host diapause (normally occurring in the third stadium) has been averted and parasitoid development continues. But *A. quadridentata* must delay commitment to further development until its host's program to pupate or diapause is communicated through the hemolymph endocrine milieu. *Cydia pomonella* diapauses as a mature larva. This occurs after it has exited the food source, wandered in search of a hibernation site, purged its gut, and spun a cocoon. If *A. quadridentata* initiated its first larval–larval molt in response to its host's molt into the ultimate stadium, as does *A. reticulatus,* then adult parasitoids could eclose when no suitable hosts were available for this egg–larval parasitoid (i.e., when unparasitized larvae were in diapause and no eggs had been produced).

The pattern of ecdysteroids in the hemolymph of nondiapausing codling moth larvae is similar to that reported in other lepidopteran species destined to pupate (Riddiford and Kiely, 1981), but larvae programmed to enter diapause have no such midstadium release of the commitment peak of ecdysteroids (Fig. 8-4). In healthy males destined to diapause, the absence of ecdysteroids at the time of wandering (Benz and Ren 1987) programs the testes to cease development and remain small throughout diapause. The same is true for codling moth larvae parasitized by *A. quadridentata.* The host exits its diet, wanders in search of a hibernaculum site, and spins a cocoon, but the parasitoid larva remains in its first stadium until an ecdysteroid pulse signals that host dormancy is terminating.

VIII. Postdiapause Development

In male codling moth larvae which are destined to diapause, the postwandering peak of ecdysteroids does not occur until after diapause termination and the testes remain small throughout the winter. Diapausing codling moth larvae chilled at 4°C, 0L:24D for 100 days readily respond to diapause terminating conditions (DTC) of 25°C, 16L:8D by pupating (Brown et al., 1990). However, if diapausing larvae are ligated behind the head within 48 h of being transferred to DTC, the thorax and abdomen will remain in the larval stage. An injection of 20-hydroxyecdysone (Benz, 1991) or topical treatment with an ecdysone-agonist (RH-5992 = tebufenozide) to ligated abdomens will stimulate metamorphosis in the integument (Friedländer and Brown, 1995). This indicates that ecdysteroids or cephalic factors that stimulate ecdysteroid synthesis are released in the first 48 h following ligation.

Testicular development in postdiapause codling moth larvae is inhibited by ligation of the head region, but growth resumes in response to 20-hydroxyecdysone, and spermatogenesis is renewed after a pulse of 20-hydroxyecdysone (Friedländer and Benz, 1982) or an ecdysone-agonist (Friedländer and Brown, 1995) appears in the hemolymph. In the spring, following six or more months of diapause, the most noticeable difference in the physiology of male codling moth larvae is the growth and resumption of spermatogenesis in the testes. However, in the absence of testes in an *A. quadridentata*-parasitized codling moth, it is the parasitoid larva's growth and molt to the second instar that is most apparent (Fig. 8-5).

Codling moth larvae parasitized by *A. quadridentata* overwinter as fourth stadium rather than fifth stadium larvae. Host exposure to DTC after 30 days of chilling causes the parasitoid to resume growth and molt to a second, and eventually ectoparasitic third instar larva that then consumes the host (Brown et al., 1990). If these chilled hosts are ligated behind their heads within 48 h of transfer to DTC, the *Ascogaster* larvae will die as first stadium larvae without molting to the second instars. A head-factor from the host released during the first 48 h after transfer to DTC is essential to renewed development of a diapausing *Ascogaster* larva (Brown et al., 1988). An injection of 20-hydroxyecdysone (Brown et al., 1990) or an ecdysone-agonist (Brown, 1994) will cause the parasitoid larva to resume growth and in some cases (~ 23%) the parasitoid will molt (Fig. 8-5). The diapausing *Ascogaster* larva is completely dependent upon its host's endocrine system to interpret the changing environmental conditions. The presence of host ecdysteroids is required for *Ascogaster* larvae to initiate a larval-larval molt. One could hypothesize that the parasitoid larva has replaced the host testes as a target receptor for hemolymph ecdysteroids. In healthy larvae, the testes respond to a pulse of ecdysteroid in the hemolymph in the spring with rapid growth, fusion and spermatogenesis; in their absence, such as in an *Ascogaster*-parasitized larva, the parasitoid larva responds to the same pulse of ecdysteroids with its own rapid growth and development (Fig. 8-6).

In the spring, adult *A. quadridentata* eclose shortly after eclosion of the unparasitized moths. Therefore, host eggs are available within a few days of parasite emergence. Host castration has maximized the endoparasitoid's overwintering

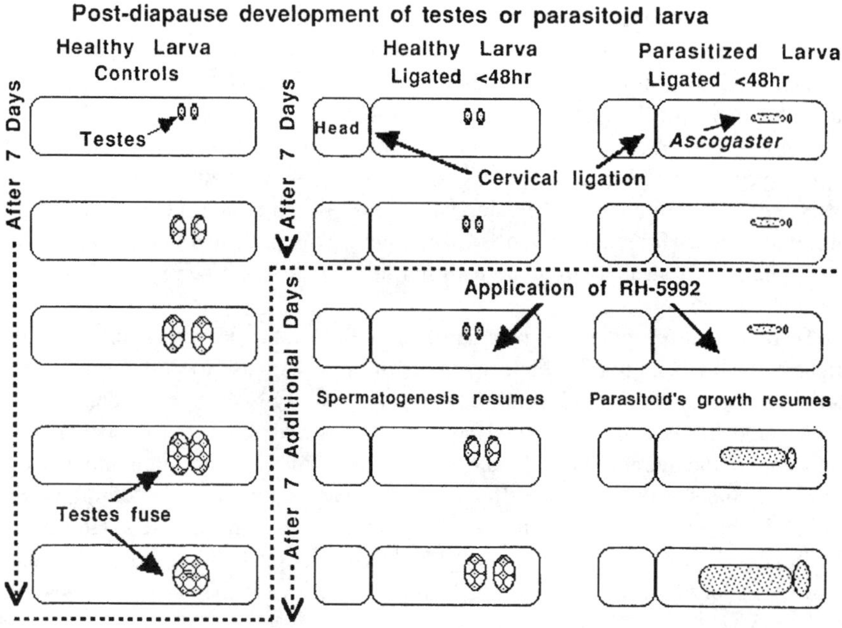

Figure 8-5. Postdiapause development of testes of an *Ascogaster quadridentata* larva in response to ligation within 48 h of transfer to diapause-terminating conditions (Brown et al., 1990). The application of RH–5992, an ecdysone agonist can, in one week, initiate spermatogenesis in nonparasitized larvae (Friedländer and Brown, 1995) or cause the parasitoid larva to resume its growth and development in parasitized codling moth abdomen (Brown, 1994).

habitat in the hemocoel of the host; not only through energy diversion, but the absence of host gonads also allows the parasitoid larva to accurately monitor fluctuations in the host endocrine system associated with diapause termination.

IX. Parasitoid Egression

The parasitoid's postdiapause molt from the first to second instar is in response to host ecdysteroids, but what about the second postdiapause molt, when the parasitoid molts from a second to a third instar and egresses from the apolysing host? The change from an endoparasitic to an ectoparasitic existence occurs just prior to the parasitoid egression from its host. The timing of this molt is very critical, a newly ecdysed third stadium larva has developed a tracheal system for acquiring oxygen directly, rather than via diffusion through the integument. It has shed its anal vesicle (the evaginated hindgut) and will soon initiate feeding on a solid, rather than liquid diet. A newly ecdysed third stadium larva would

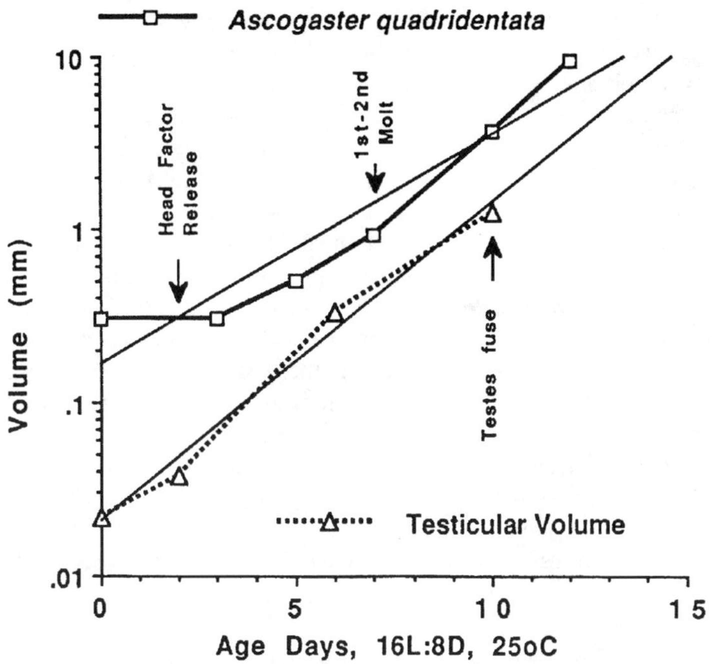

Figure 8-6. Comparison of postdiapause growth rates between testes in healthy larvae [$y = 2.0349^{-2} \times 10^{(0.18372x)}$, $r^2 = 0.988$] and *Ascogaster quadridentata* larvae [$y = 0.16291 \times 10^{(0.13229x)}$, $r^2 = 0.906$] in parasitized *Cydia pomonella* larvae, 0 to 15 days after being transferred from diapause-maintaining conditions (0L:24D, 4°C), to diapause terminating conditions (16L:8D, 25°C) (Adapted from Brown et al., 1990).

soon suffocate inside the hemocoel of its host. Its egression must accompany its own molt to a third instar, and yet the mandibles of a newly ecdysed larva are not sclerotized enough to forcibly chew through the host's integument.

Pseudoparasitized hosts are unable to produce ecdysteroids and generally die as larvae. Hemolymph ecdysteroid titers monitored in three pseudoparasitized host species parasitized by three chelonine species are very low (Brown and Reed-Larsen, 1991; Jones et al., 1992; Brown et al., 1993); and yet, in truly parasitized hosts at the moment of parasitoid egression, there are enough ecdysteroids to cause apolysis of the host integument, forcing it to weaken and thus, facilitating the parasitoid's escape. In some braconid and ichneumonid systems, there is documented degeneration of the host prothoracic glands (Johnson, 1965) due to the venom and calyx fluid (Tanaka and Vinson, 1991) or the polydnavirus delivered during parasitoid attack (Dover et al., 1988). In the absence of functional prothoracic glands and testes, the host tissues most capable of producing ecdysteroids are eliminated (Jarvis et al., 1994), allowing the parasitoid to be in control of when the host integument will initiate apolysis (Fig. 8-7). Natural selection pressures would strongly favor parasitoids capable of timing host apolysis with their own molt and egression from the host hemocoel.

In vitro incubations of late second stadium parasitoids that exhibit "fat body spots" have demonstrated an ability of chelonine parasitoid larvae to release ecdysteroids into their environment (Brown et al., 1993). Incubations involving the removal and replenishment of *in vitro* media over time revealed that the parasitoids were indeed capable of continual release of ecdysteroids (Brown and Reed-Larsen, 1991). Parasitoids appear to release large quantities of the active form of molting hormone, 20-hydroxyecdysone, at the end of the second stadium (Brown et al., 1993). The molting hormone is used not only for its own ecdysis, but the excess is released and may act on the host integument causing apolysis, thus facilitating the parasitoid's escape from the host hemocoel. This ability to cause a breach in the integrity of the host's integument may be more crucial after a long winter's diapause, when the success of the parasitoid may be determined by its ability to develop and emerge as quickly as possible after surviving six or more months of severe conditions. In the absence of host testes, the parasitoid-released ecdysteroids can maximize their effect on the integument rather than

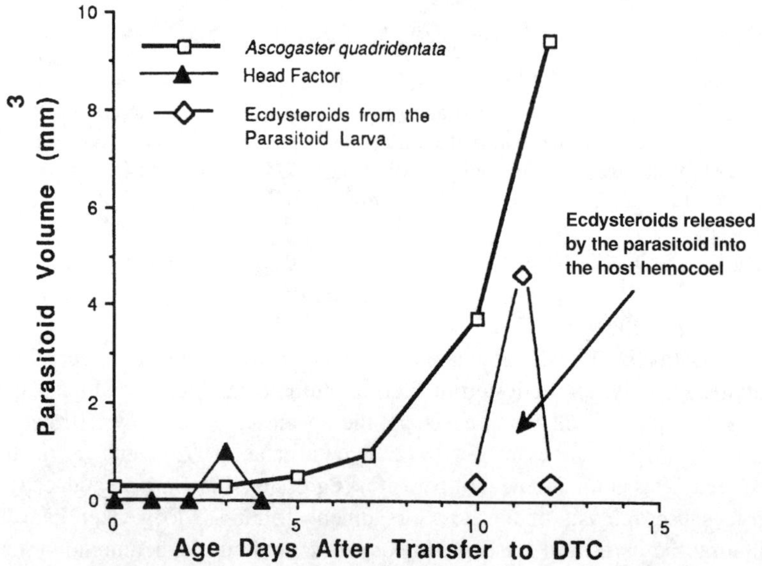

Figure 8-7. Growth of an *Ascogaster* larva in codling moth host following transfer of the host from diapause-maintaining (0L:24D; 4°C) to diapause-terminating conditions (DTC, 16L:8D; 25°C). Ligation of the host in the cervical area within 48 h of host transfer to DTC will prevent growth and development of the parasitoid larva (Brown et al., 1988; 1990). Just prior to egression from the host, the parasitoid larva releases large quantities of ecdysteroids (Brown and Reed-Larsen, 1991).

being lost to testicular uptake. If ecdysteroids were taken up by the testes, it might lessen the hormone's effect on the integument, thereby leaving the integument impermeable to any attempt at egression by the newly ecdysed parasitoid larva. Natural selection pressures, in the form of severe winter conditions and diapausing hosts have rewarded parasitoids capable of castrating their hosts by favoring the castrators with successful egression through the hosts' integument in the spring.

X. Summary

Insect parasitoids benefit from castrating their hosts by (1) eliminating competitive tissues for vital host nutrients, which may be particularly crucial during periods of high stress, such as diapause; (2) eliminating a tissue which may confuse or conflict with the hormonal cues and development of the parasitoid; and (3) preventing host resistance. As a long term benefit to the parasitoid population, castration would prevent hosts from developing resistance to parasitoid genes, gene products, or polydnavirus products. Any mutant host capable of resisting the lethal attack of the parasitoid larva would still not be able to transfer resistance genes to the next generation. Wülker (1975) identified host castration as the most important practical aspect and hope for biological control of insect pests, claiming that a parasite's ability to prevent host reproduction was more important than its ability to kill the individual pest. We must add to this the fact that even pseudoparasitized individuals are castrated and killed. This means that the true effect of parasitoids on the pest population is often underestimated if researchers assume that the number of parasitoids that emerge is equal to the number of hosts eliminated. In the future, a parasitoid's ability to castrate and prevent reproduction may again be recognized as its most valuable contribution to pest control, but the actual parasitoid may not have a direct involvement in the castration. We may be able to develop and exploit generic delivery systems, such as recombinant viruses, that carry several lethal factors plus a castration factor to prevent or delay development of pest resistance to the lethal factor(s). Of course, any effort toward developing this recombinant strategy must be tempered by the realization that the evolutionary battle will continue, regardless of the presence of the castration factor. The target pest can still develop resistance to the delivery system. Evidence for this possibility has been documented in the case history (Fuxa, 1993) of *Spodoptera frugiperda* developing resistance to nuclear polyhedrosis virus (NPV).

XI. Acknowledgments

This research was made possible by several grants to John J. Brown from the Washington Tree Fruit Research Commission and a National Science Foundation travel grant to Darcy A. Reed.

XII. References

Baudoin, M. (1975). Host castration as a parasitic strategy. *Evolution* **29:**335–352.

Benz, G. (1991). Physiology and genetics. In *Tortricid pests: Their biology, natural enemies and control* (L.P.S. van der Geest and H. H. Evenhuis, eds.), vol. 5 of World Crop Pests, pp. 89–147. Amsterdam: Elsevier.

Benz G., and Ren, S. X. (1987). The ecdysone titer in last instar larvae of *Cydia pomonella* L.) (Lep, Tortricidae). *Mitt. Schweiz. Entomol. Ges.* **60:**227–232.

Brown, J. J. (1991). Diapause. In *Tortricid Pests: Their biology, natural enemies and control* (L.P.S. van der Geest, and H. H. Evenhuis, eds.), vol. 5 of World Crop Pests, pp. 175–186. Amsterdam: Elsevier.

Brown, J. J. (1994). Effects of a nonsteroidal ecdysone agonist, tebufenozide, on host/parasitoid interactions. *Arch. Insect Biochem. Physiol.* **26:**235–248.

Brown, J. J., and Kainoh, Y. (1992). Host castration by *Ascogaster* spp. (Hymenoptera: Braconidae). *Ann. Entomol. Soc. Am.* **85:**67–71.

Brown, J. J., Ahl, J., and Reed-Larsen, D. (1988). Endocrine communication between a host and its endoparasitoid in relationship to dormancy. In Endocrinological frontiers in physiological insect ecology" (F. Sehnal, A. Zabza, and D. L. Denlinger, eds.), vol. 1, pp. 443–447. Wroclaw, Poland: Wroclaw Technical University Press.

Brown, J. J., Kiuchi, M., Kainoh, Y., and Takeda, S. (1993). *In vitro* release of ecdysteroids by an endoparasitoid, *Ascogaster reticulatus* Watanabe. *J. Insect Physiol.* **39:**229–234.

Brown, J. J., and Reed-Larsen, D. (1991). Ecdysteroids and insect host/parasitoid interactions. *Biological Control* **1:**136–143.

Brown, J. J., Reed-Larsen, D., and Ahl, J. (1990). Physiological relationship between a diapausing endoparasitoid *(Ascogaster quadridentata)* and its dormant host *(Cydia pomonella). Arch. Insect Biochem. Physiol.* **13:**229–238.

Caiger, K. M., and Alexander, A. J. (1973). The control of dominance in brachyuran crustacean, *Cyclograpsus punctatus.* Mil. Edw. *Zool. Afr.* **8:**138–140.

Clausen, C. P. (1940). *Entomophagous insects*, New York: McGraw-Hill.

Coudron, T. (1991). Host-regulating factors associated with parasitic Hymenoptera. *ACS Symp. Ser.* **449:**41–65.

Dahlman, D. L., and Vinson, S. B. (1993). Teratocytes: Developmental and biochemical characteristics. In *Parasites and pathogens of insects* (N. E. Beckage, S. N. Thompson, and B. A. Federici, eds.), vol. I, pp. 145–166. San Diego: Academic Press.

Delbeque, J. P., Weidner, K., and Hoffman, K. H. (1990). Alternative sites for ecdysteroid production in insects. *Invertebr. Reprod. Dev.* **18:**29–42.

Doutt, R. L. 1963. Pathologies caused by insect parasites. In *Insect pathology* (E. A. Steinhaus, ed.), vol. 2, pp. 393–422. New York: Academic Press.

Doutt, R. L., Annecke, D. P., and Tremblay, E. (1976). Biology and host relationships of parasitoids. In *Theory and practice of biological control* (C. B. Huffaker and P. S. Messenger, eds.), pp. 143–168. New York: Academic Press.

Dover, B. A., Davies, D. H., and Vinson, S. B. (1988). Degeneration of last instar *Heliothis virescens* prothoracic glands by *Campoletis sonorensis* polydnavirus. *J. Invert. Pathol.* **51:**80–91.

Drea, J. J. (1968). Castration of male alfalfa weevils by *Microctonus* spp. *J. Econ. Entomol.* **61:**1291–1295.

Ferro, D. N., and Harwood, R. F. (1973). Interspecific larval competition by the codling moth, *Laspeyresia pomonella. Environ. Entomol.* **2:**783–789.

Fleming, J. G. W., and Krell, P. J. (1993). Polydnavirus genome Organization. In *Parasites and pathogens of insects* (N. E. Beckage, S. N. Thompson, and B. A. Federici, eds.), vol. 1, pp. 189–226. San Diego: Academic Press.

Friedländer, M., and Benz, G. (1982). Control of spermatogenesis resumption in post-diapausing larvae of the codling moth. *J. Insect Physiol.* **28:**349–355.

Friedländer, M., and Brown, J. J. (1995). Tebufenozide (Mimic®), a non-ecdysteroidal ecdysone agonist, induces spermatogenesis reinitiation in isolated abdomens of diapausing codling moth larvae *(Cydia pomonella). J. Insect Physiol.* **41:**403–411.

Fuxa, J. R. (1993). Insect resistance to viruses. In *Parasites and pathogens of insects* (N. E. Beckage, S. N. Thompson, and B. A. Federici, eds.), vol. II, pp. 197–209. San Diego: Academic Press.

Gelman, D. B., and Woods, C. W. (1985). Metabolism of ecdysone by testes of the European corn borer, *Ostrinia nubilalis* (Hübner). *Am. Zool.* **25:**737.

Gelman, D. B., Woods, C. W., Loeb, M. J., and Borkovec, A. B. (1989). Ecdysteroid synthesis by testes of 5th instars and pupae of the European corn borer, *Ostrinia nubilalis* (Hübner). *Invertebr. Reprod. Dev.* **15:**177–184.

Girardie, J., and Girardie, A. (1977). Intervention des cellules neurosecretrices medianes dans la castration parasitaire *D'Anacridium aegyptium* (Orthoptere). *J. Insect Physiol.* **23:**461–467.

Grossniklaus-Burgin, C., Connat, J. L., and Lanzrein, B. (1989). Ecdysone metabolism in the host-parasitoid-system *Trichoplusia ni/Chelonus* sp. *Arch. Insect Biochem. Physiol.* **11:**79–92.

Hawlitzky, N. (1972). Mode de penetration d'un parasite ovo-larvaire *Phanerotoma flavitestacea* Fish. [Hym.: Braconidae] dans son hote embryonnaire, *Anagasta keuhniella* Zell. [Lep.: Pyralidae]. *Entomophaga* **17:**375–389.

Hawlitzky, N. (1979). Devenir de l'euf et comportement de la larve de *Phanerotoma flavitestacea* [Hym.: Braconidae] lorsque la femelle pond danse des eufs d'*Anagasta kuehniella* Zell. [Lep.: Pyralidae] ayant atteint des stades de developpement varies. *Entomophaga* **24:**237–245.

Hunter, K. W., and Stoner, A. (1975). *Copidosoma truncatellum:* Effect of parasitism on food consumption of larval *Trichoplusia ni. Environ. Entomol.* **4:**381–382.

Hurd, H. (1990a). Physiological and behavioral interactions between parasites and invertebrate hosts. *Adv. Parsitol.* **29:**271–318.

Hurd, H. (1990b). Parasite induced modulation of insect reproduction. *Adv. Invertebr. Reprod.* **5:**163–168.

Hurd, H. (1993). Reproductive disturbances induced by parasites and pathogens of insects. In *Parasites and pathogens of insects* (N. E. Beckage, S. N. Thompson, and B. A. Federici, eds.), vol. 1, pp. 125–144. San Diego: Academic Press.

Jackson, D. J. (1924). Insect parasites of the pea-weevil. *Nature* **113:**353–354.

Jans, P., Benz, G., and Friedländer, M. (1984). Apyrene-spermatogenesis-inducing factor is present in the haemolymph of male and female pupae of the codling moth. *J. Insect Physiol.* **30:**495–497.

Jarvis, T. D., Earley, F. G. P., and Rees, H. H. (1994). Ecdysteroid biosynthesis in larval testes of *Spodoptera littoralis. Insect Biochem. Mol. Biol.* **24:**531–537.

Johnson, B. (1965). Premature breakdown of the prothoracic glands in parasitized aphids. *Nature* **206:**958–959.

Jones, D. (1985). The endocrine basis for developmentally stationary prepupae in larvae of *Trichoplusia ni* pseudoparasitized by *Chelonus insularis. J. Comp. Physiol.* **155B:** 235–240.

Jones, D. (1989). Protein expression during parasite redirection of host *(Trichoplusia ni)* biochemistry. *Insect Biochem.* **19:**445–455.

Jones, D. and Coudron, T. (1993). Venoms of parasitic Hymenoptera as investigatory tools. In *Parasites and pathogens of insects* (N. E. Beckage, S. N. Thompson, and B. A. Federici, eds.), vol. 1, pp. 227–244. San Diego: Academic Press.

Jones, D., Gelman, D., and Loeb, M. (1992). Hemolymph concentrations of host ecdysteroids are strongly suppressed in precocious prepupae of *Trichoplusia ni* parasitized and pseudoparasitized by *Chelonus* near *curvimaculatus. Arch. Insect Biochem. Physiol.* **21:**155–165.

Jones, D., Jones, G., and Hammock, B. D. (1981a). Growth parameters associated with endocrine events in larval *Trichoplusia ni* (Hubner) and timing of these events with developmental markers. *J. Insect Physiol.* **27:**779–788.

Jones, D., Jones, G., and Hammock, B. D. (1981b). Developmental and behavioral responses of larval *Trichoplusia ni* to parasitization by an imported braconid parasite *Chelonus* sp. *Physiol. Entomol.* **6:**387–394.

Jones, D., Jones, G., Van Steenwyk, R., and Hammock, B. D. (1982). Effect of the parasite *Copidosoma truncatellum* on development of its host *Trichoplusia ni. Ann. Entomol. Soc. Am.* **75:**7–11.

Jones, D., Taylor, T., Farkas, R., Chelliah, J., Haene, B., Brown, J., and Reed-Larsen, D. (1990). Intercession of parasitic wasps (Cheloninae) in host developmental and biochemical pathways. *Ad. Invertebr. Reprod.* **5:**157–162.

Junnikkala, E. (1985). Testis development in *Peiris brassicae* parasitized by *Apanteles glomeratus. Entomol. Exp. Appl.* **37:**283–288.

Krishnan, A., Nair, P. N., and Jones, D. (1994). Isolation, cloning, and characterization of new chitinase stored in active form in chitin-lined venom reservoir. *J. Biol. Chem.* **269:**20971–20976.

Kuris, A. M. (1974). Trophic interactions: similarity of parasitic castrators to parasitoids. Q. Rev. Biol. **49:**129–148.

Lawrence, P. O. (1986). Host-parasite hormonal interactions: an overview. *J. Insect Physiol.* **32:**295–298.

Loan, C., and Holdaway, F. G. (1961). *Pygostolus falcatus* (Nees) (Hymenoptera: Braconidae), a parasite of *Sitona* species (Coleoptera: Curculionidae). *Bull. Entomol. Res.* **52:**473–488.

Loeb, M. J., Woods, C. W., Brandt, E. P., and Borkovec, A. B. (1982). Larval testes of the tobacco budworm: A new source of insect ecdysteroids. *Science* **218:**896–898.

Milinski, M. (1990). Parasites and host decision-making. In *Parasitism and host behavior* (C. J. Barnard and J. M. Behnke, eds.), pp. 95–116. London: Taylor & Francis.

Read, A. F. (1990). Parasites and the evolution of host sexual behavior. In *Parasitism and host behavior* (C. J. Barnard, and J. M. Behnke, eds.), pp. 117–158. London: Taylor & Francis.

Reed-Larsen, D., and Brown, J. J. (1990). Embryonic castration of the codling moth, *Cydia pomonella* by an endoparasitoid, *Ascogaster quadridentata. J. Insect Physiol.* **36:**111–118.

Riddiford, L. M., and Kiely, M. L. (1981). The hormonal control of commitment in the insect epidermis-cellular and molecular aspects. In *Regulation of insect development and behavior* (F. Sehnal, A. Zabza, J. J. Menn, and B. Cymborowski, eds.), pp. 485–496. Wroclaw, Poland: Wroclaw Technical University Press.

Salt, G. (1968). The resistance of insect parasitoids to the defence reactions of their hosts. *Biol. Rev.* **43:**200–233.

Stoltz, D. B. (1993). The polydnavirus life cycle. In *Parasites and pathogens of insects* (N. E. Beckage, S. N. Thompson, and B. A. Federici, eds.), vol. I, pp. 167–188. San Diego: Academic Press.

Strambi, C., Strambi, A., and Augier, R. (1982). Protein levels in the haemolymph of the wasp *Polistes gallicus* L. at the beginning of imaginal life and during overwintering. Action of the Strepsipteran parasite *Xenos vesparum* Rossi. *Experientia* **38:**1189–1191.

Strand, M. R. (1986). The physiological interactions of parasitoids with their hosts and their influence on reproductive strategies. In *Insect parasitoids* (J. Waage and D. Greathead, eds.), pp. 97–136. London: Academic Press.

Strand, M. R. (1989). Development of the polyembryonic parasitoid *Copidosoma floridanum* in *Trichoplusia ni. Entomol. Exp. Appl.* **50:**37–46.

Tamaki, Y. (1991). Tortricids in Tea. In *Endocrinological frontiers in physiological insect ecology* (F. Sehnal, A. Zabza, and D. L. Denlinger, eds.), vol. 1, pp. 541–552. Amsterdam: Elsevier.

Tanaka, T. and Vinson, S. B. (1991). Depression of prothoracic gland activity of *Heliothis virescens* by venom and calyx fluids from the parasitoid, *Cardiochiles nigriceps. J. Insect Physiol.* **37:**139–144.

Tanaka, T., Tagashira, E., and Sakurai, S. (1994). Reduction of testis growth of *Pseudaletia separata* larvae after parasitization by *Cotesia kariyai. Archs. Insect Biochem. Physiol.* **26:**111–122.

Thompson, S. N. (1990). Physiological alterations during parasitism and their effects on host behavior. In *Parasitism and host behavior* (C. J. Barnard, and J. M. Behnke, eds.), pp. 64–94. London: Taylor & Francis.

Thompson, S. N. (1993). Redirection of host metabolism and effects on parasite nutrition. In *Parasites and pathogens of insects* (N. E. Beckage, S. N. Thompson, and B. A. Federici, eds.), vol. 1, pp. 125–144. San Diego: Academic Press.

Wülker, W. (1975). Parasite-induced castration and intersexuality in insects. In *Intersexuality in the animal kingdom* (R. Reinboth, ed.), pp. 121–134. New York: Springer-Verlag.

Yagi, S. and Tanaka, T. (1992). Retardation of testis development in the armyworm, *Pseudaletia separata,* parasitized by the braconid wasp, *Cotesia kariyai. Invertebr. Reprod. Dev.* **22:**151–157.

9

The Role of Endocrinological Versus Nutritional Influences in Mediating Reproductive Changes in Insect Hosts and Insect Vectors

Hilary Hurd and Tracey Webb

In this chapter we review the adverse effects that parasites and parasitoids have on insect fecundity. The potential changes in host resource management that arise from a reduction in the amount of nutrients required for egg production are discussed. The role of two possible mediators are evaluated: nutrient competition between parasite and host, and parasite-induced modulation of the endocrine control of reproduction. Attention is focused on the similarity of the mechanisms underlying fecundity reduction in three model systems, *Hymenolepis diminuta* infections in *Tenebrio molitor, Onchocerca lienalis* infections in *Simulium ornatum* and *Plasmodium yoelii nigeriensis* infections in anopheline mosquitoes.

Evidence is presented for the presence of a parasite-induced inhibitor which interferes with the binding of juvenile hormone to the ovaries of infected beetles.

I. Introduction

Parasite-host coevolution implies the imposition of strong selection pressures by each partner upon the other which may result in changes in the biological functioning of the host (Price, 1980; Dawkins, 1982; Dobson, 1988). Of prime importance in these associations is the management of resources, with some degree of withdrawal from the host being associated with the enhancement of parasite fitness (See Fig. 9-1). The energetic cost of host reproduction is high but this is not an essential metabolic process for individual organisms. Thus, diversion of resources away from sexual reproduction will not decrease individual survival and may increase the life span of the host-parasite complex (Read, 1990). This may be why some degree of reduction in host reproductive success is demonstrable in many associations between parasites and both their vertebrate and invertebrate hosts.

Parasite-induced reduction in host fecundity is a strategy that has been discussed

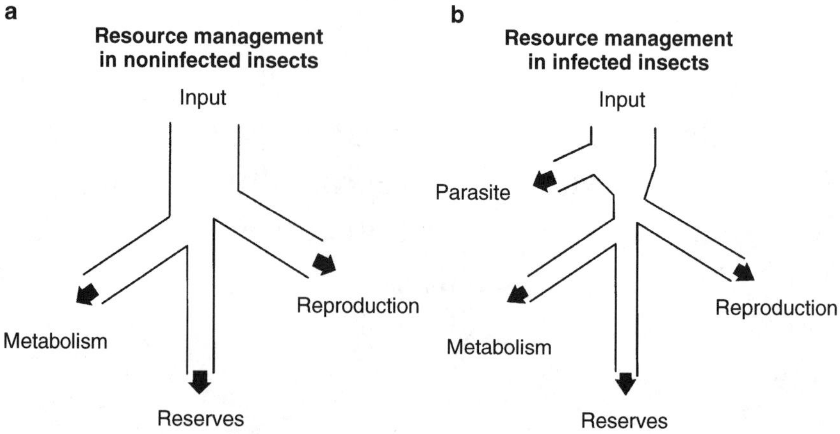

Figure 9-1. A diagrammatic representation of possible nutrient partitioning in noninfected and infected female insects: (*a*) a noninfected insect; (*b*) an infected insect in which a parasite competes directly with the host for the products of digestion, thus making less nutrients available for reproduction, other metabolic processes and storage; (*c*) an infected insect in which the parasite competes directly with the host ovaries for specific nutrients (e.g., vitellogenin); (*d*) an infected insect in which the parasite has an indirect influence on the development of oocytes via the host endocrine system, thereby releasing metabolites that may be used by the parasite immediately, or at a later stage; (*e*) an infected insect in which the parasite directly affects the ovary by inhibiting the tissue response to hormones, thereby releasing metabolites that will not be utilized by developing oocytes.

by many authors (e.g., McClelland and Bourns, 1969; Baudoin, 1975; Hurd, 1990a) and in particular its role in host life history and population dynamics (e.g., Dobson, 1988; Holmes and Zohar, 1990; Read 1990). In this chapter, our discussion will focus on the biochemical and physiological mechanisms underlying parasite-induced reductions in female insect fecundity. In particular, we wish to present evidence that indicates that this pathophysiology may not always be caused by nutrient competition between parasite and insect host. In some cases, a reduction in reproductive output may result from subtle changes in the functioning of the host endocrine system.

II. Fecundity Reduction: Definitions

Many parasitic infections of insects engender some degree of host reproductive loss. However, true castration, defined as an irreversible destruction of reproductive tissue, is rare. What we observe is a spectrum of effects ranging from male and female sterility to a slight reduction in host fecundity or fertility. It is important to distinguish here between the terms fecundity and fertility. Fecundity is a measure of the number of gametes matured (i.e., reproductive potential),

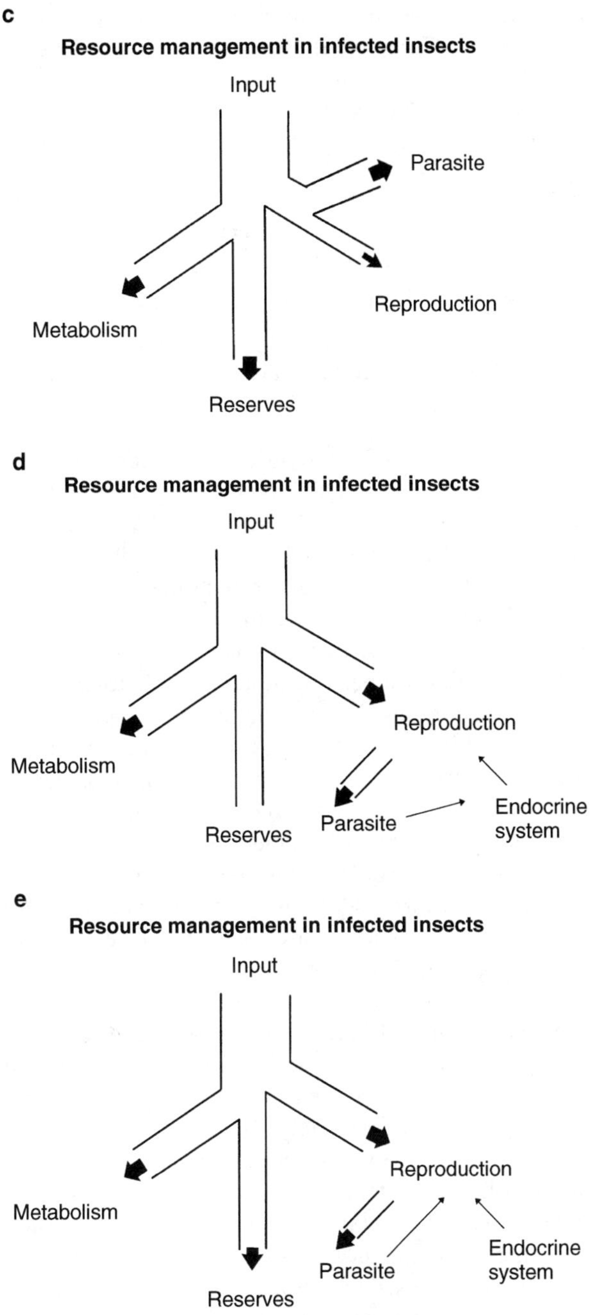

Figure 9-1. Continued

whereas fertility relates to the number of viable offspring produced (see discussions in Clements, 1992; Hurd, 1993; Hurd, et al., 1995). Thus, determination of egg output will give information about fecundity whereas measurements of egg hatch rate or larvae production can be used to assess fertility.

Parasites may affect host egg development in a number of ways:

1. Fully formed eggs may develop in the host ovary by the normal process of oogenesis, but the oviposition of these eggs may be delayed or inhibited.
2. The development of a varying number of terminal follicles may be completely inhibited.
3. The growth of oocytes may be retarded such that, during a particular time period, egg production will be reduced.

An example of the first scenario is provided by the tachinid parasitoid, *Ormia ochracea* which induces complete inhibition of oviposition in the field crickets, *Gryllus integer* and *G. rubens*, although eggs are fully formed. (Adamo, et al., in press). Ovarian damage is usually slight and it is possible that the destruction observed in the host fat body may interfere with the production of a factor(s) necessary to initiate oviposition. Thus, fecundity reduction may be a result of the pathology induced by the infection. In this association it would appear that a large store of metabolites has been made unavailable to the parasitoid larvae (as it has been deposited in host eggs) or the next generation of hosts (as eggs do not develop). It is difficult to envisage this form of fecundity reduction as adaptive, although Adamo is of the opinion that retained eggs may represent a nutrient store which is available to parasitoid larvae when the infestation is high and host tissues destroyed (personal communication).

Suppressed development of some oocytes results in reduced fecundity as observed in several mosquito-*Plasmodium* associations (Hacker and Kilama, 1974; Frier and Friedman, 1976; Hogg and Hurd, 1995a, b). Oogenesis is also inhibited in some follicles of blackflies infected with filarial nematodes of the genus *Onchocerca*, again giving rise to reduced reproductive output (Cheke et al., 1982, Ham and Banya, 1984; Ham and Gale, 1984; Renshaw and Hurd, 1994). *Mermis nigrescens* infections of the locust, *Schistocerca gregaria*, result in the complete resorption of all terminal and penultimate follicles by 3 weeks post-infection, thus inhibiting egg production completely (Gordon et al., 1973).

Similarly, in the association between metacestodes of the rat tapeworm, *Hymenolepis diminuta*, and the coleopteran, *Tenebrio molitor,* the resorption of some follicles contributes to a reduction in host reproductive output (Hurd and Arme, 1987). However, more than one mechanism is in operation in this case because fecundity reduction is also caused by a retardation in the development of oocytes that are not resorbed. This pathology is most evident early in the infection, when metacestodes are undergoing rapid growth and maturation (Webb and Hurd,

1995a). Recent work in our laboratory has demonstrated that infection with the rodent malaria parasite, *Plasmodium yoelii nigeriensis*, may also have a duel effect on the ovaries of anopheline mosquitoes. When infected, *Anopheles gambiae* contain at least 50% more resorbing ovarian follicles than noninfected females. We also see a retardation in the maturation of those follicles that are developing. For example, 24 h post blood feeding, all follicles in ovaries from noninfected females have developed to stage III of vitellogenesis (stages described by Christophers, 1911 and revised by Clements and Boocock, 1984), however 48% have remained at stage II in ovaries from infected females (Carwardine and Hurd, personal observations).

III. Underlying Mechanisms

There is a paucity of information concerning the mechanisms by which parasites are able to modulate or curtail egg production in insects. As mentioned previously, direct gonadal damage, or mechanical castration, is rarely observed in parasitic infections although some microsporidia invade insect ovaries and testes, causing reduced fecundity (reviewed in Hurd, 1993). It is thus apparent that reproductive suppression usually occurs indirectly, via a humoral mechanism and falls under the umbrella of what has been termed "chemical castration" (Cheng et al., 1973).

Many authors have concluded that developing parasites divert nutrients away from host oocyte development for their own usage (the "food robbery" discussed by Von Brand, 1979). If true, then changes in egg production could result from ovaries being starved of resources and the degree of fecundity reduction is likely to be correlated with the intensity of infection (see Figs. 9-1*b* and 9-1*c*).

Studies of allantonematid infections of Scolytidae provide an example of the relationship between parasite density and insect fecundity. Histological examination of host gonads led to the conclusion that the degree of reproductive curtailment was related to the abundance of juvenile nematodes in the bark beetle hemocoel and the timing of their release. In *Polygraphus rufipennis* infected with *Sulphuretylenchus pseudoundulatus* castration of males occurred, whereas, in *Ips pertubatus* infected with *Neoparasitylenchus ipinius* only slight reductions in oviposition occurred and spermatogenesis was unaffected (Tomalak et al., 1990). Although fecundity reduction was related to intensity of infection in these associations, no studies of hemolymph metabolites or other nutrient reserves were reported. Thus, links between putative nutrient depletion and ultimate fecundity reduction remain to be demonstrated.

Reduced host reproductive output will almost certainly decrease the metabolic demands imposed upon reproducing females and may increase the pool of metabolites available to the parasite. To date, there is little direct evidence to support the suggestion that competition for nutrients between symbionts is the *cause* of chemical castration; the end effect, however, may be to increase the nutrients available to support parasite growth. Whatever the mechanism, the potential for parasite fitness may be increased and, ultimately, the success of the parasite-host relationship enhanced.

It is possible that host reproduction may not be affected by direct competition for nutrients, but indirectly, for example, via a perturbance in the host neuroendocrine control mechanism that regulates reproductive physiology. This mechanism could operate directly, by inhibition of endocrine gland functioning, or indirectly by secretion of mediator substances (by host or parasite), which may alter gonadal sensitivity to hormones (see Fig. 9-1*d*, *e* and the discussion in Holmes and Zohar, 1990). These mechanisms are not mutually exclusive and may well act synergistically in some associations.

IV. The Role of Host Nutrients

The large parasite biomass often associated with helminth or parasitoid infections in insects will inevitably impose a burden upon host metabolic resources. Although hosts could compensate by increasing food intake, we are not aware of data that supports the use of this strategy and, indeed, there are several examples of social insects which cease foraging when infected (see Baudoin, 1975), and thus no longer compete for nutrients with the uninfected members of the colony (Gabrion et al., 1976).

It has been suggested that parasite-induced depletion of hemolymph metabolites (Fig. 9-1*b*), and in particular the yolk protein, vitellogenin (Fig. 9-1*c*), may impair oogenesis and result in a reduction in host fecundity. Lieutier (1984) suggested that the delay in ovarian maturation seen in nematode-infected *Ips sexdentatus* (Coleoptera) was due to a reduction in vitellogenin sequestration. Fat body protein content was depleted in infected beetles, but concentrations of hemolymph vitellogenins were not measured (Licutier, 1982a, b, 1984). However, an overall reduction in hemolymph proteins was demonstrated in another nematode-coleopteran association, namely *Dendroctonus pseudosugae* beetles infected with nematodes (Thong and Webster, 1975). In female grasshoppers parasitized by *Metacemyia calloti* (Diptera, Tachnidae), a depletion in hemolymph protein concentration was also associated with an inhibition of egg production (Giardie, 1977).

With the exception of the above examples, evidence to support the hypothesis that a parasite-induced delay in host ovarian development or reduction or cessation of oogenesis is due to lack of sufficient circulating vitellogenins is largely circumstantial. Indeed, experiments that we have designed to test this hypothesis have not provided supporting evidence.

Attempts to elucidate the mechanisms underlying parasite-induced fecundity reduction are few and conclusions are mostly tentative and often speculative. However, it is our opinion that the few systems that have been studied in any detail are demonstrating many common features, despite the diverse nature of both the insect hosts and the parasites involved. In the examples described below, it would appear that direct nutrient competition between parasite and host may not occur. Endocrine-regulated changes in reproductive physiology, and in particular vitellogenesis, take place very early in infection, when parasite biomass is low, and may be modulated by parasite-derived signals.

V. Experimental Models

Three experimental models have been used by the principal author to investigate the mechanism underlying fecundity reduction. In all cases the definitive host is a vertebrate. In one model, passive transfer occurs between the definitive host and the insect intermediate host, and in the other two models a dipteran insect acts as a vector for the parasite.

The rat tapeworm, *Hymenolepis diminuta,* has a metacestode stage that infects a variety of stored grain pests. Much is known of the biology of the adult worm (Arai, 1980) and several interactions between the parasite and intermediate host reproductive processes have been explored over the last decade (Hurd, 1990b). *Tenebrio molitor* (Coleoptera) was employed as intermediate host in these studies. Embryonated parasite eggs are ingested by the beetle and the oncospheres released following mechanical and enzymatic disruption of the shell and embryonic membranes. Once the oncosphere has passed through the midgut wall no further tissue invasion takes place. Development takes 10–12 days at 26°C and occurs in the hemocoel, the parasite developing through five stages as defined by Voge and Heyneman (1957). Although host castration does not occur, infection induces modulation of several aspects of reproduction. This results in a delay of the onset of oviposition, a reduction in the number of eggs laid per beetle 13–16 days postinfection (the time at which the parasites reach maturity), and a decrease in egg viability (reviewed by Hurd, 1990b).

Onchocerca lienalis is a filarial nematode that infects cattle in the UK. Microfilariae are ingested by blackflies of the genus *Simulium*, while taking a blood meal. They penetrate the midgut and migrate through the hemocoel to the thoracic muscles where they undergo development to L3 larvae before migrating to the salivary glands. From the salivary gland the L3 larvae invade the definitive host when the fly takes the next blood meal. Parasite-induced fecundity reduction is known to occur in both the UK parasite-host association (Ham and Gale, 1984; Ham and Banya, 1984; Renshaw and Hurd, 1994) and in *S. damnosum*, a blackfly of sub-Saharan Africa which acts as vector for *O. volvulus,* the causative agent of river blindness (Cheke et al., 1982).

Gametocytes of *P. y. nigeriensis*, imbibed by anopheline mosquitoes within the erythrocytes of the blood meal, undergo fertilization in the mosquito midgut after exflagellation has occurred. The resultant ookinetes penetrate the midgut and migrate to a position below the basal lamina of the midgut. Here they develop into oocysts and undergo sporogony. Sporozoites are released into the hemocoel and invade the salivary gland from whence they are injected into a vertebrate host during blood feeding. Egg production is significantly reduced in females that have fed on infected blood (Hogg and Hurd, 1995a) and in those containing oocysts or sporozoites (Hogg and Hurd, 1995b).

These three associations provide very different systems with which to investigate the mechanisms underlying reduced reproductive output in parasitized insects. The mode of parasite development is different in each case. The malaria parasite undergoes both sexual and asexual reproduction within the insect, whereas the metacestode and the microfilaria have a phase of growth and development prior to transmission. Each parasite is located in a different site, although none invade reproductive tissues, and all spend some time in the hemolymph. Almost nothing is known about the metabolic requirements of the parasite stages in vector insects. However, we do know that metacestodes are able to utilize monosaccharides and amino acids (Arme, 1988) and *Plasmodium*-infected midguts utilize more glucose than noninfected midguts (see Maier, et al., 1987).

The reproductive physiology of dipterans differs from coleopterans in several ways. *S. ornatum, An. stephensi* and *An. gambiae* are anautogenous and require a blood meal to complete a gonotrophic cycle and produce a batch of eggs. In contrast, oocyte maturation in *T. molitor* is not directly linked to a protein meal, egg production being continuous and asynchronous. Despite these differences, we find that some of the mechanisms underlying parasite-induced fecundity reduction in each system may be similar.

VI. Host Vitellogenesis

A decrease in egg production has been linked to a reduction in the protein content of ovaries from infected insects in all three associations. Ovaries from blackflies, infected with microfilariae by intrathoracic injection immediately after blood feeding, were compared with ovaries from sham-injected females and a significant reduction in protein content was detected as early as 24 h post infection or feeding (Renshaw and Hurd, 1994). Similarly, the total protein content of ovaries from *P. y. nigeriensis*-infected *An. stephensi* was reduced by 21 h post feeding when mosquitoes were fed on gametocytaemic blood and also, when mosquitoes were infected with developing oocysts and/or sporozoites fed on the same mouse as the noninfected mosquitoes (Hogg, 1995). A comparison in the uptake *in vivo* of radiolabelled proteins by ovaries from infected and control *T. molitor* revealed a 51.5% decrease in protein sequestration in females 12 days post infection (Hurd and Arme, 1986). Using ELISA, Webb and Hurd (1995a) have shown that terminal follicles at all stages of development contain significantly smaller amounts of vitellin than same-sized follicles from noninfected females. This depletion occurs as early as 3 days post infection with *H. diminuta* but the reduction declines with time post infection and has disappeared when metacestodes are fully developed (i.e., by stage V) (see Fig. 9-2).

It is unlikely that the impairment of ovarian sequestration in these systems is due to lack of circulating vitellogenins resultant upon nutrient deprivation. In

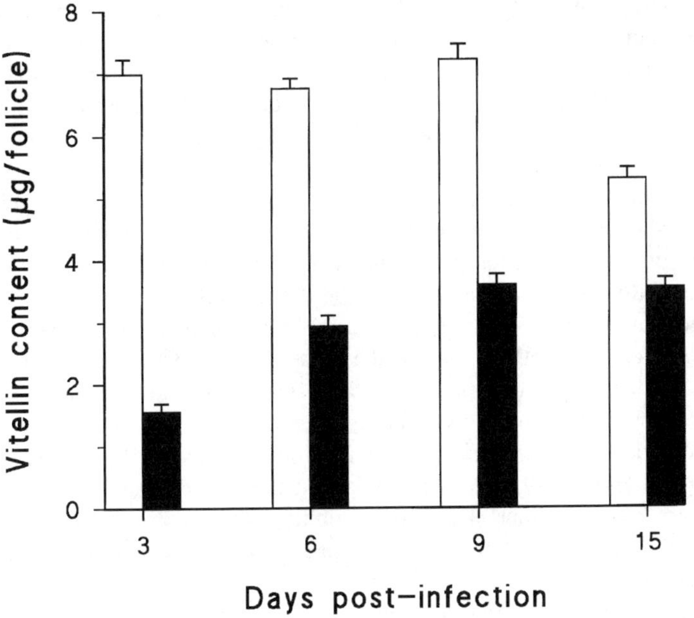

Figure 9-2. The vitellin content of *Tenebrio molitor* follicles of less than 200 µm in length, at various days post infection, detected using an ELISA. Data for both control (open bars) and *H. diminuta*-infected beetles (solid bars) were collected, the bars represent the mean of 8 separate experiments, each carried out in duplicate, ± S.E.M. values.

fact, vitellogenin appears to accumulate in the hemolymph at later stages of infection and gonotrophic cycles. The hemolymph protein content of female *T. molitor* is significantly elevated by 12 days post infection. Electrophoretic separation of hemolymph proteins (Hurd and Arme, 1984) and more recently by ELISA (Major and Hurd, personal observations), has demonstrated that it is the vitellogenin fractions of hemolymph that are elevated. This is also the case in *O. lienalis*-infected blackflies, where, after an initial decrease in vitellogenin, circulating titers are significantly elevated by 24 h post infection. Recent work in our laboratory indicates that this elevation of vitellogenin titers may also occur in malaria-infected anophelines (Hogg, 1995). It appears that, although vitellogenin proteins are in abundant supply later in infection, the ability to sequester them has been affected. It is interesting to note that this accumulation of vitellogenin is unlikely to benefit the parasites directly. Developing metacestodes are not able to take up radiolabelled vitellogenin (Webb and Hurd, 1996) and we were not able to detect vitellogenin on the surface of *O. lienalis* microfilaria by immunofluorescence (Renshaw and Hurd, personal observations).

Vitellogenin, the hemolymph form of the yolk protein, vitellin, passes to the oolemma through spaces that develop between the cells of the follicular epithelium and is taken up by receptor-mediated endocytosis (Wang and Davey, 1992). The development of these spaces is known as patency. Using Evans' Blue dye to visualize interfollicular spaces, Hurd and Arme (1987) demonstrated that a parasite-induced reduction in ovarian patency occurred and proposed that this could be the cause of the observed reduction in yolk sequestration. This has not yet been investigated in the vector insects.

In addition to the parasite-induced changes in ovarian uptake of vitellogenin we have shown that fat body synthesis and secretion of vitellogenin is markedly reduced in infected beetles (Hurd and Arme, 1986, Webb and Hurd, 1996) and blackflies (Renshaw and Hurd, 1994). The status of fat bodies from *Plasmodium*-infected mosquitoes has not yet been examined. When fat bodies from noninfected *T. molitor* are cultured *in vitro* in the presence of metacestodes, inhibition of vitellogenin synthesis occurs with as little as two parasites per fat body. These effects take place within 4 h of culture and are produced by stage I–II parasites that are undergoing rapid growth and development, but not by mature parasites (Webb and Hurd, 1996). It is possible that the metacestodes are producing a modulator molecule that is directly inhibiting vitellogenin synthesis. Alternatively, inhibition may occur indirectly via an unknown signalling system in the fat body.

It would appear that more than one mechanism may be operating to decrease or retard egg production in parasitized insects. In the vectors, it is possible that a reduction in synthesis, and hence circulating titers of vitellogenin, early in the gonotrophic cycle may precipitate changes in the developing follicles that render them unable to continue development, despite the fact that, later in the cycle, vitellogenin is not a limiting factor. The asynchronous nature of egg development in *Tenebrio*, and the fact that egg production has been initiated prior to infection, makes it difficult to assess the effect of downward regulation of vitellogenin synthesis on ovarian sequestration. Some resorption of follicles occurs but the majority of oocytes continue to grow and to accumulate vitellogenin although they may never contain the same amount as control follicles.

In all three models, ovaries are affected early in infection, when parasite biomass is extremely low. Renshaw and Hurd (1994) demonstrated that a single microfilaria was sufficient to significantly reduce ovarian protein content, and a mean oocyst burden of 4.4 per midgut produced fecundity reduction in *An. stephensi* (Hogg and Hurd, 1995b). It would seem unlikely that the biomass of these parasites could have a marked impact on vector nutrient reserves. There is evidence from the *H. diminuta–T. molitor* model that demonstrates a parasite-induced change in carbohydrate reserves. In male beetles, fat body glycogen reserves are significantly depleted 3 days post infection, however this depletion is delayed until 5 days post infection in females (Kearns et al., 1994). It is possible that a switch of resources away from egg production in infected females has maintained reserves a little longer.

The direct utilization of some products of digestion en route to the fat body

(Fig. 9-1b) is likely to occur in all three associations. However, it is difficult to reconcile what must be small demands early in infection relative to host reserves with such significant changes in fat body vitellogenin synthesis and the decrease in oocyte development. The possibility that specific or nonspecific changes in host reproductive physiology are induced indirectly via endocrine control mechanisms must be considered.

VII. Endocrine Interactions

Juvenile hormone (JH) is synthesized in the corpora allata of insects, secreted into the hemolymph and transported in association with specific binding proteins. JH is known to regulate both fat body anabolism of vitellogenin and the development of ovarian patency in many insects. However, it would appear that, in the Diptera, titers of JH and ecdysteroids may be in antiphase during a gonotrophic cycle. Ecdysone, produced largely by the ovary, is converted to 20-hydroxyecdysone which controls the transcription of vitellogenin genes in the dipteran fat body. The role of JH after a blood meal is under debate (Clements, 1992). Some authors have attempted to link parasite-induced fecundity reduction in insects with real or putative changes in these endocrine systems.

Mermithid infections (*Mermis nigrescens*) of female locusts result in cessation of egg production, possibly due to the inability of the host ovaries to sequester vitellogenin (reviewed by Gordon, 1981). The nematodes affect host fat body protein turnover. However, depletion of hemolymph protein reserves occurred *after* the impairment of ovarian functioning (Gordon et al., 1973), and the authors suggested that interaction between parasite and host endocrine system may induce changes in both fat body protein synthesis and ovarian sequestration of yolk protein. In particular, they suggest that the inability of the host to continue vitellogenesis, which began normally, was due to reduced corpora allata activity and subsequent lowering of juvenile hormone titers (Gordon et al., 1973). Another mermithid, *Neomermis flumenalis*, has been shown to increase host corpora allata volume and DNA/RNA synthesis (measured in terms of nuclear/cytoplasmic ratios) and to increase stored neurosecretory substances in its blackfly host, *Simulium venustum* (Condon and Gordon, 1977). Thus, a tenuous link has been made between impaired ovarian functioning and endocrinological interactions between a parasite and host in these systems.

Some parasitoids may affect host reproduction via corpora allata development. For example, although no change occurred in the median neurosecretory cells, the observed decrease in volume and degeneration of corpora allata in *Eurygaster* (Pentotomidae) infected with Tachinidae indicated a reduction of JH titer (Panov et al., 1972). In *Bombus terrestris* queens parasitized by the nematode *Sphaerularia bombi*, a decrease in corpora allata volume was associated with a lack of ovarian development. The hemolymph vitellogenin fraction was found to be much less concentrated in parasitized queens than in control queens with maturing eggs, and parasite-induced corpora allata injury was linked to suppressed yolk protein synthesis (Röseler and Röseler, 1973).

Infection of the wasp *Polistes gallicus* by the strepsipteran *Xenos vesparum* suppresses the maturation of host oocytes, a syndrome known as stylopization (Hoffmann, 1914). In parasitized female wasps, a large increase in the size of the corpora allata was noted at the beginning of the host pupal phase. In contrast, at the end of the diapause, the corpora allata growth that precedes oocyte development in uninfected wasps was much reduced (e.g., Strambi and Strambi, 1973). Synthesis of JH by corpora allata from stylopized wasps that had overwintered was significantly reduced (Strambi, 1981). This deficiency could explain the parasite-induced inhibition of ovarian maturation. Implantation of active corpora allata stimulated vitellogenesis and induced ovarian development, demonstrating that this was not an irreversible effect (Strambi and Girardie, 1973). The mechanism whereby *Xenos* affects host JH production is unknown. Cytological evidence suggests that material was discharged from neurosecretory cells of infected hosts, whereas, in control wasps, large quantities of material were present in the lateral neurosecretory cells (Strambi, 1966, 1967). The parasite may act directly on the neurosecretory cells via a chemical signal or, indirectly, via other aspects of pathophysiology (Strambi et al., 1982).

Laverdure (1970) demonstrated a link between corpora allata products and vitellogenesis in *T. molitor*, JH III being the only homologue produced (Weaver et al., 1980). Although key events in vitellogenesis were adversely affected by metacestode infection, the level of JH synthesis, degradation and titer in the hemolymph appeared unaffected (Hurd and Weaver, 1987; Hurd et al., 1990). Work in the authors' laboratory has therefore focused on the action of JH at the tissue-binding level, and, particularly, on the binding of JH to sites in the follicular epithelium.

The precise mechanism of JH action in the insect ovary is unclear. Although there is evidence for JH acting as a classical steroid- or thyroid-type hormone (i.e., promoting DNA synthesis (Koeppe and Wellman, 1980)), other reports indicate a role in the regulation of a membrane-bound Na^+- K^+ ATPase on the follicle cell membrane (Ilenchuk and Davey, 1983; Davey, 1994). Activation of the ATPase causes the follicle cell to lose water and shrink; hence, intercellular spaces develop (the development of patency), allowing the passage of vitellogenin to the surface of the oocyte, where it is taken up by receptor-mediated endocytosis (Wang and Davey, 1992).

Patent follicles of *T. molitor* (i.e., those actively accumulating vitellogenin) were selected by measurement of their length (600–1000 μm, Hurd and Arme, 1987). Following homogenization of the follicles, subcellular fractionation was achieved by differential centrifugation (Webb and Hurd, 1995c). The kinetics of JH III binding were examined in the presence of a general or specific esterase inhibitor and nonspecific and total binding quantified (Webb and Hurd, 1995b).

The pattern of JH III binding in the various subcellular fractions reflected the dichotomy in the current paradigm for JH action in the ovary. In *Tenebrio* follicle cells, specific JH III binding was demonstrated in the postmicrosomal supernatant (containing the cytosol), the nuclei and microsomes. Infection with *H. diminuta*, however, only resulted in reduced binding in the latter (Webb and Hurd, 1995b). Intriguingly, the reduction in binding was not uniform throughout the infection (Fig. 9-3), but was most acute in the early stages: by day 15 post infection, when metacestode development is complete and the parasite virtually dormant, JH-binding is not significantly different from that of control insects.

Since little was known of JH-binding proteins in *Tenebrio* ovaries (although Schmialek et al., [1975] had shown that JH-binding proteins existed in post-centrifugation (30,000 g) supernatants of *Tenebrio* ovaries) it was necessary to demonstrate specific, high-affinity sites in the follicle cell microsomes, before examining the effects of infection. Using Scatchard analysis (Scatchard, 1949), two specific binding sites were detected in noninfected insects, having affinity

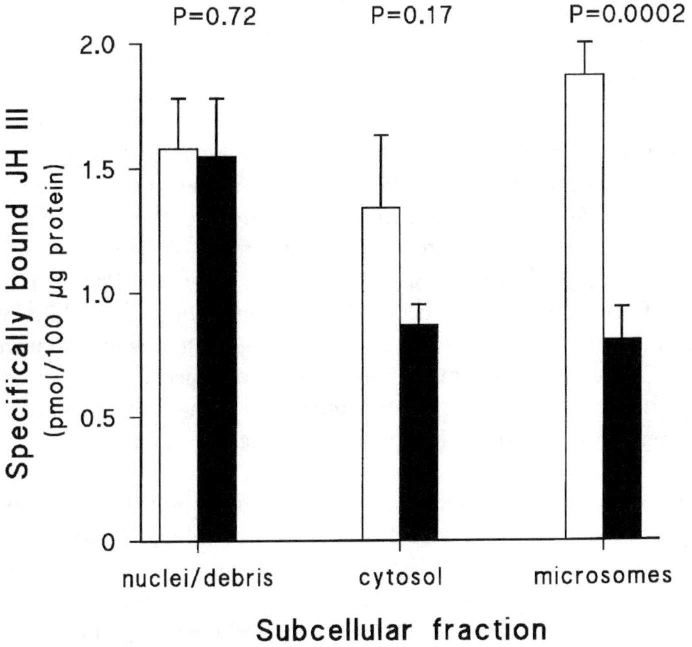

Figure 9-3. Specific JH III binding in subcellular fractions from follicle cells of *Tenebrio molitor* for control (open bars) and *Hymenolepis diminuta*-infected insects (solid bars), measured at day three post infection. Each bar is the mean of eight separate investigations performed in duplicate ± S.E.M. values; significance (P) values were calculated using ANOVA.

constants (K_d values), of 1.0×10^{-8}M and 4.3×10^{-7}M, and the concentration of binding sites (B_{max}) values being 1.4×10^{-7}M/mg protein and 9.9×10^{-7}M/mg protein, respectively. Using microsomal preparations from day three postinfection insects, it was clear that the B_{max} values for both the higher and lower affinity sites were unchanged by infection, but that the K_d values for the higher affinity site was increased approximately fivefold in infected beetles. These data were suggestive of a binding inhibitor being induced by the presence of *H. diminuta* which was capable of competing with JH III for the high affinity site.

These results reinforced the findings of an earlier study, in which a topical application of the JH analogue, methoprene, could reverse the infection-associated accumulation of vitellogenin in the hemolymph (Hurd et al., 1990). Presumably, the high dose of the analogue (1–10 ng/beetle) successfully competed with the binding inhibitor for the binding site.

The nature and source of this inhibitor is unknown. Hemolymph from infected female beetles is able to induce fecundity reduction when injected into noninfected recipient females (Major, Webb, and Hurd, personal observations), which is a finding that suggests the presence of a circulating factor.

VIII. Conclusion

There are many examples of parasite-host associations in which the parasite is known to influence the behavior and/or physiology of the host (reviewed by Hurd, 1990a,b). Dawkins (1990) suggested that there are two alternative mechanisms whereby this may occur; i.e., directly, via an extension of the parasite phenotype or, indirectly, as a by-product of infection (The Extended Phenotype Theory versus the Boring By-product Theory). The former theory requires the existence of a modulator molecule(s), produced by the parasite, that is able to manipulate the host. The latter theory implies an indirect influence upon the host which could be triggered by such factors as stimulation of the host immune response, competition for host metabolites, or the presence of toxins produced by the parasite (See Fig. 9-4).

Examples of parasites that are able to influence their hosts directly via excretory or secretory products are provided elsewhere in this volume. Work has yet to be done to demonstrate the direct effect of a parasite-derived molecule upon insect host reproductive physiology. However, our *in vitro* studies of the effect of developing metacestodes of *H. diminuta* on vitellogenin synthesis appear to provide preliminary evidence for the existence of such a modulator molecule.

Our understanding of fecundity reduction in dipteran vectors is still rudimentary. It will be interesting to see whether future studies reveal a common mechanism whereby both helminths and protozoa affect dipteran vitellogenesis, and to what extent parasite manipulation of host endocrine systems occurs in these associations.

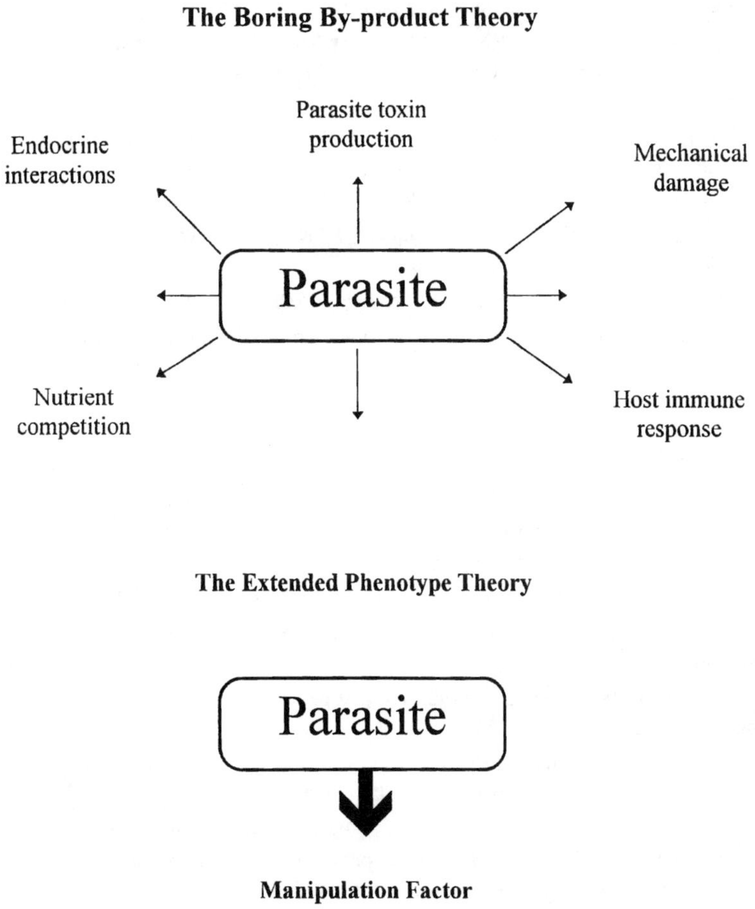

Figure 9-4. Two possible mechanisms whereby a parasite could manipulate its host, described as the "boring by-product theory" and the "extended phenotype theory" (Dawkins, 1990). In the former, one or more stress factors could induce pathology whereas the latter, directed, approach would require a specific factor to be produced by the parasite.

IX. References

Adamo, S. A., Robert D., and Hoy, R. R. (1995). Effects of a Tachinid parasitoid, *Ormia ochracea,* on the behaviour and reproduction of its male and female field cricket hosts (*Gryllus* spp.). *J. Insect Physiol.* (in press).

Arai, H. P. (1980). In: Biology of the tapeworm *Hymenolepis diminuta.* London and New York: Academic Press.

Arme, C. (1988). Ontogenic changes in helminth membrane function. *Parasitology* **96:**S83–S104.

Baudoin, M. (1975). Host castration as a parasitic strategy. *Evolution* **29:**335–352.

Cheke, R. A., Garms, R., and Kerner, M. (1982). The fecundity of *Simulium damnosum* s.l. in northern Toga and infections with *Onchocerca* spp. *Ann. Trop. Med. Parasitol.* **54:**561–568.

Cheng, T. C., Sullivan, J. T., and Harris, K. R. (1973). Parasitic castration of the marine prosobranch gastropod *Nassarius obsoletus* by the sporocyst of *Zoogonus rubellus* (Trematoda): Histopathology. *J. Invertebr. Pathol.* **42:**42–50.

Christophers, S. R. (1911). The development of the egg follicles in anophelines. *Paludism,* **2:**73–8.

Clements, A. N. (1992). *The biology of the mosquitoes.* London: Chapman & Hall.

Clements, A. N., and Boocock, M. R. (1984). Ovarian development in mosquitoes: Stages of growth and arrest, and follicular resorption. *Physiol. Entomol.* **9:**1–8.

Condon, W. J., and R. Gordon. (1977). Some effects of mermithid parasitism on the larval blackflies *Prosimulium mixtum fuscum* and *Simulium venustum. J. Invertebr. Pathol.* **29:**56–62.

Davey, K. G. (1994). Insect vitellogenesis: Putting the Diptera in context. In *Perspectives in comparative endocrinology* (K. G. Davey and S. S. Tobe, eds.), pp. 299–303. Montreal: National Research Council of Canada.

Dawkins, R. (1982). *The extended phenotype.* Oxford: Oxford University Press.

———. (1990). Parasites, desiderata lists and the paradox of the organism. *Parasitology,* **100:**S63–S75.

Dobson, A. P. (1988). The population biology of parasite-induced changes in host behaviour. *Q. Rev. Biol.* **63:**139–165.

Freier, J. E., and Friedman, S. (1976). Effect of host infection with *Plasmodium gallinaceum* on the reproductive capacity of *Aedes aegypti. J. Invertebr. Pathol.* **28:**161–166.

Gabrion, D., Plateaux, L., and Quintin, C. (1976). *Anomotaenia brevis* Clere, 1902) Fuhrmann 1908 Cestoda Cyclophyllidea, parasite of *Leptothorax nylandier* (Förster), Hymenoptera, Formicidae. *Ann. Parasitol.* **51:**407–420.

Girardie, J. (1977). Contrôle neuroendocrine des protéines sanguines vitellogénes d'*Anacridium aegyptium* sains et parasités. *J. Insect Physiol.* **23:**569–577.

Gordon, R. (1981). Mermithid nematodes: Physiological relationships with their insect hosts. *J. Nematol.* **13:**266–274.

Gordon, R., Webster, J. M., and Hislop, T. G. (1973). Mermithid parasitism, protein turnover and vitellogenesis in the desert locust, *Schistocerca gregaria. Comp. Biochem. Physiol.* **46B:**575–593.

Hacker, C. S., and Kilama, W. L. (1974). The relationship between *Plasmodium gallinaceum* density and the fecundity of several strains of *Aedes aegypti. J. Invertebr. Pathol.* **23:**101–105.

Ham, P. J., and Banya, A. J. (1984). The effect of experimental *Onchocerca* infections on the fecundity and oviposition of laboratory reared blackflies. (Diptera, Simuliidae). *Tropenmed. Parasitol.* **35:**211–216.

Ham, P. J., and Gale, L. (1984). Blood meal enhanced *Onchocerca* development and its correlation with fecundity in laboratory reared blackflies (Diptera, Simuliidae). *Tropenmed. Parasitol.* **35:**61–66.

Hoffman, R. W. (1914). Die embryonalen Vorğange bei der Strepsiteren und ihre Deutung. *Verh. Dtsch. Zool. Ges.,* **24:**192–216.

Hogg, J. C. (1995). Effects of *Plasmodium* infection on anopheline mosquito fecundity. Ph.D. Thesis, University of Keele, UK.

Hogg, J. C., and Hurd, H. (1995a). Malaria induced fecundity reduction during the first gonotrophic cycle of *Anopheles stephensi. Med. Vet. Entomol.* **9:**176–180.

———. (1995b). *Plasmodium yoelii nigeriensis:* the effects of high and low incidence of infection upon the egg production and blood-meal size of *Anopheles stephensi* during three gonotrophic cycles. *Parasitology.* (in press).

Holmes, J. C., and S. Zohar. (1990). Pathology and host behaviour. In *Parasitism and host behaviour,* (C. J. Barnard and J. M. Behnke, eds.), pp. 34–64. London: Taylor and Francis, Ltd.

Hurd, H. (1990a). Physiological and behavioural interactions between parasites and invertebrate-hosts. In *Advances in parasitology,* vol. 29 (J. R. Baker and R. Muller, eds.), pp. 271–317, London: Academic Press.

———. (1990b). Parasite induced modulation of insect reproduction. In *Advances in invertebrate reproduction,* vol. 5 (M. Hoshi & H. Yamashita, eds.), pp. 163–169. B. V.: Elsevier Science Publishers.

———. (1993). Reproductive disturbance induced by parasites and pathogens of insects. In *Parasites and pathogens of insects* (N. E. Beckage, S. N. Thompson, and B. A. Federici, eds.), pp. 87–105. San Diego: Academic Press.

Hurd, H. and Arme, C. (1984). Pathology of *Hymenolepis diminuta* infections in *Tenebrio molitor:* Effect of parasitism on haemolymph proteins. *Parasitology* **89:**253–262.

———. (1986). *Hymenolepis diminuta:* Influence of metacestodes upon synthesis and secretion of fat body protein and its ovarian sequestration in the intermediate host, *Tenebrio molitor. Parasitology,* **93:**111–20.

———. (1987). *Hymenolepis diminuta:* Effect of infection upon the patency of the follicular epithelium of the intermediate host *Tenebrio molitor. J. Invertebr. Pathol.* **49:**227–234.

Hurd, H., and Weaver, R. J. (1987). Evidence against the hypothesis that metacestodes of *Hymenolepis diminuta* inhibit corpora allata functioning in the intermediate host, *Tenebrio molitor. Parasitology,* **95:**93–97.

Hurd, H., Strambi, C., and Beckage, N. E. (1990). *Hymenolepis diminuta.* An investigation of juvenile hormone titre, degradation and supplementation in the intermediate host, *Tenebrio molitor. Parasitology* **100:**445–452.

Hurd, H., Hogg, J. C., and Renshaw, M. (1995). Interactions between bloodfeeding, fecundity and infection in mosquitoes. *Parasitology Today,* **11:**411–416.

Ilenchuk, T. T., and Davey, K. G. (1983). Juvenile hormone increases ouabain binding capacity of microsomal preparations from vitellogenic follicles. *Can. J. Cell Biol.* **61:**826–831.

Kearns, J., Hurd, H., and Pullin, A. S. (1994). The effect of metacestodes of the rat tapeworm, *Hymenolepis diminuta,* on storage and circulating carbohydrates in the intermediate host, *Tenebrio molitor. Parasitology* **108:**473–478.

Koeppe, J. K., and Wellman, S. E. (1980). Ovarian maturation in *Leucophaea maderae:* Juvenile hormone regulation of thymidine uptake into follicle cell DNA. *J. Insect Physiol.* **26:**219–228.

Laverdure, A. M. (1970). Action de l'ecdysone et de l'ester méthylique du farnesol su l'ovaire nymphal de *Tenebrio molitor* (Coléoptere) cultiré *in vitro. Ann. Endocrinol.* **31:**516–524.

Lieutier, F. (1982a). Les variations pondérales du tissu adipeux et des ovaries, et les variations de longueur des ovocytes, chex *Ips sexdentatus* Boern (Coleoptera: Scolytidae); relations avec le parasitisme par les nématodes. *Ann. Parasitol. Hum. Comp.* **57:**407–418.

———. (1982b). Action des nématodes endoparasites sur la ponte du scolytide *Ips sexdentatus* Boerner (Insecta: Coleoptera). *Acta Oecol. Oecol. Appl.* **3:**191–204.

———. (1984). Ovarian and fat body protein concentrations in *Ips sexdentatus* (Coleoptera: Scolytidae) parasitized by nematodes. *J. Invertebr. Pathol.* **43:**21–31.

Maier, W. A., Becker-Feldman, H., and Seitz, H. M. (1987). Pathology of malaria-infected mosquitoes. *Parasitology Today* **3:**216–218.

McClelland, G., and Bourns, T. K. R. (1969). Effects of *Trichobilharzia ocellata* on growth, reproduction and survival of *Lymnea stagnalis. Exp. Parasitol.* **24:**36–41.

Panov, A. A., Bassurmanova, O. K., and Balyaeva, T. G. (1972). Ultrastructural changes in the corpus allatum of the bug *Eurygaster* infected by the larvae of *Clytiomyia helluo. J. Insect Physiol.* **18:**1787–1792.

Price, P. W. (1980). *Evolutionary Biology of Parasites.* Princeton, New Jersey: Princeton University Press.

Read, A. F. (1990). Parasites and the evolution of host sexual behaviour. In *Parasitism and host behaviour* (C. J. Barnard and J. M. Behnke, eds.), pp. 117–158. London: Taylor and Francis.

Renshaw, M. and Hurd, H. (1994). The effect of *Onchocerca* infection on the reproductive physiology of the British blackfly, *Simulium ornatum. Parasitology* **109:**337–345.

Röseler, I., and Röseler, P. F. (1973). Änderungen im Muster der Haemolymphproteine von Adulten Königinen der Hummelart *Bombus terrestris. J. Insect Physiol.* **19:**1741–1752.

Scatchard, G. (1949). The attraction of proteins for small molecules. *Ann. N. Y. Acad. Sci.* **51:**660–672.

Schmialek, P., Geyer, A., Misoga, V., Nündel, M., and Zapf, B. (1975). Juvenilhormonbindende Substanzen mit allosterischen Eigenschafen in den Ovarian von *Tenebrio molitor* L. *Z. Naturforsch.* **30C:**730–733.

Strambi, A. (1966). Action de *Xenos vesparum* Rossi (Strepsiptére) sur la neurosécrétion des fondatrices filles de *Polistes gallicus* L. (Hyménoptére, Vespide) en diapause. *C.R. Hebd. Seances Acad. Sci.* **263:**533–535.

———. (1967). Effets de la disparition du parasite *Xenos* (Strepsiptére) sur la neurosécrétion protocérébrale de son hôte *Polistes* (Hyménoptère, Vespide). *C.R. Hebd. Seances Acad. Sci.* **264:**2646–2648.

———. (1981). Some data obtained by radioimmunoassay of juvenile hormone. In *Juvenile hormone biochemistry* (G. E. Pratt and G. T. Brooks, eds.), pp. 59–63. North Holland: Elsevier Biomedical Press.

Strambi, A., and Girardie, A. (1973). Effect de l'implantation de corpora allata actifs de *Locusta migratoria* (Orthoptére) dans des femelles de *Polistes gallicus* L. (Hyménoptère) saines et parasitées par *Xenos vesparum* Rossi (Insecte, Strepsiptère). *C.R. Hebd. Seances Acad. Sci.* **276:**3319–3322.

Strambi, A., and Strambi, C. (1973). Influence du développement du parasite *Xenos vesparum* Rossi (Insecte, Strepsiptère) sur le système neuroendocrinien des femelles de *Polistes* (Hyménoptère, Vespide) au début de leur vie imaginale. *Arch. Anat. Microsc. Morphol. Exp.* **62:**39–54.

Strambi, C., Strambi, A., and Augier, R. (1982). Protein levels in the haemolymph of the wasp *Polistes gallicus* L. at the beginning of imaginal life and during overwintering. Action of the strepsipteran parasite *Xenos vesparum* Rossi. *Experientia* **38:**1189–1191.

Thong, C. H. S., and Webster, J. M. (1975). Effects of *Contortylenchus reversus* (Nematoda: Sphaerulariidae) on haemolymph composition and oocyte development in the beetle *Dendroctonus pseudotsugae* (Coleoptera: Scolytidae). *J. Invertebr. Pathol.* **26:**91–98.

Tomalak, M., Welch, H. E., and Galloway, T. (1990). Pathogenicity of allantonematidae (Nematoda) infecting bark beetles (Coleoptera: Scolytidae) in Manitoba. *Can. J. Zool.* **68:**89–100.

Voge, M., and Heyneman, D. (1957). Development of *Hymenolepis nana* and *Hymenolepis diminuta* (Cestoda: Hymenolepididae) in the intermediate host *Tribolium confusum. Univ. Cal. Publ. Zool.* **59:**549–580.

Von Brand, T. (ed.) (1979). Pathophysiology of the host. In *Biochemistry and physiology of endoparasites,* pp. 321–390. North Holland: Elsevier Biomedical Press.

Wang, Z., and Davey, K. G. (1992). Characterisation of yolk protein and its receptor on the oocyte membrane in *Rhodnius prolixus. Insect Biochem.* **2:**757–767.

Weaver, R. J., Pratt, G. E., Hamnett, A. F., and Jennings, R. C. (1980). The influence of incubation conditions on the rates of juvenile hormone biosynthesis by corpora allata isolated from adult females of the beetle *Tenebrio molitor. Insect Biochem.* **10:**245–254.

Webb, T. J., and Hurd, H. (1995a). The use of a monoclonal antibody to detect parasite induced reduction in vitellogenin content in the ovaries of *Tenebrio molitor. J. Insect Physiol.* **41:**745–751.

———. (1995b). *Hymenolepis diminuta*-induced fecundity reduction in *Tenebrio molitor* may be caused by changes in hormone binding to host ovaries. *Parasitology* **110:**565–571.

Webb, T. J., and Hurd, H. (1995c). Microsomal juvenile hormone III binding proteins in the follicle cells of *Tenebrio molitor. Insect Biochem. Mol. Biol.* **25:**631–637.

———. (1996). *Hymenolepis diminuta:* Metacestode-induced reduction in the synthesis of the yolk protein, vitellogenin, in the fat body of *Tenebrio molitor. Parasitology* (in press).

PART 3

Parasites, Pathogens, and Host Behavior

10

Behavioral Abnormalities and Disease Caused by Viral Infections of the Central Nervous System

Carolyn G. Hatalski and W. Ian Lipkin

I. Introduction

Viral infections of the central nervous system (CNS) can produce a wide variety of behavioral consequences ranging from subtle modifications in behavior to coma or death. The determinants of the outcome of infection include the age of the host, intrinisic properties of the virus, and nature of the immune response. Mechanisms for behavioral disturbances include cell death or damage due to viral replication, cell damage second to the host immune responses, and impairment of differentiated cellular function without cytopathology (reviewed in Oldstone, 1989).

The nature and duration of the behavioral disturbance are determined by the lineage and anatomic distribution of the infected cells. For example, whereas infection of oligodendrocytes may lead to deficits in neural conduction (motor or sensory disturbances), infection of neurons may cause seizures or neural degeneration. Furthermore, astrocytes and oligodendrocytes have the capacity to replicate, but neurons do not. Thus, the potential for repair after neuronal loss is more limited.

The expression of cytokines due to systemic infection can mediate changes in behavior through interaction with cytokine receptors on neurons and glia (reviewed in Benveniste, 1992). The hypothalamus, an area of the CNS controlling homeostasis and endocrine functions, does not have an intact blood–brain barrier (BBB) and is sensitive to some systemic cytokines. Fever, for example, is a result of cytokine binding to temperature-regulating cells in the hypothalamus. However, the majority of CNS neurons are not exposed to systemic cytokines unless the BBB has been compromised by direct CNS infection or trauma. Although behavioral effects due to systemic infection are intriguing in their own right, this chapter will focus on manifestations of CNS infection.

II. Changes in Behavior with Acute Viral Infections of the CNS

Many viruses have the potential to cause transient, acute infections of the nervous system with prominent inflammatory features. Viruses commonly associated with acute human CNS disease include flaviviruses, alphaviruses, rhabdoviruses (rabies virus), bunyaviruses, enteroviruses, and herpesviruses (Specter et al., 1992). Behavioral manifestations directly due to the viral infection can be difficult to distinguish from those resulting from the host immune response. In viral meningitis, although neither the virus nor the immune cells penetrates the brain parenchyma, patients usually have a syndrome of fever, headache, and possibly cognitive impairment, presumably through exposure of neural cells to cytokines. Fortunately these symptoms typically resolve without clinically significant sequellae (Johnson, 1982). In parenchymal infection (infection of neurons, glia, microglia, or epithelial cells of the choroid plexus) immune cells recruited to the brain may effect viral clearance at considerable cost to the host.

Acute viral infections cause a wide range of behavioral abnormalities including the transient effects of immune infiltrates and long-term changes due to cellular damage. In human subjects, Herpes simplex type I (HSV-1) can infect the frontal and temporal lobes to cause a destructive encephalitis with personality changes, hallucinations, and terrifying delusions (Drachman et al., 1962) as well as seizures, hemiparesis, and stupor. Introduction of specific antiviral therapy has greatly reduced the morbidity and mortality of CNS HSV-I infection; however, many of these patients have persistent neurological deficits such as aphasia and amnesia as well as sensory and motor deficits or psychosis (Mohammed et al., 1993). These lasting symptoms are likely due to either loss of specific neurons or persistence of the virus. Rabies virus infection results in behavioral changes such as anxiety, agitation, delirium, aggression, and hypersexuality (Wunner et al., 1989), but these profound disturbances are not associated with significant cytophatology. In the context of immunosuppression, cytomegalovirus can disseminate into the nervous system to cause cellular dysfunction resulting in motor abnormalities, disturbances in consciousness, confusion, and psychotic behavior (Dorfman, 1973).

Permanent lesions of the CNS resulting in long-term behavioral changes are also a consequence of some viral infections. In the early 1900s, Von Economo described a complex syndrome known as encephalitis lethargica (Wilkins et al., 1968; Von Economo, 1929) thought to result from an infection of the CNS. Patients had a febrile illness characterized by abnormal sleep cycles, psychotic behavior, and movement disorders reminiscent of Parkinsonism. Although pathology was present throughout the brains of these patients, damage was particularly marked in the substantia nigra, a dopamine-rich structure in the midbrain important to motor control. Movement disorders tended to persist in these patients, giving rise to the term postencephalitic Parkinsonism.

III. Chronic Viral Infections and Gradual Changes in Behavior with Progressive Disease

Some viruses cause persistent infection and slowly progressive disease. The infection may be subtle and escape detection for years or even decades until signs of disease emerge. Infection with measles virus can result in subacute sclerosing panencephalitis, a relentlessly progressive disease that begins in childhood with intellectual deterioration and later manifests as incoordination and motor abnormalities (myoclonic jerks, ataxia, and dystonia) and culminates in coma and death (Moench et al., 1989). Another chronic infection causing slowly progressive disease is rubella. Long-term rubella infection results in the chronic inflammatory syndrome of progressive rubella panencephalitis (PRP). The clinical signs of PRP begin with changes in behavior and progress to global dementia with ataxia, spasticity, and myoclonus. The underlying pathology of PRP is loss of cortical neurons and white matter that correlates with CNS inflammation (Wolinsky, 1990).

Disease is determined by the host's immune system in some persistent viral infections. The papovavirus JC virus, for example, does not cause disease as long as the immune system is functioning properly. If the host is immunocompromised, there is a productive infection of astroglia and oligodendrocytes resulting in the demyelinating disease of progressive multifocal leukoencephalopathy (PML). The clinical features of PML reflect the anatomical distribution of the demyelinating lesions. Frequent presentations include disturbances in personality, sensory abnormalities, motor paralysis, ataxia, cortical blindness, and dementia (Johnson et al., 1975; Major et al., 1992; Greenlee 1989).

IV. Changes in Behavior Due to Cytopathic Effects of Infection

Viruses may cause behavioral changes through both direct and indirect destruction of neural cells. Some viruses directly damage cells without generating an immune response. For example, the spongiform encephalopathy of mice caused by infection with Moloney murine leukemia virus (Gardner et al., 1994) results in generalized tremors and progressive hind-limb paralysis in the absence of inflammation. In contrast, the cell-mediated immune response to Theiler's virus in mice causes destruction of oligodendrocytes and demyelinating lesions in the spinal cord resulting in paralysis (Lipton, 1975). In addition to direct damage caused by the virus and indirect damage caused by cell-mediated immunity, behavioral changes and disease may also reflect indirect cytopathic mechanisms due to soluble factors. Without infecting neurons and glia, human immunodeficiency virus (HIV-1) can lead to damage of these cells causing AIDS dementia complex, a subcortical dementia characterized by disturbances of cognition, motor performance, and behavior (Grant et al., 1990). Although the pathogenesis of this disorder remains an enigma, cytotoxic soluble mediators including HIV gp120, TNFα, and quinolinic acid have been implicated (Georgsson, 1994; Spencer et al., 1992).

V. Changes in Behavior due to Persistent Infection

Noncytolytic viruses that evade the immune response can cause persistent infection for the life of the host. Such persistent infections are more readily established in the CNS than in other organ systems due to unique features that abrogate or ameliorate an efficient immune response, namely absence of lymphatic drainage and reduced MHC class I expression. In addition, some viruses (e.g., HIV-1, cytomegalovirus, hepatitis B virus, and lymphocytic choriomeningitis virus (LCMV)) may infect cells of the immune system resulting in selective immunosuppression.

Animals with incompetent immunity due to immunosuppression or tolerance may become persistently infected and may have only mild neurologic disease. Neonatally infected rats (Borna disease virus, BDV) and mice (LCMV) have persistent neuronal infection with subtle signs of CNS dysfunction: altered locomotive and exploratory behaviors, substandard performances in learning and memory tasks, and altered taste preferences (Hotchin et al., 1977; Bautista et al., 1994; Dittrich et al., 1989). Behavioral dysfunction in these animals has been mapped to architectural abnormalities in the hippocampus and cerebellum (Baldridge et al., 1993; Carbone et al., 1991), which are two structures in the rodent CNS that continue to develop after birth.

VI. Changes in Behavior due to Alterations in the Levels of Neurochemicals or Their Receptors as a Result of Infection

Behavioral changes are often correlated with imbalances in the normal levels of neurochemicals. Treatment of some neurologic and psychiatric diseases (Parkinson's disease, Huntington's disease, schizophrenia, and depression) can be understood in terms of the effects of drugs on activity in specific catecholamine circuits. Brain levels of neurotransmitters or their receptors are reported to be abnormal in both acute and persistent CNS viral infections. These disturbances may reflect viral destruction of particular neural pathways, alteration in normal cell function due to persistent viral infection, or disruption of neuronal function due to soluble mediators released by infiltrating cells.

An example of an acute infection altering neurotransmitter levels and behavior is temperature-sensitive vesicular stomatitis virus (ts-VSV) infection of young rats. The ts-VSV causes acute inflammation and necrosis. Viral antigen is cleared from the brain within 2 wk of infection, however, 4 mo thereafter the animals are hyperactive and have learning deficits in association with a marked reduction in CNS levels of the neurotransmitter serotonin and its metabolite (5-HIAA) (Mohammed et al., 1990).

Although simian immunodeficiency virus infection of rhesus monkeys does not cause neuronal infection, it results in profound cognitive and motor impairment. These behavioral disturbances have been linked to increased mRNA for the neurotransmitter precursor preprosomatostatin in layer IV of the cortex (Da Cunha et al., 1995). Mice infected with avirulent Semliki Forest virus have decreased levels of norepinephrine and epinephrine in specific brain regions (Mehta et al., 1993). Interestingly, treatment of these mice with a tricyclic antidepressant (amitriptyline) prevents this alteration in brain catecholamines. Rabies virus infection causes alterations in the levels of gamma-amino-butyric acid (GABA) (Ladogana et al., 1994) and opiates (Munzel et al., 1981) in cultured cells and the levels of serotonin and its receptor in the infected rat brain (Bouzamondo et al., 1993; Ceccaldi et al., 1993).

Persistent LCMV infection of mice results in decreased levels of mRNA for one peptide neurotransmitter (somatostatin) but not another (cholecystokinin) (Lipkin et al., 1988). In this model, altered neurotransmitter levels are associated with decreased motor activity and increased sensitivity to stimuli (Hotchin et al., 1977). BDV-infected adult rats causes transient encephalitis and abnormalities in movements and behaviors including hyperactivity, hypersensitivity, aggression, and stereotypies (Narayan et al., 1983; Solbrig et al., 1994). The movement and behavioral disorders seen in these animals is due at least in part to increased activity in selected dopamine circuits (Solbrig et al., 1996).

VII. Speculations on Viral Control of Host Behavior to Enhance Transmission

The evolution of some neurotropic viruses may have involved selection for the ability to cause specific host behaviors that enhance the probability of viral transmission. Rabies virus infects neurons of the mammalian limbic system including the hippocampus, hypothalamus, amygdala, and cingulate cortex. Lesions in these structures can result in hyperexcitability, hypersexuality, and aggression. Rabies virus concentrates in the saliva of infected animals. As rabid animals become agitated, irritable, and aggressive (Wunner et al., 1989), they are more likely to approach, bite, and scratch other animals of their own and different species, facilitating virus transmission to new hosts.

Another example may be HSV-2, which causes a latent infection of the sacral sensory ganglia innervating sex organs. At intervals, the virus travels centrifugally from the sensory ganglia to the skin releasing virus to initiate new cycles of infection in sex partners. It is intriguing to speculate that ganglion infection may modulate sensory input to sex organs leading to increased sexual activity and enhanced probability of virus transmission.

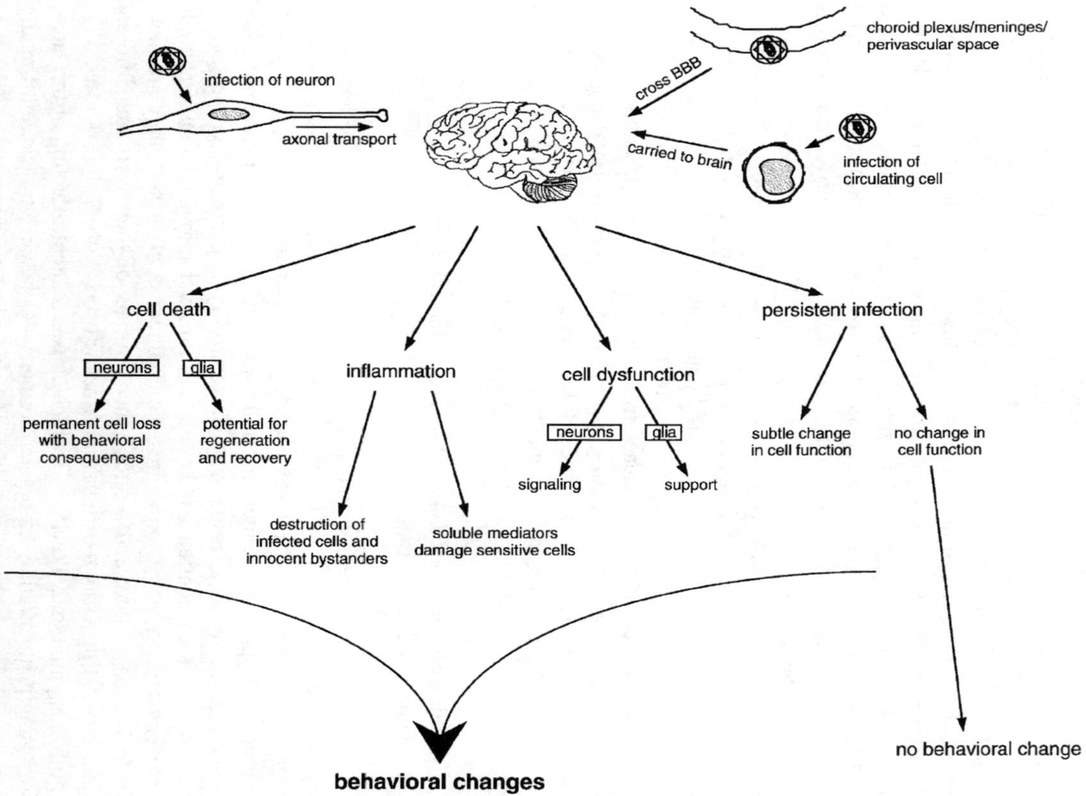

Figure 10-1. Mechanisms for behavioral changes resulting from neurotropic viral infections.

VIII. Summary

The mechanisms by which CNS viral infection may result in behavioral disorders are summarized in Figure 10-1. Viruses enter the brain either through neuronal processes (axons or dendrites) or the vascular system. Once inside the brain some viruses effect direct lysis of neurons and/or glia. Because neurons are postmitotic, their loss is permanent and the only possibility for compensation is through new neural connections. In contrast, glia are not terminally differentiated. Thus, lysed glia may be replaced offering enhanced potential for recovery of function. The host immune response to the virus may itself cause disease either through destruction of infected cells (and innocent bystanders) or cell damage second to soluble factors (e.g., cytokines, free radicals, excitatory amino acids).

Neural cells may be functionally impaired even in the absence of cytopathology. Signaling (neurotransmission) may be disrupted by altered neurochemical levels, disturbances in ion permeability, or changes in the extracellular environment due to glial dysfunction. Neurologic disturbances due to infection need not be dramatic. Some persistent CNS infections can cause subtle alterations in neural functions that are only detected with sensitive behavioral tests.

The increasing interest in pathogenesis of neurotropic viral infections in part reflects the recognition that mechanisms for neurologic dysfunction are more varied than previously appreciated. Given that classical hallmarks of viral infection such as cytopathology may not be present with some neurotropic viruses, neurovirologists have extended their focus beyond diseases known to have a viral basis to idiopathic CNS diseases that may be infectious in origin. If, as predicted from animal models, links can be established between viruses and complex behavioral syndromes, it is likely that even more intricate virus–host relationships will be found than those presented here.

IX. References

Baldridge, J. R., Rearce, B. D., Parekh, B. S., and Buchmeier, M. J. (1993). Teratogenic effects of neonatal arenavirus infection on the development rat cerebellum are abrogated by passive immunotherapy. *Virology* **197:**669–677.

Bautista, J. R., Schwartz, G. J., de la Torre, J. C., Moran, T. H., and Carbone, K. M. (1994). Early and persistent abnormalities in rats with neonatally acquired Borna disease virus infection. *Brain Res. Bull.* **34:**31–40.

Benveniste, E. N. (1992). Inflammatory cytokines within the central nervous system: Sources, function, and mechanism of action. *Am. J. Physiol.* **263:**C1–16.

Bouzamondo, E., Ladogana, A., and Tsiang, H. (1993). Alteration of potassium-evoked 5-HT release from rabies virus-infected rat cortical synaptosomes. *Neuroreport* **4:**555–558.

Carbone, K. M., Park, S. W., Rubin, S. A., Waltrip, R. W., and Vogelsang, G. B. (1991). Borna disease: Association with a maturation defect in the cellular immune response. *J. Virol.* **65**(11):6154–6164.

Ceccaldi, P. E., Fillion, M. P., Ermine, A., Tsiang, H., and Fillion, G. (1993). Rabies virus selectively alters 5-HT-1-receptor subtypes in rat brain. *Eur. J. Pharm.* **245:**129–138.

Da Cunha, A., Rausch, D. M., and Eiden, L. E. (1995). An early increase in somatostatin mRNA expression in the frontal cortex of rhesus monkeys infected with simian immunodeficiency virus. *Proc. Natl. Acad. Sci. USA* **92:**1371–1375.

Dittrich, W., Bode, L., Ludwig, H., Kao, M., and Schneider, K. (1989). Learning deficiencies in Borna disease virus–infected but clinically healthy rats. *Biol. Psych.* **26:**818–828.

Dorfman, L. J. (1973). Cytomegalovirus encephalitis in adults. *Neurology* **23:**136–144.

Drackman, D. A., and Adams, R. D. (1962). Herpes simplex and acute inclusion body encephalitis. *Arch. Neurol.* **33:**362–367.

Gardner, M. B., and Wiley, C. A. (1994). Murine retroviral spongiform polioencephalopathy. In *Slow infections of the central nervous system* (J. Björnsson, R. I. Carp, A. Löve and H. M. Wisniewski, eds.), pp. 385–398. New York: The New York Academy of Sciences.

Georgsson, G. (1994). Neuropathologic aspects of lentiviral infections. In *Slow infections of the central nervous system* (J. Björnsson, R. I. Carp, A. Löve and H. M. Wisniewski, eds.), pp. 50–67. New York: The New York Academy of Sciences.

Grant, I., and Atkinson, J. H. (1990). Neurogenic and psychogenic behavioral correlates of HIV infection. In *Immunologic mechanisms in neurologic and psychiatric disease.* (B. H. Waksmans, eds.), pp. 291–304. New York: Raven Press, Ltd.

Greenlee, J. E. (1989). Papovavirus infections of the nervous system. In *Clinical and molecular aspects of neurotropic virus infection.* (D. H. Gilden and H. L. Liptons, eds.), pp. 319–342. Boston: Kluwer Academic Publishers.

Hotchin, J., and Seegal, R. (1977). Virus-induced behavioral alterations in mice. *Science* **196:**671–674.

Johnson, R. T. (1982). *Viral infections of the nervous system.* New York: Raven Press.

Johnson, R. T., Narayan, O., Weiner, L. P., and Greenlee, J. E. (1975). Progressive multifocal leukoencephalopathy. In *Slow virus infections of the central nervous system.* (V. ter Meulen and M. Katzs, eds.), pp. 91–100. Berlin: Springer-Verlag.

Ladogana, A., Bouzamondo, E., Pocchiari, M., and Tsiang, H. (1994). Modification of tritiated gamma-amino-n-butyric acid transport in rabies virus–infected primary cortical cultures. *J. Gen. Virol.* **75:**623–627.

Lipkin, W. I., Battenberg, E. L. F., Bloom, F. E., and Oldstone, M. B. A. (1988). Viral infection of neurons can depress neurotransmitter mRNA levels without histologic injury. *Brain Res.* **451:**333–339.

Lipton, H. L. (1975). Theiler's virus infection in mice: An unusual biphasic disease process leading to demyelination. *Infect. Immun.* **11:**1147–1155.

Major, E. O., Vacante, D. A., and Houff, S. A. (1992). Human papovaviruses: JC virus, progressive multifocal leukoencephalopathy, and model systems for tumors of the central nervous system. In *Neuropathogenic viruses and immunity.* (S. S. Specter, M. Bendinelli and H. Friedmans, eds.), pp. 207–228. New York: Plenum Press.

Mehta, S., Parsons, L. M., and Webb, H. E. (1993). Effect of amitriptyline on neurotransmitter levels in adult mice following infection with the avirulent strain of Semliki Forest virus. *J. Neurol. Sci.* **116:**110–116.

Moench, T. R., and Johnson, R. T. (1989). Measles. In *Clinical and molecular aspects of neurotropic virus infection.* (D. H. Gilden and H. L. Liptons, eds.), pp. 203–229. Boston: Kluwer Academic Publishers.

Mohammed, A. H., Norrby, E., and Kristensson, K. (1993). Viruses and behavioral changes: A review of clinical and experimental findings. *Rev. Neurosci.* **4:**267–286.

Mohammed, A. K., Magnusson, O., Maehlen, J., Fonnum, F., Norrby, E., Schultzberg, M., and Kristensson, K. (1990). Behavioral deficits and serotonin depletion in adult rats after transient infant nasal viral infection. *Neuroscience* **35**(2):355–363.

Munzel, P., and Koschel, K. (1981). Rabies virus decreases agonist binding to opiate receptors of mouse neuroblastoma–rat glioma hybrid cells 108-CC-15. *Biochem. Biophys. Res. Com.* **101:**1241–1250.

Narayan, O., Herzog, S., Frese, K., Scheefers, H., and Rott, R. (1983). Behavorial disease in rats caused by immunopathological responses to persistent Borna virus in the brain. *Science* **220:**1401–1403.

Oldstone, M. B. A. (1989). Viruses can cause disease in the absence of morphological evidence of cell injury: Implication for uncovering new diseases in the future. *J. Infect. Dis.* **159:**384–389.

Solbrig, M. V., Koob, G. F., Fallon, J. H., and Lipkin, W. I. (1994). Tardive dyskinetic syndrome in rats infected with Borna disease virus. *Neurobiol. of Dis.* **1:**111–119.

Solbrig, M. V., Koob, G. F., Fallon, J. H., Reid, S., and Lipkin, W. I. (1996). Prefrontal cortex dysfunction in Borna disease virus–infected rats. *Biol. Psych.* **40:**629–636.

Specter, S., Bendinelli, M., and Friedman, H. (1992). Viruses and neuropsychiatric disorders. In *Neuropathogenic viruses and immunity.* (S. Specter, M. Bendinelli and H. Friedmans, eds.), pp. 1–12. New York: Plenum Press.

Spencer, D. C., and Price, R. W. (1992). Human immunodeficiency virus and the central nervous system. *Annu. Rev. Microbiol.* **46:**655–693.

Von Economo, C. (1929). *Encephalitis lethargica.* Berlin: Urban and Schwarzenberg.

Wilkins, R. H., and Brody, I. A. (1968). Encephalitis lethargica. *Arch. Neurol.* **18:**324–328.

Wolinsky, J. S. (1990). Subacute sclerosing panencephalitis, progressive rubella panencephalitis, and multifocal leukoencephalopathy. In *Immunologic mechanisms in neurologic and psychiatric disease.* (B. H. Waksmans, ed), pp. 259–268. New York: Raven Press, Ltd.

Wunner, W. H., and Koprowski, H. (1989). Clinical and molecular aspects of rabies virus infections of the nervous system. In *Clinical and molecular aspects of neurotropic virus infection.* (D. H. Gilden and H. L. Lipton, eds.), pp. 269–302. Boston: Kluwer Academic Publishers.

11

Effects of Hormones on Behavioral Defenses Against Parasites

Benjamin L. Hart

I. Host Behavioral Defenses against Parasites

Most animals live in environments that are, at times, teeming with parasites. Yet animals survive, and even thrive, in such environments, to some degree because behavioral defenses against fitness-compromising parasites have been enhanced by natural selection. Animals, ranging from the smallest insects and worms to the largest land and sea mammals, are at risk of parasites and are likely to have parasite defensive behaviors. Most is known about the defensive behaviors of vertebrates, so the emphasis in this chapter is on vertebrate species. Behavioral patterns that vertebrates may use to defend against helminths and arthropods (macroparasites) and pathogens (microparasites) have been categorized into a number of strategies (Hart, 1990). The first strategy, and the one for which most information is available, comprises those behaviors that enable animals to avoid, remove, destroy, or minimize their exposure to parasites. Various aspects of foraging, grooming, microhabitat seeking, grouping, maternal, and sexual behavior fall into this category. A second strategy is controlled exposure, in which animals may expose themselves or their offspring to small samples of particular parasites or pathogens to facilitate development of the body's immunological competence. A third strategy is that of the behavior of sick animals and relates to the adaptive value of anorexia and depression that accompany a febrile response in enabling animals to recover from an acute microparasite infection. A fourth strategy is helping sick groupmates or kin survive a microparasitic infection. The fifth strategy, which has received a good deal of attention in the recent research literature as well as in this volume, is the selection of mates to provide offspring with the genetic basis for resistance to parasites. Actually, this strategy might be more accurately defined as rejection of mates with evidence of low levels of resistance to parasites.

Behavioral defenses against parasites are complementary to the immune system which, along with other biochemical and physiological forms of resistance, are extremely important in enabling wild animals, domestic animals, and humans to survive and reproduce in a parasite-contaminated environment. The array of behavioral defenses for which there is some evidence of effectiveness may be thought of as a first line of defense against parasites, with physiological and immunological forms of resistance as a second barrier (Fig. 11-1).

A coevolutionary analysis reveals that parasites have sometimes evolved behavioral or physiological mechanisms to evade or penetrate these defenses (Hart, 1994). The movement of lice to parts of the feathers where they are not so easily preened off when they detect preening movements by their host is an example. The feather louse *Columbicola columbae*, when disturbed by preening, moves from the surface of the feather to insert itself between two barbs, grabbing a barbule firmly between its mandible, or it may run rapidly to the base of the feather where it is hidden under wing coverts (Clayton, 1991). Other examples are discussed later in this chapter. Parasites have also evolved ingenious ways

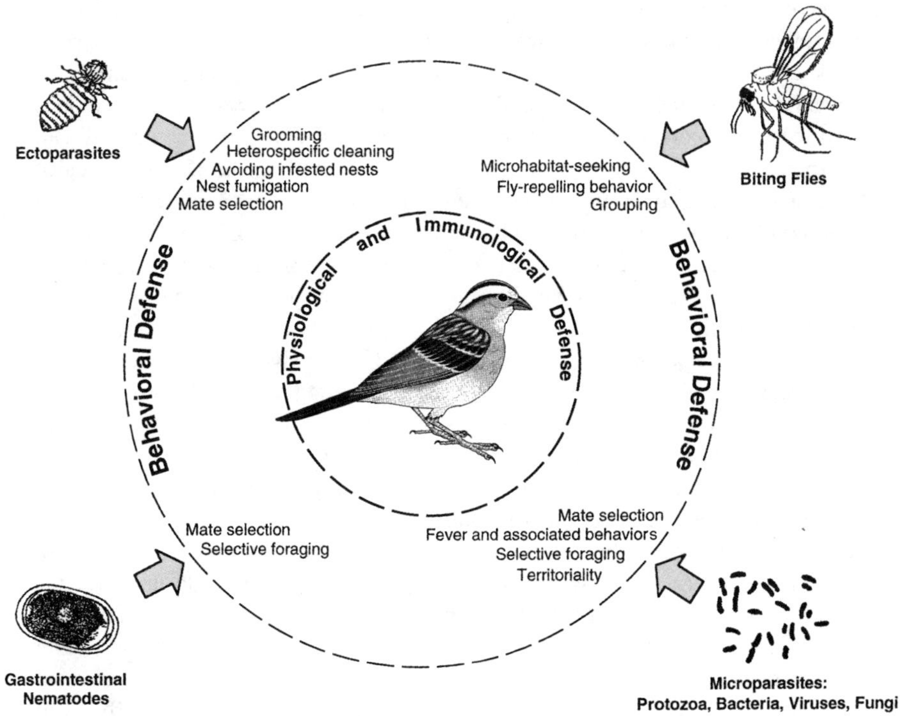

Figure 11-1. Schematic illustration of some behavioral patterns that have proven effective as a first line of defense against macroparasites and microparasites with physiological and immunological responses represented as a second line of defense. Not all behavioral defenses are seen in any one host species. From Hart, 1994.

of locating hosts. One of the widely recognized means by which flies, as well as ticks, locate or move towards hosts is by reacting to gradients of carbon dioxide in expired air (Garcia, 1962; Nicolas and Sillans, 1989; Steullet and Guerin, 1992a). Finally, there are many examples in which parasites physiologically alter the behavior of hosts to facilitate their survival or transmission. Brief exposure of mice to mosquitoes (Colwell and Kavaliers, 1990) or stable flies (Colwell and Kavaliers, 1992) results in an opioid-mediated reduction in pain sensitivity. Assuming that this opioid enhancement is the result of bites and injection of some substance from the flies, the flies may be decreasing host pain sensitivity so that if their first feeding attempts are only partially successful, subsequent bites may not be felt as much and the flies will eventually succeed at feeding. Alteration of host behavior is particularly notable when an intermediate host is involved and the parasite is transmitted when a definitive host predator eats the intermediate host. The increased vulnerability of pill bugs to predation by starlings (bugs seek shelter and dark, humid places less frequently) caused by thorny-headed worms is well-documented (Moore, 1983).

The complexity of host–parasite interactions sets the stage for host hormones to influence the interaction in ways that are advantageous either to the host or parasite. The purpose of this chapter is to explore, both theoretically and from the standpoint of the limited available empirical evidence, the role of hormones from these two perspectives: (1) hormones secreted by the host may coordinate host parasitic defenses with other essential host activities; and (2) hormones of the host may be used as cues or physiological modulators by parasites to increase their likelihood of reproduction and/or transmission. This latter topic is discussed to only a limited degree. Hormones, of course, may also alter the host–parasite interactions through effects on the immune system and this topic is covered elsewhere in this volume.

This first section presents some examples of defensive behaviors to lay the background for discussion of the alteration of some of these behaviors by hormonal influences. However, something should be said first about the criteria for accepting a particular behavioral pattern as having a parasite defense function. As outlined previously (Hart, 1990), two principal criteria are: (1) that the parasites in question be shown to have a detrimental effect on the host's fitness, and (2) the behavior in question actually has the effect of avoiding, removing, or otherwise controlling the parasite. The cost in fitness of an infectious bacterium or virus that kills an animal or weakens it to the point where it is susceptible to starvation or predation is self-evident. Fitness costs of blood-sucking or tissue-consuming macroparasites such as lice, fleas, ticks, and gastrointestinal nematodes that are not generally carried in overwhelming numbers are more subtle and difficult to demonstrate. Typical of such costs is the annualized decrement in weight gain in growing calves of about 0.6 kg per engorging tick (Norval et al., 1988, 1989; Sutherst et al., 1983). Carrying several ticks, while not life-threatening, results in a loss of blood proteins and other nutrients. Loss of such body resources can compromise ability to escape from predators, compete with conspecifics in producing milk for nursing young, and fight for territory by breeding males (Hart, 1990). It is these subtle costs of parasites, rather than the effects of an overwhelming number of parasites, that would appear to be responsible for the evolution of behavioral patterns that maintain a parasite load at low levels (reviewed elsewhere, Hart, 1990, 1995).

The second requirement for accepting a particular behavior as having a parasite defense function is that the behavior in question be shown to be effective in helping the animal, its offspring, and sometimes groupmates avoid, remove, or destroy the parasite in question. Defensive behaviors usually carry some costs. With grooming behavior, for example, there may be distraction from vigilance over predators (FitzGibbon, 1989; Mooring and Hart, 1995a). Grouping, which sometimes offers protection from biting flies, may compromise foraging efficiency in grazing herbivores. Fly-repelling behaviors may carry costs in terms of energy used as illustrated by a time and motion study of slapping flies by howler monkeys. On a typical day when flies were numerous, the energy expended in fly slapping (acceleration and deceleration of the forelimb) accounted for an average 4.6 percent (ranging up to 24 percent) of metabolic costs of living after basal metabolic rate was discounted (Dudley and Milton, 1990).

When animals are studied in clean laboratories or zoological parks, the interplay between host and parasite is not obvious. In some ways host-parasite interactions are not unlike those envisioned for predator–prey interactions. Animals in parasite-free environments may show antiparasite behaviors such as grooming, grouping, selective foraging, and infant care just as they show predator avoidance behavior even when there are no predators around. Grooming, for example—a defense against ectoparasites—occurs in tick-free environments (Hart et al., 1992). In nature, animals may be almost parasite-free even though parasites are present. This does not mean that a particular defense in unnecessary; the relative freedom from a parasite may mean that the behavioral defense has led to a standoff between the parasite's invasiveness and the animal's resistance in the host's favor. A let-up in the host's defenses could lead to acquisition of a substantial parasite load just as a let-up in predator vigilance may result in greater likelihood of predation.

A. *Selective Foraging*

One of the primary means of parasite infection of vertebrates is by ingestion of food while the host is feeding. For a number of intestinal helminths of herbivores, infection occurs when the animals ingest larval forms that have developed from parasite eggs that have passed out with feces and have contaminated forage that the animals graze upon. The adult parasites may reside in either the gastrointestinal tract or in other organ systems from which the eggs can find their way into the feces. Adult lung worms, for example, reside in the bronchial tree and lay eggs that usually hatch in the tracheal trunks into larvae that are swallowed and passed out with the feces.

It is well-known that cattle, sheep, and horses, and presumably other wild and domestic ungulates, using olfactory cues (Ödberg and Francis-Smith, 1992), tend to avoid grazing upon forage that is near recently dropped feces (even though the forage may be quite lush). One study that quantitatively examined the effectiveness of this avoidance behavior was conducted by Michel (1955) on the transmission of the cattle lung worm *Dictyocaulus viviparus*. He found that the larval count in areas that had just been grazed was only one-third that of random sampling. Despite the fact that more quantitative studies are necessary, it is quite logical that animals that avoid grazing near recently dropped feces would ingest fewer parasites than they would if they were indifferent to the presence of fecal contamination. However, some fecal-borne parasites have evolved behavioral strategies to counteract host defensive selective foraging. In general, larvae of fecal-borne parasites are quite mobile and tend to migrate away from the fecal pad, thereby increasing their chances of being consumed. However, in migrating away from fecal pads, larvae increase their susceptibility to desiccation. Distances traveled by larvae under moist conditions reportedly range from a mean of 0.3 m (English, 1979) to a mean of 3.0 m (Durie, 1961).

In the case of larval forms of *D. viviparus,* the parasite has a rather dramatic means of escape from its fecal tomb which relates to the life cycle of a fungus that grows on fecal pads. The larvae of the cattle lung worm parasite migrate to the surface of fecal pads at roughly the same time that the fungus starts growing on the pad. The fungus, *Philobolus* spp., which is a common growth on cattle feces, produces conspicuous sporangiophores loaded with spores. The spores cannot undergo development until they pass through the gut of an herbivore. By avoiding feces, herbivores also avoid ingesting the fungal spores. However, sporangiophores spread their ripe spores through a violent discharge to the surrounding herbage. The spores may be thrown up to 2 m into the air, reaching a horizontal distance of 3.0 m. Lung worm larvae tap into this system for a free ride. The larvae are activated by light to move actively about the surface of the fecal pad, occasionally adopting an erect posture as though searching for an upright support. When a sporangiophore is contacted the larvae attach themselves to it and migrate to the upper part where they remain coiled. The sporangiophores, which also react to illumination, discharge spores towards the source of the light and the larvae along with the spores are catapulted as far as 3.0 m away from the fecal pad (Robinson, 1962). Recent work illustrated that at least 17% of *D. viviparus* larvae are transported in this manner (Eysker, 1991).

B. Avoiding Biting Flies

Biting flies and blood-sucking flies represent another group of fitness-compromising parasites for mammals as small as mice to as large as elephants. There are a variety of defensive behaviors used by various species to eliminate or reduce the intensity of attacks by flies. These include fly-repelling behaviors, microhabitat seeking, and grouping. Small mammals and birds have excellent mosquito-repelling behaviors in the form of tail flipping, ear flipping, face rubbing, and foot stamping. These behaviors undoubtedly protect them from mosquito-borne malaria and other serious diseases as well as loss of blood. The frequency with which fly-repelling behaviors are performed varies with exposure to mosquitoes. When woodrats were exposed to mosquitoes, fly-repelling movements increased from around 1,400 to almost 4,000 movements per hour (Edman and Kale, 1971). Evidently such behaviors are quite protective. Mosquito host-blood meal identification procedures are well-established (Washino and Tempelis, 1983) and analysis of blood of wild mosquitoes reveals that they rarely feed on small mammals and birds (Edman, 1971). In ungulates, ear twitching, head tossing, leg stamping, muzzle flicking, muzzle twitching, and tail switching with a bushy tail increase in the presence of flies (Espmark, 1967; Okumura, 1977; Harvey and Launchbaugh, 1982).

Even elephants are pestered by flies, but neither the trunk nor tail is well-suited for fly switching. Fortunately for the elephant, what it lacks in the way of a bushy tail, it makes up for in highly evolved intelligence. Using their prehensile trunk, Asian elephants grab branches from shrubs or trees as fly switches to repel pesky biting flies (*Tabanus* spp., *Haematopota* spp.) that can apparently bite through the thick skin. This example of tool use by Asian elephants was initially alluded to by Darwin (1871) as an example of intelligence in animals. Recently, we have documented that captive Asian elephants switched branches in correspondence to the number of flies on or around them (Hart and Hart, 1994). Also, an experiment conducted when fly intensity was high revealed that the median number of flies on and around the elephants was reduced by 43 percent when the animals were able to switch with a branch compared to when no branch was available.

Birds and mammals that are pestered by flies may seek a habitat where there are fewer flies. Caribou, in summer, avoid the intense mosquito activity at lower elevations by moving to higher elevations (Downes et al., 1986). An example in native Hawaiian birds points to a microhabitat interplay between parasite and host that has just recently evolved. In the early part of this century, many species of native Hawaiian birds were eliminated by the introduction of non-native birds carrying malaria to which the non-native birds were resistant but native birds were susceptible. For those native species that survived, malaria became a limiting factor in habitat availability. Non-native bird species took over the mesic forest areas, which had high numbers of insects and nectar-producing plants but in which malaria became endemic. Native bird species that did survive relocated to higher mountainous areas where mosquitoes, and hence malaria, were absent. To continue to feed on nectar-producing plants, the native birds adopted a modification of a previous feeding strategy. Before the introduction of malaria, native birds followed an elevation-related flowering sequence throughout a season by gradually moving from lower to higher elevations as plant flowering and nectar production progressed up the mountains. With the introduction of malaria, this seasonal migration changed to a daily migration pattern in which the birds, now roosting at higher elevations, leave the overnight roosting early in the morning, when mosquito vectors in the lower areas are becoming inactive, to work their way downhill to nectar-producing plants. They begin their return the same day up the mountain in time to reach the upper elevations before the mosquitoes are active again (Van Riper et al., 1986).

Grouping as a method of avoiding biting flies has been quantitatively documented in ungulates. The tendency of animals to seek the company of conspecifics has attracted the attention of behaviorists for decades, probably because such animal grouping often brings with it interesting social behaviors. It is only recently that grouping has been shown to offer protection from some types of flying parasites. There are certainly disadvantages to grouping, including increased competition for food and increased likelihood of infection from contact parasites such as lice, bacteria, and viruses (Poulin, 1991). It has long been realized that grouping reduces the risk of predation because the ability of predators to encounter individual prey does not increase linearly with the increase in group size. Animals may gain a within-group protection by getting closer to other animals which usually means moving towards the center of the group; this is referred to as the selfish herd effect (Hamilton, 1971). There is now ample evidence that the encounter effect protects animals from attacks by flies just as with predation, and in fact there are more quantitative studies supporting the fly-protection aspect of grouping than protection from predators (Mooring and Hart, 1993).

C. Protection from Nest-Borne Ectoparasites

One of the places where ectoparasites are a continuous and serious threat is the dark, warm, stable environment of the nest. A bird or mammalian nest offers an abundant food supply, forming an excellent breeding ground for a variety of ectoparasites including mites, fleas, and ticks. Long-term occupancy of nests may lead to an accumulation of nest-borne parasites and, in fact, it is thought that many ectoparasitic arthropods have evolved specifically in such nest environments (Marshall, 1981; Waage, 1979). Not surprisingly, investigators have pointed to an array of behavioral defenses against nest-borne ectoparasites, including nest selection and desertion. Both birds and mammals face a dilemma with regard to the avoidance or reuse of old nests. By building a new nest or constructing a new burrow for each group of young, much of the build-up of nest-borne parasites is avoided. However, if the old nest can be reused, energy that might go otherwise into building a nest can be utilized in caring for and provisioning the new young (Barclay, 1988). In birds that are fairly mobile and able to select a nest at the beginning of the breeding season, there is an opportunity to avoid nests in which they detect a high number of ectoparasites left over from the previous season. This behavior has been reported in cavity-nesting birds, namely barn swallows (Barclay, 1988; Møller, 1991), great tits (Christe et al., 1994), and cliff swallows (Loye and Carroll, 1991).

Another type of defense against nest-borne ectoparasites is the use of plants containing substantial quantities of secondary plant compounds as nest fumigants against ectoparasites. The European starling, which reuses nests over several seasons, was the subject of studies by Clark and Mason (1985) in examining nest fumigation in birds. Starlings, which were offered nest boxes at the beginning of the breeding season, were observed to weave fresh green plants into the matrix of old dry grass in constructing the nest. Green material was continually added until the eggs hatched. Of 66 plant species available to the starlings, 9 were selected more frequently than would have been expected by chance. The preferred plants retarded the hatching of louse eggs in comparison to non-preferred plants (Clark and Mason, 1985) and when the nests from starling boxes were removed and placed in plastic bags with preferred plant leaves (wild carrot or fleabane) the emergence of immature nest mites was delayed in comparison to nests placed in bags with non-preferred plant material (Clark and Mason, 1988). Burrow selection and desertion and the use of plants as fumigants to reduce nest-borne ectoparasites have not been investigated in mammalian species, but clearly this could be a promising area of research.

D. Grooming and Control of Ectoparasites

Grooming, another parasite defense, is one of the most frequently and regularly performed behavioral patterns of rodents, felids, ruminants, and primates. Laboratory rats may spend up to one-third of their waking time grooming (Bolles, 1960), and depending upon species, antelope of east African plains deliver 80–200 oral grooming bouts consisting of 600–2000 grooming episodes per 12-h day (Hart et al., 1992). The best evidence for the effectiveness of grooming in removing ectoparasites comes from manipulations in which grooming is restricted or prevented. In mice, Murray (1961; 1987) showed that louse infestation increases about 60-fold above baseline following temporary restraint of oral grooming. Some data on cattle, which groom with their tongues, revealed that when they were infested with larval stages of the one-host tick *Boophilus microplus*, restraint of self oral grooming resulted in an increase in ticks surviving through the engorgement stage to about 4 times that of baseline (Snowball, 1956; Bennett, 1969). Impala (*Aepyceros melampus*) and other small or medium-sized antelope groom with their lateral incisors and canine teeth, and in a recent study we found that the application of neck harnesses to impala, which partially prevented self oral grooming, resulted in these animals carrying 20 times more adult female ticks than those wearing control harnesses that allowed grooming (Mooring et al., 1996a).

Along with colleagues, I have proposed that in antelope and some other ungulate species, grooming is delivered on a regular basis to an animal's body in response to a loosely running, species-specific biological clock (Hart et al., 1992). Thus, ectoparasites tend to be removed before they attach and begin to feed and even before they produce any cutaneous stimulation. The parameters of such a grooming clock reflect the costs of grooming (water loss, distraction from vigilance over predators, attrition of grooming incisors) balanced against the costs of ectoparasites. Some predictions from this model have been explored in field studies. One prediction, reflecting the body size principle, is that species of smaller body size cannot tolerate the same number of blood-sucking ticks per unit of body surface area as species of larger body size because of the greater body surface to body mass ratio. Correspondingly, Thomson's gazelle, typically weighing 22 kg, were found to groom twice as frequently as wildebeest (214 kg) which inhabit the same grassland plains and are frequently seen side-by-side with Thomson's gazelle (Hart et al., 1992. When adult ticks (primarily *Rhipicephalus pulchellus*) were collected from the skins of the antelope culled at a game ranch, Thomson's gazelle typically had about one-third the number of adult ticks per unit of body surface area as wildebeest (Olubayo et al., 1993).

Another prediction stemming from the programmed grooming model is that of species of similar body size, those that have evolved in habitats with relatively few ticks should groom at a rate that is lower than the grooming rate of size-matched counterparts that have evolved in habitats subjecting them to higher numbers of ticks. These differences should be apparent even when members of the species are examined in tick-free environments. This prediction, reflecting the habitat principle, was supported by our finding that steinbok, which inhabit tick-dense grassland areas, were found to groom about 10 times as much as size-matched klipspringers, which inhabit tick-sparse rocky outcroppings (Hart et al., 1996). Diagrammatic representations of the body size and habitat principles are presented in Fig. 11-2.

As alluded to above, parasites sometimes have evolved behavioral patterns to evade host defenses. With regard to self-grooming, the head, ears, and neck cannot be reached by oral grooming and the perineal area cannot be reached by oral grooming or hind-leg scratching. Unlike fleas and lice, which can move out of the area being groomed, ticks move very slowly and are more susceptible to being groomed off. In surveys done on the distribution of ticks on domestic and wild hoofstock, the protected areas of head, neck, and perineum are indeed where one finds most adult and even immature forms of ticks. The best evidence for selective migration of ticks to protected areas, as opposed to being groomed off unprotected areas, is the attraction of different tick species to different areas. For example, *Rhipicephalus appendiculatus* is concentrated on the head and ears of antelope and is rarely found in the perianal area, whereas *R. evertsi* is found almost exclusively in the perianal region (Howell et al., 1978). Studies on the detection of chemical cues by ticks showing that they will move towards a variety of biologically relevant volatiles (Yunker et al., 1992) provides the basis for

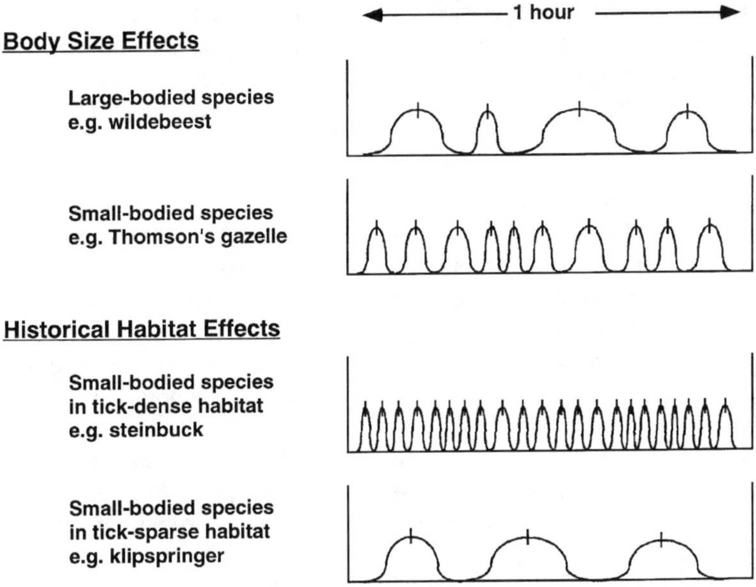

Figure 11-2. Representation of a model of programmed grooming that occurs on a periodic basis, as in African antelope. The model assumes that grooming bouts are regulated by an oscillating biological clock with a species-specific, but somewhat variable, periodicity (loosely running clock). Shown are species differences in frequency of grooming bouts in an ectoparasite-free environment as a function of: 1) body size (smaller-bodied species groom more frequently), and 2) evolutionary (historical) exposure to tick-dense versus tick-sparse habitat when body size is similar (species typically found in tick-dense habitat groom more frequently). The marks at the top of the waves represent the occurrence of a grooming bout. Data from Hart et al., 1992; Hart et al., 1996.

suggesting that ticks follow chemical cues in heading towards these preferred areas. A tick could follow the carbon dioxide gradient produced by expired air inasmuch as they are not only sensitive to small changes in ambient carbon dioxide (important in finding hosts), but can respond to higher concentrations ranging up to 5 percent (Steullett and Guerin, 1992a). Other components of expired air including hydrogen sulphide may play a role in orientation to the head and ears (Steullett and Guerin, 1992b).

II. Hormonal Influences on Host Defensive Behaviors

Hormones influence a variety of behavioral patterns of vertebrates including social interactions, sexual behavior, and maternal behavior, and are mediators of seasonal influences bringing on sexually receptive behavior at an appropriate time of year for breeding and various aspects of parental behavior for raising young (Nelson, 1995). Given the well-studied pervasive effects of hormones on a variety of behavioral systems, one would expect to find hormonal regulation of some behavioral defenses against parasites as well, provided that such hormonal regulation proved to be adaptive in integrating parasite defenses with other critical behavioral systems. In this section I will present evidence suggesting a logical role of hormones in modulating some of the behavioral defenses discussed in the previous section that would work to an animal's advantage. Unfortunately, in contrast to other areas of behavioral endocrinology, this area has received little attention from investigators. One purpose of this section is to suggest fruitful areas of research. The following areas of hormonal control are discussed: (1) mediation of seasonal occurrence of defenses against nest-borne ectoparasites, (2) down-regulation of grooming by testosterone in rutting male antelope to enhance vigilance over females, (3) possible mediation of acceleration of grooming in antelope as a result of an increase in ectoparasite exposure, (4) mediation of enhanced grooming of the nipples and anogenital area of mammalian mothers to protect young from pathogens during birth and suckling after being born, and (5) mediation of a pheromone that attracts young rat pups to consume maternal feces high in deoxycholic acid which protects young rats from bacterial enteritis.

A. Nesting and Control of Nest-Borne Ectoparasites

It is well-known that hormone systems responsive to changes in day length bring on behavioral patterns in birds and mammals that are involved in the onset of seasonal sexual and parental behavior. Because some behavioral defenses relate to protection of vulnerable newly born or hatched young from parasites, one might expect to find seasonal influences on these behavioral patterns. A common behavior of birds at the beginning of the breeding season, after moving into a previously used nesting cavity, is to remove old nest material, and, in fact, this turns out to be a very effective procedure for reducing nest-borne mites (Clark and Mason, 1988). Nest cleaning, as well as the use of fresh green material for fumigation of ectoparasites, are undoubtedly under hormonal control; as soon as the eggs hatch, males and females no longer engage in the behavior of weaving green vegetation into the nest matrix (Clark and Mason, 1985). Because little is known about the defenses in burrowing mammals with regard to nest-borne ectoparasites, one can only speculate that if such use of plants occurs, the behavior would also be under hormonal control. Consistent with this viewpoint, Neal (1986) observed that in the spring, European badgers clean out the parts of their burrows devoted to raising young before bringing in new dried forage from the field for replacement.

B. Male Rutting and Grooming Behavior

A prominent aspect of testosterone regulation of seasonal behaviors is the stimulation of rutting behavior in males. Rutting ungulates, such as North American cervids or southern African antelope, characteristically compete with other males for territory, guard and herd females, and remain vigilant to fend off conspecific males and acquire reproductively active females (Elgar, 1989; Murray, 1982). In southern Africa, rutting in impala occurs in February–April after the rainy season and just when ticks are becoming a threat (Mooring, 1995). Inasmuch as the behavioral defense against ticks, namely oral grooming, can distract from rutting behavior (Mooring and Hart, 1995), this is an area where one might look for hormonal regulation of both grooming and rutting activities, but in different degrees (females only need a few seconds to bolt from a territory or be highjacked by a competing male). In keeping with the prediction that territorial males should thus groom less than females, observations of four species of East African antelope—Thomson's gazelle, Grant's gazelle, wildebeest, and impala—revealed that males groomed at only about half the rate of females (Hart et al., 1992). In another study focusing on differences in grooming and adult tick load between territorial male and female impala in Zimbabwe, we found that the territorial males, which self oral groomed at one-third the rate of females and engaged in no allogrooming, were found to carry about six times more adult ticks than female impala (Mooring and Hart, 1995b). In a follow-up study where grooming rate in territorial males was compared with bachelor male impala, as well as with females, and all stages of ticks were examined, territorial male impala engaged in significantly less self oral grooming than both bachelor males and females and carried significantly more ticks of all stages (Mooring et al., 1996b).

The above findings suggest that reduction in grooming in territorial male impala may be due to testosterone suppression of grooming, a concept supported by the finding that testosterone levels are higher in territorial male impala than in bachelor males (Bramley and Neaves, 1972). The hypothesis that grooming is actively suppressed by testosterone as opposed to grooming being reduced because of competing behavioral demands associated with rutting comes from observations in the above studies on Thomson's gazelle, wildebeest, and impala, showing that territorial males, whether they have females on their territories or are solitary (at the time observed), have reduced grooming rates. If the solitary territorial males, which are just sitting around with no females close enough to attract their attention, still exhibit the slow grooming rate typical of more active territorial males, this argues for testosterone suppression of grooming as opposed to reduction of grooming by competing demands of rutting behavior.

In other taxa of ungulates comparative grooming rates among females, rutting males, and non-rutting males have yet to be determined. However, comparisons of tick loads between males and females point to such behavioral differences. Observations on male white-tailed deer reveal they had about 3 times as many adult ticks as females (Main et al., 1981). These authors mention that the sex differences may reflect behavioral differences such as grooming. Likewise, in moose during the rutting season, males seem to carry twice as many winter ticks per individual as females (Drew and Samuel, 1985). Admittedly, testosterone has been recognized as having a suppressive effect on the immune system (Folstad and Karter, 1992), and because resistance to ticks is known to be influenced by the immune system (Wakelin, 1984), it is possible that males carry higher tick loads than females partially because of suppression of the immune system. A reasonable conclusion is that the differences in ectoparasites between males and females is a function of testosterone down-regulation of grooming and lowered immune responses in males. Interestingly, it has been found that in all of the antelope species studied, scratch grooming is not nearly as sexually dimorphic and, in fact, tends not to significantly differ among territorial males, bachelor males, and females (Hart et al., 1992; Mooring and Hart, 1995b; Mooring et al., 1996b). This is not unreasonable because the posture adopted when animals scratch groom does not block the visual field, and theoretically would not interfere with vigilance, as does self oral grooming. A schematic illustration of the possible effect of testosterone on programmed grooming is illustrated in Figure 11-3.

Even though species-specific and gender-specific rates of grooming may be set by the body size, habitat, and vigilance principles, within the age–sex classes of a species, grooming is accelerated by an increase in ectoparasite load. This was evident in comparing grooming rates in antelope across high-tick and low-tick seasons (Mooring, 1995), between tick-free and tick-dense environments (Hart et al., 1992; Mooring and Hart, 1997), and when ticks were directly applied to impala (Mooring et al., 1996a). A possible model for modulation of grooming rate by stimuli arising from tick bites is the systemic absorption of tick saliva or some host substance such as histamine into the systemic circulation, which then acts on the central nervous system to accelerate the grooming rate. A number of endogenous hormonal systems, notably vasopressin and oxytocin, are known to increase grooming rate in rodents (Colbern and Gispen, 1988), and it is possible that tick saliva and/or histamine (or some other substance related to tick bites) increases grooming through activating one or more endogenous hormonal sys-

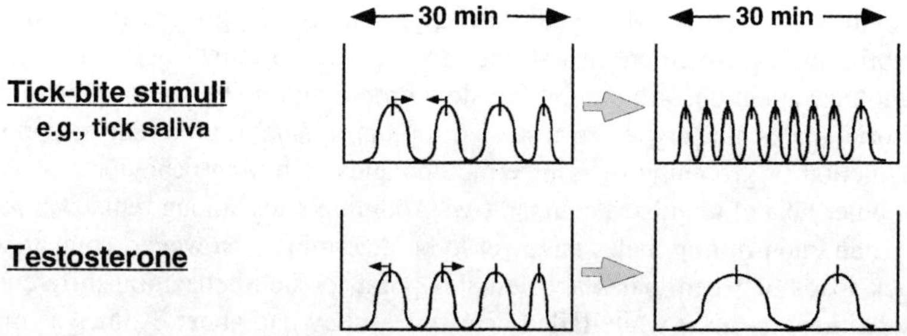

Figure 11-3. Representation of the hypothesized effects of tick-bite stimuli, such as tick saliva, in increasing the rate of grooming through presumably central, hormone-mediated effects and the effect of testosterone in some male ungulates (e.g., impala) in decreasing the rate of grooming. Data from Hart et al., 1992; Mooring and Hart, 1995; Mooring et al., 1996b.

tems. A schematic representation of this hypothetical mechanism is shown in Figure 11-3.

C. Coordination of Defenses against Parasites with Maternal Behavior

Maternal behavior, which is activated and coordinated by hormones, has an important role in protecting the young from pathogens and parasites. Although not investigated, one would expect, based upon the demonstrated effectiveness of grooming in controlling ectoparasites of adult animals, that grooming of newborns would be important in protecting them from excessive ectoparasitic infestations especially in species that use nests repeatedly and where the young are altricial and incapable of self-grooming.

In addition to the threat of ectoparasites, the young of altricial mammalian species, such as rodents and carnivores, are particularly susceptible to nest-borne bacteria to which adults are resistant because the young are born with an undeveloped immune system, have a sterile gut, and lack the intestinal protective bacterial flora present in adult animals. Even the relatively benign *E. coli* can be a threat to survival of newborns if they consume these bacteria prior to the ingestion of protective colostrum, as they would during the birth process or in initially attaching to nipples prior to suckling. As a defensive strategy, females late in pregnancy generally engage in enhanced licking of the mammary and anogenital areas (Rheingold, 1963). Systematic observations on rats revealed that it is in the last few days just before parturition that licking of the anogenital and mammary areas peaks while licking of areas such as the shoulder and back decline (Roth and Rosenblatt, 1967). Saliva of rats and dogs, and presumably other mammals, is bactericidal (Hart and Powell, 1990; Hart et al., 1987) and this, along with the physical licking-off of bacteria, undoubtedly helps clean the nipples of bacteria. This change in grooming behavior would appear to result from the pattern of estrogen, progesterone, and prolactin changes of the mother that occur during this latter stage of pregnancy (Rosenblatt et al., 1985). As a fail-safe mechanism assuring that pups will not attempt to suckle contaminated nipples or nipple-like protrusions, it is reported that rat pups will not attach to nipples that have been experimentally washed, but attachment is induced when maternal saliva is applied to the nipples (Blass and Teicher, 1980).

Young animals of altricial species are also susceptible to environmental opportunistic pathogens just after they are weaned and start taking solid food, which could easily be contaminated with bacteria from the immediate environment. *E. coli* enteritis is a serious problem in weanling rats that may be subjected to stress. One of the things that protects adult rats from *E. coli* is the presence of deoxycholic acid, a derivative of bile secretion (Moltz, 1984). However, weanling rodents do not secrete deoxycholic acid. A hormone-driven pheromone plays a role in protection of rat pups from *E. coli* enteritis by leading the rat pups to consume maternal feces, which are especially rich in deoxycholic acid. As illustrated in

Figure 11-4, rat pups are attracted by the fecal-borne pheromone to consume maternal feces 14–27 days after the onset of lactation, which is just when the rats start to take solid food. Through the action of prolactin, the pheromone is produced by bacteria in the colon. About day 28, when prolactin levels drop off and pheromone production stops, attraction to maternal feces wanes and the rat pup's secretion of deoxycholic acid also reaches adult levels (Kilpatrick et al.,).

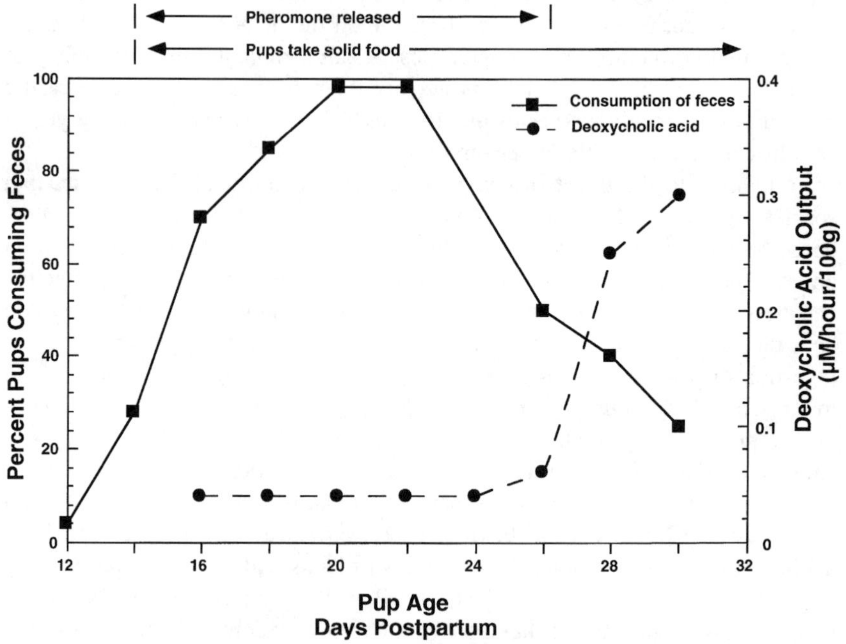

Figure 11-4. Relation between pup consumption of maternal feces (a source of the protective deoxycholic acid), release of the prolactin-dependent pheromone in maternal feces, when pups take solid food, and level of deoxycholic acid in the bile of pups as a function of pups age. From Kilpatrick et al., 1983.

III. Exploitation of Host Hormonal Secretions by Parasites

Although the exploitation of host hormone signals by parasites is not the primary subject of this chapter, one classic example is useful to mention because of its relevance to the topic of maternal care of newborns just discussed. As mentioned, the young of altricial species are vulnerable to nest-borne ectoparasites not only because of the concentration of such parasites in the nest but because they are not capable of self-grooming. Newborns in most altricial species are groomed extensively by their mothers throughout the day. One species of mammals, however, the European rabbit (*Oryetolagus cuniculus*), does not receive this protective grooming from the mother and the newborns would be quite vulnerable to uninterrupted attack by ectoparasites. Presumably as a reflection of predator-protection behavior, mother rabbits, which must forage for a considerable period of time each day, visit the burrows of their newborns only once a day and rabbit pups are not groomed even during this visit (Hudson and Distel, 1982). During this brief visit, the newborn suckle for a few minutes, then the mother rabbit immediately leaves the burrow, replacing the grass over the entrance.

In classic studies on the rabbit flea (*Spilopsyllus cuniculi*), Rothschild and associates (Rothschild, 1965; Rothschild and Ford, 1966; 1973) seem to have uncovered a system by which the rabbit flea apparently exploits the rather helpless newborn rabbits. The rabbit flea has evolved a reproductive cycle that is entrained by hormone changes in the pregnant female rabbit. The fleas mature and move to the doe's face just before birth, presumably for easy transfer to newborn pups, as a function of adrenal corticosteroid and estrogen changes in the blood imbibed by the fleas from the pregnant female rabbit. At the time of birth of pups, the fleas, which have just reached maturity, are stimulated to migrate from the doe to the pups where the fleas feed voraciously, copulate, and lay eggs. The newly laid eggs will give rise to adult fleas some 30 days later, after the rabbit pups have left the nest. The fleas may then transfer back to the doe where they remain reproductively quiescent until the next pregnancy when the fleas reproductive maturation is again synchronized with hormonal changes.

Although I have put this naturalistic functional interpretation on this work of Rothschild and associates, I should mention that Rothschild recognized that this coupling of the cycle of the flea with the rabbit guaranteed the flea a breeding habitat and food supply (Rothschild, 1965). The emphasis on the vulnerability of rabbit pups to fleas is my own interpretation and it suggests that the lack of grooming of rabbit pups, which is unique among altricial polytocous species, may be the reason the rabbit flea seems to be unique in having its reproductive cycle so completely entrained to hormones of the host.

IV. Conclusion

In the discussion above, I have taken a general concept from the field of hormones and behavior, namely that of coordination or integration of several behavioral events with changing physiological processes, to illustrate some possibilities for hormonal control of behavioral defenses against parasites. Although the examples I have given are logical and supported by some lines of evidence, the specific hypotheses have yet to be fully explored through experimental or field studies involving manipulation. This topic represents a new area of hormones and behavior and certainly a new area from the standpoint of host-parasite interactions. There are probably many examples of hormonal regulation of host behavioral defenses that will become evident in future studies. The converse side of the picture is that one can expect parasites, particularly those that have access to changing levels of hormones in the blood of their hosts, to capitalize on these changing hormone levels as cues for reproductive or other processes that would allow them to more effectively exploit hosts. Hopefully, I have pointed to some areas of fruitful exploitation by other investigators that may add to the perspective and depth of this relatively novel area.

V. Acknowledgment

Preparation of this review was supported in part by the NSF-BNS 9109039 grant to B. L. Hart and the NSF-DIR 9113287 Research Training Group in Behavior.

VI. References

Barclay, R. M. R. (1988). Variation in the costs, benefits, and frequency of nest reuse by barn swallows (*Hirundo rustica*). *Auk* **105:**53–60.

Bennett, G. F. (1969). *Boophilus microplus* (Acarina: *Ixodidae*): Experimental infestations on cattle restrained from grooming. *Exp. Parasitol.* **26:**323–328.

Blass, E. M., and Teicher, M. H. (1980). Suckling. *Science* **210:**15–21.

Bolles, R. C. (1960). Grooming behavior in the rat. *J. Comp. Physiol. Psychol.* **53:**306–310.

Bramley, P. S., and Neaves, W. B. (1972). The relationship between social status and reproductive activity in male impala, *Aepyceros melampus. J. Reprod. Fert.* **31:**77–81.

Christe, P., Oppliger, A., and Richner, H. (1994). Ectoparasite affects choice and use of roost sites in the great tit, *Parus major. Anim. Behav.* **47:**895–898.

Clark, L., and Mason, J. R. (1985). Use of nest material as insecticidal and antipathogenic agents by the European starling. *Oecologia* **67:**169–176.

———. (1988). Effect of biologically active plants used as nest material and the derived benefit to starling nestlings. *Oecologia* **77:**174–180.

Clayton, D. H. (1991). Coevolution of avian grooming and ectoparasite avoidance. In *Bird–parasite interactions: Ecology, evolution, and behavior* (J. E. Loye and M. Zuk, eds.), pp. 258–289. New York: Oxford University Press.

Colbern, D. L., and Gispen, W. H., eds. (1988). *Neural mechanisms and biological significance of grooming behavior.* Ann. N.Y. Acad. Sci.

Colwell, D. D., and Kavaliers, M. (1990). Exposure to mosquitoes, *Aedes togoi* (Theo.), induces and augments opioid-mediated analgesia in mice. *Physiol. and Behav.* **48:**397–401.

———. (1992). Evidence for activation of endogenous opioid systems in mice following short exposure to stable flies. *Med. and Vet. Entomol.* **6:**159–164.

Darwin, C. (1871). *The descent of man and selection in relation to sex,* Vol. 1. London: Murray.

Downes, C. M., Theberge, J. B., and Smith, S. M. (1986). The influence of insects on the distribution, microhabitat choice, and behavior of the Burwash caribou herd. *Can. J. Zoo.* **64:**622–629.

Drew, M. L., and Samuel, W. M. (1985). Factors affecting transmission of larval winter ticks, *Dermacentor albipictus* (Packard), to moose, *Alces alces* L., in Alberta, Canada. *J. Wild. Dis.* **21:**274–282.

Dudley, R., and Milton, K. (1990). Parasite deterrence and the energetic costs of slapping in howler monkeys, *Alouatta palliata*. *J. Mammol.* **7:**463–465.

Durie, P. H. (1961). Parasitic gastroenteritis of cattle: The distribution and survival of infective strongyle larvae on pasture. *Aust. J. Agr. Res.* **12:**1200–1211.

Edman, J. D. (1971). Host-feeding patterns of Florida mosquitoes I. *Aedes, Anopheles, Coquillettidia, Mansonia* and *Psorophora*. *J. Med. Entomol.* **8:**687–695.

Edman, J. D., and Kale II, H. W. (1971). Host behavior: Its influence on the feeding success of mosquitos. *Ann. of Ent. Soc. of America* **64**(2):513–516.

Elgar, M. A. (1989). Predator vigilance and group size in mammals and birds: A critical review of the empirical evidence. *Biol. Rev.* **64:**13–33.

English, A. W. (1979). The epidemiology of equine strongylosis in southern Queensland. *Aust. Vet. J.* **55:**299–305.

Espmark, Y. (1967). Observations of defence reactions to oestrid flies by semidomestic forest reindeer (*Rangifer tarandus* L.) in Swedish Lapland. *Zoologische Beitrage* **14:**155–167.

Eysker, M. (1991). Direct measurement of dispersal of *Dictyocaulus viviparous* in sporangia of *Philobolus* species. *Res. Vet. Sci.* **50:**29–32.

FitzGibbon, C. D. (1989). A cost to individuals with reduced vigilance in groups of Thomson's gazelles hunted by cheetahs. *Anim. Behav.* **37:**508–510.

Folstad, I., and Karter, A. J. (1992). Parasites, bright males, and the immunocompetence handicap. *Amer. Nat.* **139:**603–622.

Garcia, R. (1962). Carbon dioxide as an attractant for certain ticks (*Acarina argasidas* and *Ixodidas*). *Ann. Entomol. Soc. of America* **14:**605–606.

Hamilton, W. D. (1971). Geometry for the selfish herd. *J. Theor. Biol.* **31:**295–311.

Hart, B. L. (1990). Behavioral adaptations to pathogens and parasites: Five strategies. *Neurosci. Biobehav. Rev.* **14:**273–294.

Hart, B. L. (1994). Behavioral defense against parasites: Interaction with parasite invasiveness. *Parasitol.* **109:**S139–S151.

———. (1995). Behavioral defenses against parasites. In (D. H. Clayton and J. Moore, eds.). *Coevolutionary ecology of birds and their parasites* Oxford University Press (in press).

Hart, B. L., and Hart, L. A. (1994). Fly switching by Asian elephants: Tool use to control parasites. *Anim. Behav.* **48:**35–45.

Hart, B. L., Hart, L. A., Mooring, M. S., and Olubayo, R. (1992). Biological basis of grooming behavior in antelope; the body-size, vigilance, and habitat principles. *Anim. Behav.* **44:**615–631.

Hart, L. A., Hart, B. L., and Wilson, V. J. (1996). Grooming rates in klipspringer and steinbok reflect environmental exposure to ticks. *Afr. J. Ecol.* **34:**79–82.

Hart, B. L., Korinek, E., and Brennan, P. (1987). Postcopulatory genital grooming in male rats: Prevention of sexually transmitted infections. *Physiol. and Behav.* **41:**321–325.

Hart, B. L., and Powell, K. L. (1990). Antibacterial properties of saliva: Role in maternal periparturient grooming and in licking wounds. *Physiol. and Behav.* **48:**383–386.

Harvey, T. L., and Launchbaugh, J. L. (1982). Effect of horn flies on behavior of cattle. *J. Econ. Entomol.* **75:**25–27.

Howell, C. J., Walker, J. B., and Nevill, E. M. (1978). Ticks, mites, and insects infesting domestic animals in South Africa. Part 1. Descriptions and biology. *Sci. Bull. Depart. Agr. Tech. Serv. RSA.*

Hudson, R., and Distel, H. (1982). The pattern of behavior of rabbit pups in the nest. *Behavior* **79:**255–271.

Kilpatrick, S. J., Lee, T. M., and Moltz, H. (1983). The maternal pheromone of the rat: Testing some assumptions underlying a hypothesis. *Physiol. Behav.* **30:**539–432.

Loye, J. E., and Carroll, S. P. (1991). Nest ectoparasite abundance and cliff swallow colony site selection, nestling development, and departure time. In *Bird–parasite interactions: Ecology, evolution, and behavior* (J. E. Loye and M. Zuk, eds.), pp. 222–241. New York: Oxford University Press.

Main, A. J., Sprance, H. E., Kloter, K. O., and Brown, S. E. (1981). *Ixodes dammini* (Acari: *Ixodidae*) on white-tailed deer *Odocoileus virginianus* in Connecticut, USA. *J. Med. Entomol.* **18:**487–492.

Marshall, A. G. (1982). *The ecology of ectoparasitic insects.* London: Academic.

Michel, J. F. (1955). Parasitological significance of bovine grazing behaviour. *Nature* **175:**1088–1089.

Møller, A. P. (1991). Ectoparasite loads affect optimal clutch size in swallows. *Funct. Ecol.* **5:**351–359.

Moltz, H. (1984). Of rats and infants and necrotizing enterocolitis. *Perspect. Biol. Med.* **27:**327–335.

Moore, J. (1983). Responses of an avian predator and its isopod prey to an acanthocephalan parasite. *Ecology* **64:**1000–1015.

Mooring, M. S. (1995). The effect of tick challenge on grooming rate by impala. *Anim. Behav.* **50:**377–392.

Mooring, M. S., and Hart, B. L. (1993). Animal grouping for protection from parasites; selfish herd and encounter/dilution effects. *Behavior* **123:**173–193.

———. (1995a). Costs of allogrooming in impala: Distraction from vigilance. *Anim. Behav.* **49:**1414–1416.

———. (1995b). Differential grooming rate and tick load of territorial male and female impala, *Aepyceres melampus. Behav. Ecol.* **6:**94–101.

———. (1997). Self-grooming in impala mothers and lambs: Testing the body size and tick challenge principles. *Anim. Behav.* (in press).

Mooring, M. S., McKenzie, A. A., and Hart, B. L. (1996a). Grooming in impala: Role of oral grooming in removal of ticks and effects of ticks in increasing grooming rate. *Physiol. Behav.* **59:**965–971.

———. (1996b). Role of sex and breeding status in grooming and total tick load in impala. *Behav. Ecol. Sociobiol.* **39:**259–266.

Murray, M. D. (1961). The ecology of the louse *Polyplax serrata* (Berm.) on the mouse *Mus musculus L. Aust. J. Zool.* **9:**1–13.

———. (1982). Home range, dispersal, and the clan system of impala. *Afr. J. Ecol.* **20:**253–269.

———. (1987). Effects of host grooming on louse populations. *Parasitol. Today* **3**(9):276–278.

Neal, E. (1986). *The natural history of badgers.* London: Croom Helm.

Nelson, R. J. (1995). *Introduction to behavioral endocrinology.* Sunderland, Mass: Sinauer Associates.

Nicolas, G., and Sillans, D. (1989). Immediate and latent effects of carbon dioxide on insects. *Ann. Rev. Entomol.* **34:**97–116.

Norval, R. A. I., Sutherst, R. W., Jorgensen, O. G., Gibson, J. D., and Kerr, J. D. (1989). The effect of the bont tick (*Amblyomma hebraeum*) on the weight gain of *Africander* steers. *Vet. Parasitol.* **33:**329–341.

Norval, R. A. I., Sutherst, R. W., Kurki, J., Gibson, J. D., and Kerr, J. D. (1988). The effect of the brown ear-tick *Rhipicephalus appendiculatus* on the growth of Sanga and European breed cattle. *Vet. Parasitol.* **30:**149–164.

Ödberg, F. O., and Francis-Smith, K. (1977). Studies on the formation of ungrazed eliminative areas in fields used by horses. *Appl. Anim. Ethol.* **3:**27–44.

Okumura, T. (1977). The relationship of attacking fly abundance to behavioral responses of grazing cattle. *Jpn. J. Appl. Entomol. Zool.* **21:**119–122.

Olubayo, R. O., Jono, J., Orinda, G., Groothenius, J. G., and Hart, B. L. (1993). Comparative differences in densities of adult ticks as a function of body size on some East African antelopes. *Afr. J. Ecol.* **31:**26–34.

Poulin, R. F. (1991). Group-living and infestation by ectoparasites in passerines. *Condor* **93:**418–423.

Rheingold, H. L. (1963). *Maternal behavior in mammals.* New York: Wiley Press.

Robinson, J. (1962). *Pilobolus* spp. and the translation of infective larvae of *Dictyocaulus viviparus* from feces to pastures. *Nature* **193:**353–354.

Rosenblatt, J. S., Mayer, A. D., and Siegel, H. I. (1985). Maternal behavior among the nonprimate mammals. In *Handbook of behavioral neurobiology* (N. Adler, D. Pfaff, and R. W. Goy, eds.), pp. 229–298. Plenum Press.

Roth, L., and Rosenblatt, J. S. (1967). Changes in self-licking during pregnancy in the rat. *J. Comp. Physiol. Psychol.* **63:**397–400.

Rothschild, M. (1965). The rabbit flea and hormones. *Endeavor* **24:**162–168.

Rothschild, M., and Ford, B. (1966). Hormones of the vertebrate host controlling ovarian regression and copulation in the rabbit flea. *Nature* **211:**261–266.

Rothschild, M., and Ford, R. (1973). Factors influencing the breeding of the rabbit flea (*Spilopsyllus cuniculi*): A spring-time accelerator and a kairomone in nestling rabbit urine (with notes on *Cediopsylla simplex* and "hormone-bound" species). *J. Zool.* **170:**87–137.

Snowball, G. J. (1956). The effect of self-licking by cattle on infestation of cattle tick *Boophilus microplus* (Canestrini). *Aust. J. Agric. Res.* **7:**227–232.

Steullet, P., and Guerin, P. M. (1992a). Perception of breath components by the tropical bout tick, *Amblyomma variegatum Fabricicus* (Ixodidae). I. CO_2-excited and CO_2-inhibited receptors. *J. Comp. Physiol.* **170**A:665–676.

———. (1992b). Perception of breath components by the tropical bout tick, *Amblyomma variegatum Fabricicus* (Ixodidae). II. Sulfide-receptors. *J. Comp. Physiol.* **170**A:677–685.

Sutherst, R. W., Maywald, G. F., Kerr, J. D., and Stegeman, D. A. (1983). The effect of cattle tick (*Boophilus microplus*) on the growth of *Bos indicus* x *B. taurus* steers. *Austral. J. Agric. Res.* **34:**317–327.

Van Riper, C., Van Riper, S. G., Goff, M. L., and Laird, M. (1986). The epizootiology and ecological significance of malaria in Hawaiian land birds. *Ecol. Monographs* **56:**(4):327–344.

Waage, J. K. (1979). The evolution of insect/vertebrate associations. *Brit. J. Linn. Soc.* **12:**187–224.

Wakelin, D. (1984). *Immunity to parasites: How animals contract parasitic infections.* London: Edward Arnold.

Washino, R. K., and Tempelis, C. H. (1983). Mosquito host-blood meal identification. *Ann. Rev. Entomol.* **28:**179–201.

Yunker, C. E., Peter, R., Norval, R. A. W., Sonenshine, D. E., Burridge, M. J., and Butler, J. F. (1992). Olfactory responses of adult *Amblyomma hebraeum* and *A. variegatum* (Acari: *Ixodidae*) to attractal chemicals in laboratory tests. *Exp. App. Acarology* **13:**295–301.

12

How Parasites Alter the Behavior of their Insect Hosts

Shelley A. Adamo

I. Introduction

An alien creature slithers into the body of its victim and with machiavellian cunning manipulates its behavior. This disturbing vision has been the premise for more than one successful horror film. As with many human fantasies, this one has a basis in biological fact: some parasites do alter their hosts' behavior (see Moore and Gotelli, 1990; Horton and Moore, 1993; Moore, 1993, 1995). What is less clear is how they do it.

This chapter focuses on the change in behavior observed in many parasitized insects for three reasons: (1) The size difference between host and parasite is usually less than that in other systems, potentially allowing for more direct interactions between the two. For example, a protozoan parasite of a large mammal would need incredible synthetic abilities to raise host hormonal levels; however an insect parasitoid that can attain 10–30 percent of its host body weight could secrete factors at behaviorally potent concentrations. (2) Insects are often parasitized by other insects. For example, insect parasitoids have a free-living adult stage, but their larvae must develop on or within the body of another insect. Because these parasites belong to the same phylogenetic class as their hosts, they can probably synthesize most, if not all, of the neuroactive and/or hormonal substances used by the host. This may give insect parasitoid species a greater opportunity to evolve specific mechanisms for altering host behavior. (3) Although not simple systems, insects have simpler endocrine and nervous systems (i.e., fewer neurons) than do vertebrates. This makes it easier to determine how and where in the central nervous system a parasite is having an effect. Moreover, insects have long been a favorite animal group of behavioral physiologists, and much is known about their behavior (e.g., Bailey and Ridsdill-Smith, 1991), neuroethology (e.g., Camhi, 1984), and endocrinology (e.g., Nijhout,

1994). This background information allows a rigorous analysis of the changes observed in the behavior and physiology of parasitized insects.

II. Why Study How Parasites Alter Host Behavior?

The question of how behavior is regulated in normal animals remains an outstanding biological problem that spans the fields of neurobiology, endocrinology, and ethology. The interconnections between different physiological systems, and the multiplicity of influences on the central nervous system (e.g., circadian rhythms, hormones, sensory stimuli, see Huber et al., 1989) make it difficult to determine the physiological mechanisms mediating behavior. Despite this complexity, some parasites have evolved at least partial solutions to the problem of how behavior is controlled. Because some parasites can induce changes in specific host behaviors (see Horton and Moore, 1993), they must have the ability to alter selectively the dynamics between the different physiological components that regulate behavior. For this reason, parasites offer the potential to be used as probes to aid us in understanding the interplay of endocrine, neural, immunological, and other factors that act in concert to control behavior. This will be true even if the change in host behavior serves no adaptive purpose to either host or parasite.

III. Possible Physiological Mechanisms Mediating Host Behavioral Changes

With the exception of mechanical damage to muscles or sensory organs, any change in behavior must involve changes within the central nervous system. Some parasites gain direct access to the host's central nervous system by residing within it. Potentially these parasites can alter behavior in a number of ways, such as by destroying neural pathways or by secreting neuroactive substances directly into specific brain areas. The liver flukes, *Dicrocoelium dendriticum* and *Dicrocoelium hospes* attack the brain of their intermediate ant host (Romig et al., 1980). Although most of the infesting flukes do not reside in the brain, one or two enter either the ventral subesophageal ganglion (*D. dendriticum*) or the dorsal part of the antennal lobe (*D. hospes*). Parasitized ants behave abnormally, and either climb to the tops of plants and fix themselves there with their mandibles (*D. dendriticum*) or they remain motionless in elevated places (*D. hospes*). The flukes induce the same behavioral change even if they are not exactly in the same place in the brain in every ant. Romig et al. (1980) suggested that the behavioral change induced by the flukes is due to something the parasites secrete and not due to specific neural damage; however, the identity of the active agent(s) or how it might cause its effect is unknown.

Most insect parasites, however, reside in the host's hemocoel and therefore do not come into physical contact with the central nervous system (CNS). These parasites can still influence host behavior using either indirect or direct mechanisms (see also Jones, 1985). Below are examples of some of the ways in which the presence of parasites has been found to alter host behavior. It should be noted that in all cases the exact mechanisms remain unknown (Hurd, 1990a; Lawrence and Lanzrein, 1993).

A. Mechanisms without Apparent Direct Effects on the CNS

For example, in some systems the parasite may damage tissue, create nutritional stress, activate the immune system, or influence the functioning of the endocrine system. These events may secondarily alter nervous system function and thus induce behavioral changes within the host.

1. Tissue Damage

Damage to host muscles and sensory organs can lead to behavioral changes (Holmes and Zohar, 1990). The fly *Glossina morsitans morsitans* can be infested by the trypanosome *Trypanosoma congolense* (Moore, 1993). The trypansomes attach themselves to the labral sensilla of the fly, probably impairing sensory functions and hence feeding behavior. Mathematical models indicate that the increased probing of the parasitized fly probably increases parasite transmission (Rossignol and Rossignol, 1988). Molyneaux and Jefferies (1986) discussed other causes for the decline in the feeding of parasitized insects such as the destruction of the host's salivary glands, obstruction of its digestive tract, and/or changes in receptor sensitivity because of changes in the ion concentration (e.g., Na^+, K^+, Ca^{2+}) of host hemolymph after parasitic infestation.

A decline in host activity levels is a common effect of parasitism (Horton and Moore, 1993). In some cases this decline appears to be caused by the destruction of muscle tissue by the parasite. For example, the microfilariae of *Brugia pahangi* invade the flight muscles of the mosquito *Aedes aegypti*. There is a dramatic decline in host flight activity when the larvae emerge from the muscle and this is coincident with major muscle damage (Rowland and Lindsay, 1986). The cricket *Gryllus integer* shows a decline in fighting and locomotion near the time of emergence of the dipteran tachinid parasitoid *Ormia ochracea* (Adamo et al., 1994). This coincides with the destruction of much of the cricket's muscle mass by the feeding *O. ochracea* larvae, and the authors suggested that this tissue damage is the likely cause of the behavioral change.

Destruction of the fat body can lead to hormonal changes (Lawrence, 1986). The fat body synthesizes juvenile hormone esterase, an enzyme that degrades juvenile hormone (JH) (Nijhout, 1994). Therefore destruction of this tissue will elevate the hemolymph JH titer. This can lead to developmental and sometimes behavioral alterations (discussed in more detail below).

2. *Nutritional Stress and Energy Metabolism*

Parasites can decrease egg laying in female hosts by altering host energy metabolism, which decreases the insect's ability to synthesize eggs (see Hurd, 1990a,b; Hurd, 1993). This mechanism will not be discussed further because it has been recently reviewed (Hurd, 1990a,b; Hurd, Chapter 9), although how nutritional stress prevents oviposition (egg laying behavior) is not known. This phenomenon does suggest that oviposition in most insect species is dependent on the physical presence of eggs and/or the female's endocrine levels and nutritional status.

Alleyne and Beckage (1997), Alleyne et al. (1997) and Thompson (1990, 1993) discussed the complex ways in which parasites can alter host metabolism and the far-reaching effects these changes can have on the host. Changes in one system, such as energy metabolism, are likely to alter the function of many other systems, such as reproduction. However, the direct links between intermediate metabolism and the nervous system are poorly understood in insects, and therefore it has not been worked out in detail how changes in protein, lipid, and sugar levels in the hemolymph can alter host behavior. Nevertheless these appear to have a direct effect on feeding, at least in some insects (Simpson and Bernays, 1983; Timmins and Reynolds, 1992; Edgecomb et al., 1994).

As suggested by the data cited above, feeding behavior in hosts might be altered by changes in the concentrations of lipids and sugars in their hemolymph. Changes in the levels of these compounds may partly explain the anorexia of *Manduca sexta* after being parasitized by the parasitoid wasp *Cotesia congregata*. At first the host feeds normally until the parasitoid larvae are about to emerge (Beckage and Riddiford, 1978). Prior to parasitoid emergence, hemolymph trehalose, but not glucose, levels are lowered (Dahlman, 1975). After the *C. congregata* larvae emerge, feeding by the host ceases, even though there is no evidence of debilitation (Beckage and Templeton, 1986; Adamo et al., 1997). This suggests that the host's anorexia is not due to sensory or motor disturbances. The cessation of feeding correlates with an increase in host hemolymph octopamine levels. These levels are higher than levels found in unparasitized animals or in hosts prior to the emergence of the *C. congregata* larvae (Adamo et al., 1997). In *M. sexta*, injecting octopamine, or its agonists, into the hemocoel has been shown to depress feeding (Ismail and Matsumura, 1992), and the authors suggested this is because octopamine induces the release of lipids and sugars into the hemolymph, and this increase inhibits feeding (Ismail and Matsumura, 1992). Sugar and lipid levels have not been measured in parasitized *M. sexta* after the larvae have emerged, but if the raised octopamine levels are increasing lipid and sugar concentrations, the prediction would be that the levels of these compounds will be elevated.

We do not know what causes the increase in octopamine levels in the hemolymph of some parasitized insects. In the common armyworm, *Mythimna separ-*

ata, parasitized by the wasp *Apanteles kariyai*, the increase in the octopamine levels in the hemolymph of the host is thought to be due to the stress of parasitism (Shimizu and Takeda, 1994). However, although octopamine is released as a neurohormone in insects under some forms of stress, this typically occurs only when the stress involves vigorous physical activity (Adamo et al., 1995). Whether the source of the increased octopamine is the same in parasitized and exercise-stressed insects needs to be determined.

3. *Activation of the Immune System*

Although some parasites evade detection by the host's immune system, others, such as tachinid larvae, elicit immune reactions in their host (Vinson, 1990). In both mammals and molluscs, it has been established that the immune, endocrine, and nervous systems are linked intricately with behavior (see Thompson and Kavaliers, 1994). Less is known about insect psycho-neuro-immunology, although preliminary evidence suggests that the immune system interacts with other physiological systems in insects, and some of these interactions may have behavioral effects (see below). The presence of parasites, then, could alter host behavior by activating the immune system.

Dunn et al. (1994) found that injecting *M. sexta* with peptidoglycans (found in bacterial cell walls) both activated the animal's immune system and induced some behavioral changes, including a depression in feeding (Dunn et al., 1994). Dunn et al. (1994) speculated that these behavioral changes may increase *M. sexta*'s chance of overcoming the infection. Were these behavioral changes caused by factors released by the insect's immune system? Insect hemocytes release a variety of factors when activated by antigens (Ratcliffe, 1993), but the effects of these substances on the central nervous system have yet to be investigated.

Prime candidates for such an investigation are compounds that are involved in the functioning of both the immune and nervous systems. For example, octopamine and serotonin are important neurohormones in insects (Nijhout, 1994), and they are also involved in mediating insect immune responses (Baines et al., 1992; Dunphy and Downer, 1994). Whether neurohormonal levels of octopamine and serotonin can influence immune function and/or whether activation of the immune system can induce behaviorally relevant changes in neurohormonal levels of these compounds remains to be tested.

4. *Effects on the Endocrine System*

Most physiological studies have examined the developmental, not behavioral, effects of parasitism (see Lawrence and Lanzrein, 1993, Table 1). Host development is usually altered by interfering with the host's endocrine system (see Lawrence and Lanzrein, 1993, and chapters by Beckage and others in this volume). For example, some parasites, or their associated tissues (e.g., teratocytes, Dahlman and Vinson, 1993), can secrete substances such as insect hormones (e.g., JH III,

Jones et al., 1990; ecdysteroids, Brown and Reed-Larsen, 1991) into their host, and thus affect host development. In most cases the site of release is unknown, although larvae of the icheumonid parasitoid *Pimpla turionellae* secrete a fungistatic factor from their anal vesicle (Führer and Willers, 1986). Below I list two types of parasite-induced host behavioral changes that, at least in part, may be due to changes in the levels of host hormones.

a. Parasitic Castration

Some parasites physically attack host gonads (see Hurd, 1990a; Brown and Reed-Larsen, 1991) but most parasites probably castrate their insect hosts by altering the endocrine functions that are necessary for gonad development. This often leads to a decline in their sexual and reproductive behavior (Hurd, 1993; Chapter 9). For example, Strambi and Girardie (1973) found that the wasp *Polistes gallicus* parasitized by the parasitoid *Xenos vesparum* shows a decline in egg laying that appears to be due to the lack of ovarian development in parasitized females. They speculated that this lack of development was due to a decline in secretions from the corpora allata, because ovarian development could be induced by implanting functional corpora allata into parasitized females. The corpora allata secretes JH which is important for egg development in insects (Nijhout, 1994). The authors did not note if these females also exhibited normal mating and egg-laying behavior after the implantation; therefore, it is unclear if this treatment also induced normal reproductive behavior.

Strambi et al. (1982) also found that parasitized *Polistes gallicus* females have less fat body and less protein in their hemolymph than unparasitized females, suggesting that parasitized females are nutritionally stressed by parasitism. According to their data, the decline in protein concentration in the hemolymph in parasitized females begins while the parasitoid is still rapidly growing. The low protein levels in the hemolymph could result in a suppression of juvenile hormone release from the corpora allata, inhibiting reproduction.

Parasitic castration has been recently reviewed (Hurd, 1990a, 1993; Brown and Reed-Larsen, 1991; Chapters 8 and 9). Despite the interest in this topic, as Hurd (1993) concludes, ". . . at the moment there is no example of parasite-induced disturbance of insect reproduction in which the underlying mechanisms have been fully elucidated (p. 100)".

b. Changes in Thermoregulation and Temperature Preference

Most infested insects have not been tested to determine if parasitism alters their preferred body temperature. Insects infected with bacteria, however, have been shown to exhibit 'behavioral fever'. During behavioral fever, insects select areas that are higher in temperature than they would normally tolerate, thus raising their body temperature (Bronstein and Conner, 1984; Louis et al., 1986; Horton and Moore, 1993; Stanley-Samuelson, 1994). In the case of crickets infected with the bacterium *Rickettsiella grylli*, animals exhibiting behavioral fever had less mortality (Louis et al., 1986); therefore, at least in some cases, raising body temperature can assist the insect in ridding itself of pathogens. To date no parasitoid has been shown to induce behavioral fever, even though it may be to a host's advantage to respond in this way. For example, the wasp parasitoid *Apanteles (Cotesia) miltaris* that parasitizes the armyworm *Pseudaletia unipuncta* can be killed by incubating the host at high temperatures (Kaya and Tanada, 1969). However, even though the parasitoids were destroyed, the host was not capable of normal development and did not pupate.

The presence of the parasitoid can decrease host temperature preferences. Müller and Schmid-Hempel (1993) have shown that bumblebees parasitized by a conopid fly larva prefer a lower temperature than do unparasitized bumblebees. Preferring the cooler temperature retards parasitoid growth and development. This increases the host's lifespan and therefore increases the amount of foraging the host can do. Because the host is typically a forager for its colony, increasing its ability to forage enhances colony success and increases the host's inclusive fitness (Müller and Schmid-Hempel, 1993).

The study on the bumblebees by Müller and Schmid-Hempel (1993) suggests that the presence of some parasitoids induces changes in temperature regulation in the host. Although it is unknown how these changes are produced, prostaglandins are involved in thermoregulation in cicadas (Toolson et al., 1994) and prostaglandin E_1 can elicit behavioral fever in leeches (Cabanac, 1989), scorpions (Cabanac and Guelte, 1980) and crayfish (Casterlin and Reynolds, 1978). How prostaglandins might affect the central nervous system to produce changes in a temperature 'set point' is unknown (see Stanley-Samuelson, 1994 for a review of prostaglandins and behavioral fever). However, some prostaglandins can act as hormones and influence behavior in insects (e.g., oviposition behavior in crickets, Stanley-Samuelson et al., 1987; Stanley-Samuelson, 1994).

B. Mechanisms with Direct Effects on the CNS

The parasite may actively secrete factors that can cross the blood-brain barrier and affect the central nervous system, and can directly influence host behavior by secreting substances that bind to receptors in the CNS, interfere with the metabolism (i.e., production, processing and/or degradation) of neuroactive substances, and/or alter the number or sensitivity of receptors for these substances on target cells.

Parasites must overcome some formidable physiological barriers to be able to directly influence their hosts (Lawrence and Lanzrein, 1993). To affect the central nervous system, substances secreted by the parasite must cross the insect blood-brain barrier. The parasite, which is smaller than its host, must also secrete enough of a substance to have a biologically significant effect after dilution in the host's hemolymph. The substance must resist degradation long enough to bind with receptors. Some of these difficulties can be lessened if the target tissue, such as a peripheral sensory receptor, is in the vicinity of the parasite, so that local concentrations of parasite-secreted factors could be high.

Some behavioral changes occur concomitantly with alterations in host development (Beckage and Templeton, 1986). Whether the same substances (e.g., JH or ecdysteroids) are responsible for both the developmental and the behavioral changes is unknown. Some insect neurons have receptors for insect hormones such as ecdysteroids (Truman et al., 1994) and the excitability of some of these neurons change as hormonal titers fluctuate during metamorphosis (Hewes and Truman, 1994). Altering neural excitability could have a wide range of behavioral effects; for example, it could change an animal's responsiveness to stimuli. Therefore some parasites that are known to alter host development (see Lawrence and Lanzrein, 1993) may also alter host behavior using the same compounds.

In some cases the presence of parasites correlates with changes in the levels of neuroactive substances such as octopamine (discussed above) and dopamine (Noguchi et al., 1995). In the armyworm *Pseudaletia separata* parasitized by the wasp *Cotesia kariyai*, the host exhibits an increase in hemolymph levels of Growth Blocking Peptide (GBP, Hayakawa, 1995). GBP is made by the host in the presence of the polydnavirus that is co-injected by the wasp with its eggs (Hayakawa, 1995). Injections of GBP result in increases in dopamine levels in the hemolymph and nerve cord of the host, but not in its brain (Noguchi et al., 1995).

It has yet to be demonstrated, however, that the changes in dopamine or GBP levels are involved in altering host behavior. Dopamine is also a tanning agent in insects (Wigglesworth, 1972) and may not always play a role in regulating behavior. Brey (1994) suggested that the increase in GBP, which occurs naturally in the unparasitized host when it is thermally stressed, is an indirect effect of the stress caused by the polydnavirus, and is not necessarily due to a specific effect of the polydnavirus on the nervous system. Beckage (1993) suggested that GBP may affect feeding behavior and development by influencing JH esterase levels.

A related hymenopteran parasitoid, the wasp *Cotesia congregata*, also injects a polydnavirus along with its eggs into its host, *M. sexta*. Dushay and Beckage (1993) speculated that the polydnavirus can affect the brain-neurosecretory axis, influencing JH and ecdysteroid levels and hence *M. sexta* development. Polydnaviruses have been detected in the nervous systems of insect hosts (Strand et al., 1992), including *M. sexta* (F. Tan and N. Beckage, unpublished data), supporting this hypothesis. However, it is the presence of the wasp larvae, not the polydnavirus, that is responsible for the increase in host hemolymph octopamine levels that occurs during parasitoid emergence (Adamo et al., 1997). It is possible that *C. congregata* affects host behavior using multiple mechanisms.

Parasitized *M. sexta* larvae also show an accumulation of neuromodulators in some of their neurons (Zitnan, Kingan, and Beckage, 1995; Zitnan, Kingan, et al.,

1995). The increase in immunohistochemical staining could be due to a decrease in neural release, a change in the rate of synthesis at the transcriptional or translational level, or alterations in posttranslational processing (Zitnan, Kingan, et al., 1995). The cause for this increase is unknown as are the effects of this increase on host behavior. The most pronounced accumulation of neural peptides (prothoracicotropic hormone, eclosion hormone, FMRF-amide and others) occurs several days after the *C. congregata* larvae emerge from their host. This is several days after the most dramatic changes in host behavior have occurred, and therefore it is unclear if these changes are causally related to the observed changes in behavior. During the time of the greatest behavioral changes, when the host stops feeding while the *C. congregata* larvae emerge, the differences in staining between the brains of parasitized and control animals are more subtle. However, there are strong increases in staining for proctolin at this time. Zitnan, Kingan, et al. (1995) suggested that proctolin (or a proctolin-like peptide) may be involved in modulating the release of neuropeptides in the brain–corpora cardiaca–corpora allata complex. If enhanced immunoreactivity corresponds to a decrease in proctolin's release, then this effect could potentially alter the release rate of a large number of insect hormones (Nijhout, 1994).

The only studies that examine the question of neural activity in parasitized insects suggest that the tachinid parasitoid, *Metacemayia calloti*, induces a decline in the activity of some of the neurosecretory cells of its host, the grasshopper *Anacridium aegyptium* (e.g. Girardie and Girardie, 1977). The female grasshopper host exhibits a decline in egg laying behavior due to an impairment of host oocyte development (Girardie, 1977). The effect on female sexual behavior was not recorded, but interestingly, male sexual behavior is unaffected (Girardie and Granier, 1974). The decline in ovarian development appears to be due to a decline in activity in the corpora allata. Less radioactive cysteine was taken up by the corpora allata in parasitized animals than in unparasitized controls (Girardie and Girardie, 1977). Implantations of corpora allata from unparasitized females increased ovarian development in hosts infested with the tachinid fly larvae (Girardie and Girardie, 1977). Furthermore Girardie and Girardie (1977) found that the median neurosecretory cells showed less fuschin staining in parasitized animals than in controls, and concluded that these cells are also less active in parasitized hosts. They increased activity in the median neurosecretory cells by electrically stimulating the pars intercerebralis. This induced some of the parasitized *A. aegyptium* females to undergo ovarian development. Electrical stimulation had no effect on females that were heavily parasitized, however, and the effects on 'lightly' parasitized females depended on the time of year they were given brain stimulation (Girardie and Girardie, 1977).

The experiments of Girardie and Girardie (1977) suggest that the presence of the tachinid fly larvae depressed neural activity in the median neurosecretory cells, and that this depression led to a decrease in the hormonal output of the corpora allata. However, alternative explanations, such as a decline in receptor sensitivity in the corpora allata, would also explain the above results. To unequivocally demonstrate that the activity of the median neurosecretory cells declined, their activity levels need to be measured directly using electrophysiological techniques.

IV. Conclusions

The study of how the presence of parasites alters host behavior remains a promising and dynamic field. The interactions among the nervous, endocrine, and immune systems can be unravelled by examining host-parasite relationships using a combination of biochemical and electrophysiological techniques. We have a much better understanding of how insect venoms affect nervous systems (and behavior) because of the extensive use of combined electrophysiological and biochemical approaches to this problem (e.g. Fóuad et al., 1994; Tipton and Dajas, 1994). The little we do know about how parasites affect behavior suggests that further study will prove fruitful, and yield insights about the regulation of behavior in 'normal' insects.

V. Acknowledgments

I would like to thank Nancy Beckage, Darcy Reed, and two anonymous reviewers for critically reading earlier versions of this chapter. The previously unpublished research on the tobacco hornworm system cited in this review was supported by USDA NRI grant 92–37302–7470 to Nancy Beckage.

VI. References

Adamo, S. A., Linn, C. E., and Beckage, N. E. (1997). Correlation between changes in host behaviour and octopamine levels in the tobacco hornworm, *Manduca sexta,* parasitized by the gregarious braconid parasitoid wasp *Cotesia congregata. J. Exp. Biol.* **200:**117–127.

Adamo, S. A., Robert, D., and Hoy R. R. (1995). The effect of a tachinid parasitoid, *Ormia ochracea,* on the behaviour and reproduction of its host, the field cricket. *J. Insect Physiol.* **41:**269–277.

Alleyne, M., and Beckage, N. E. (1997). Parasite density-dependent effects on host growth and metabolic efficiency in tobacco hornworm larvae parasitized by *Cotesia congregata. J. Insect Physiol.* **43**(4):407–424.

Alleyne, M., Chappell, M. A., Gelman, D. B., and Beckage, N. E. (1997). Effects of parasitism by the braconid wasp *Cotesia congregata* on host metabolic rate and carbon dioxide emission in host larvae of the tobacco hornworm, *Manduca sexta. J. Insect Physiol.* **43**(2):143–154.

Bailey, W. J., and Ridsdill-Smith, J. (1991). *Reproductive behavior of insects.* New York: Chapman and Hall.

Baines, D., DeSantis, T., and Downer, R. G. H. (1992). Octopamine and 5-hydroxytryptamine enhance the phagocytic and nodule formations activities of the cockroach *Periplaneta americana)* haemocytes. *J. Insect Physiol.* **38:**905–914.

Beckage, N. E. (1993). Games parasites play: The dynamic roles of proteins and peptides in the relationship between parasite and host. In *Parasites and pathogens of insects* (N.E. Beckage, S. N. Thompson, and B. A. Federici, eds.), vol. 1, pp. 25–57. San Diego: Academic Press.

Beckage, N. E., and Riddiford, L. M. (1978). Developmental interactions between the tobacco hornworm *Manduca sexta* and its braconid parasite *Apanteles congregatus. Entomol. Exp. Appl.* **23:**139–151.

Beckage, N. E., and Templeton, T. J. (1986). Physiological effects of parasitism by *Apanteles congregatus* in terminal-stage tobacco hornworm larvae. *J. Insect Physiol.* **32:**299–314.

Brey, P. T. (1994). The impact of stress on insect immunity. *Bull. Inst. Pasteur* **92:**101–118.

Bronstein, S. M., and Conner, W. E. (1984). Endotoxin-induced behavioural fever in the Madagascar cockroach, *Gromphadorhina portentosa. J. Insect Physiol.* **30:**327–330.

Brown, J. J., and Reed-Larsen, D. (1991). Ecdysteroids and insect/host parasitoid interactions. *Biol. Control* **1:**136–143.

Cabanac, M. (1989). Fever in the leech, *Nephelopsis obscura* (Annelida). *J. Comp. Physiol.* **159:**281–285.

Cabanac, M., and Le Guelte, L. (1980). Temperature regulation and prostaglandin E_1 fever in scorpions. *J. Physiol.* **303:**365–370.

Camhi, J. M. (1984). *Neuroethology.* Sunderland, Mass: Sinauer.

Casterlin, M. E., and Reynolds, W. W. (1978). Prostaglandin E_1 fever in the crayfish *Cambarus bartoni. Pharmacol. Biochem. Behav.* **9:**593–595.

Dahlman, D. L. (1975). Trehalose and glucose levels in hemolymph of diet reared, tobacco leaf-reared and parasitized tobacco hornworm larvae. *Comp. Biochem. Physiol.* **50:**165–167.

Dahlman, D. L., and Vinson, S. B. (1993). Teratocytes: Developmental and biochemical characteristics. In *Parasites and pathogens of insects* (N. E. Beckage, S. N. Thompson, and B. A. Federici, eds.), vol. 1, pp. 145–166. San Diego: Academic Press.

Dunn, P. E., Bohnert, T. J., and Russell, V. (1994). Regulation of antibacterial protein synthesis following infection and during metamorphosis of *Manduca sexta. Ann. N. Y. Acad. Sci.* **712:**117–130.

Dunphy, G. B., and Downer, R. G. H. (1994). Octopamine, a modulator of the haemocytic nodulation response of non-immune *Galleria mellonella* larvae. *J. Insect Physiol.* **40:**267–272.

Dushay, M. S., and Beckage, N. E. (1993). Dose-dependent separation of *Cotesia congregata*–associated polydnavirus effects on *Manduca sexta* larval development and immunity. *J. Insect Physiol.* **39:**1029–1040.

Edgecomb, R. S., Harth, C. S., and Schneiderman, A. (1994). Regulation of feeding behavior in the adult *Drosophila melanogaster* varies with feeding regime and nutritional state. *J. Exp. Biol.* **197:**215–235.

Fouad, K., Libersat, F., and Rathmayer, W. (1994). The venom of the cockroach-hunting wasp *Ampulex compressa* changes motor thresholds: A novel tool for studying the neural control of arousal? *Zoology* **98:**23–34.

Führer, E., and Willers, D. (1986). The anal secretion of the endoparasitic larva *Pimpla turionellae:* Sites of production and effects. *J. Insect Physiol.* **32:**361–367.

Girardie, J. (1977). Côntrole neuroendocrine des protéines sanguines vitellogenes d'*Anacridium aegyptium* sain et parasite. *J. Insect Physiol.* **23:**569–577.

Girardie, J., and Girardie, A. (1977). Intervention des cellules neurosecretrices medianes dans la castration parasitaire d'*Anacridium aegyptium* (Orthoptère). *J. Insect Physiol.* **23:**461–467.

Girardie, J., and Granier, S. (1974). Rôle des corps allates d'*Anacridium aegyptium* (Insecte Orthoptère) infesté par *Metacemyia calloti* (Insecte Diptère). *Arch. Anat. Microsc.* **63:**269–280.

Hayakawa, Y. (1995). Growth-blocking Peptide: An insect biogenic peptide that prevents the onset of metamorphosis. *J. Insect Physiol.* **41:**1–6.

Hewes, R. S., and Truman, J. W. (1994). Steroid regulation of excitability in identified insect neurosecretory cells. *J. Neurosci* **14:**1812–1819.

Holmes, J. C., and Zohar, S. (1990). Pathology and host behaviour. In *Parasitism and host behaviour* (C. J. Bernard and J. M. Behnke, eds.), pp. 34–63. New York: Taylor and Francis.

Horton, D. R., and Moore, J. (1993). Behavioral effects of parasites and pathogens in insect hosts. In *Parasites and pathogens of insects* (N. E. Beckage, S. N. Thompson, and B. A. Federici, eds.), vol. 1, pp. 107–124. San Diego: Academic Press.

Huber, F., Moore, T. E., and Loher, W. (1989). *Cricket behavior and neurobiology.* Ithaca: Cornell Univ. Press.

Hurd, H. (1990a). Physiological and behavioural interactions between parasites and invertebrate hosts. *Adv. Parasitol.* **29:**271–318.

Hurd, H. (1990b). Parasite induced modulation of insect reproduction. *Adv. Invertebr. Reprod.* **5:**163–168.

Hurd, H. (1993). Reproductive disturbances induced by parasites and pathogens of insects. In *Parasites and pathogens of insects* (N. E. Beckage, S. N. Thompson, and B. A. Federici, eds.), vol. 1, pp. 87–106. San Diego: Academic Press.

Ismail, S. M., and Matsumura, F. (1992). Studies on the biochemical mechanisms of anorexia caused by formamidine pesticides in the tobacco hornworm *Manduca sexta. Insect Biochem. Mol. Biol.* **22:**713–720.

Jones, D. (1985). Endocrine interaction between host (Lepidoptera) and parasite (Cheloninae: Hymenoptera): Is the host or the parasite in control? *Ann. Entomol. Soc. Am.* **78:**141–148.

Jones, G., Hanzlik, T., Hammock, B. D., Schooley, D. A., Miller, C. A., Tsai, L. W., and Baker, F. L. (1990). The juvenile hormone titer during the penultimate and ultimate larval stadia of *Trichoplusia ni*. *J. Insect Physiol.* **36:**77–83.

Kaya, H. K., and Tanada, Y. (1969). Responses to high temperature of the parasite *Apanteles militaris* and of its host, the armyworm *Pseudaletia unipuncta*. *Ann. Entomol. Soc. Am.* **62:**1303–1306.

Lawrence, P. O. (1986). Host-parasite hormonal interactions: An overview. *J. Insect Physiol.* **32:**295–298.

Lawrence, P. O., and Lanzrein, B. (1993). Hormonal interactions between insect endoparasitoids and their host insects. In *Parasites and pathogens of insects* (N. E. Beckage, S. N. Thompson, and B. A. Federici, eds.), vol. 1, pp. 59–86. San Diego: Academic Press.

Louis, C., Jourdan, M., and Cabanac, M. (1986). Behavioral fever and therapy in a rickettsia-infected Orthoptera. *Am. J. Physiol.* **250** (Regulatory Integrative and Comp. Physiol. 19):R991–R995.

Molyneux, D. H., and Jefferies, D. (1986). Feeding behaviour of pathogen infected vectors. *Parasitology* **92:**721–736.

Moore, J. (1993). Parasites and the behavior of biting flies. *J. Parasitol.* **79:**1–16.

Moore, J. (1995). The behavior of parasitized animals. *Bioscience* **45:**89–96.

Moore, J., and Gotelli, N. J. (1990). Phylogenetic perspective on the evolution of altered host behaviours: a critical look at the manipulation hypothesis. In *Parasitism and host behaviour* (C. J. Bernard and J. M. Behnke, eds.), pp. 193–233. New York: Taylor and Francis.

Müller, C. B., and Schmid-Hempel, P. (1993). Exploitation of cold temperature as defense against parasitoids in bumblebees. *Nature* **363:**65–67.

Nijhout, H. F. (1994). *Insect hormones.* Princeton, N.J.: Princeton Univ. Press.

Noguchi, H., Hayakawa, Y., and Downer, R. G. H. (1995). Elevation of dopamine levels in parasitized insect larvae. *Insect Biochem. Mol. Biol.* **25:**197–201.

Orchard, I., Carlisle, J. A., Loughton, B. G., Gole, J. W. D., Downer, R. G. H. (1982). In vitro studies on the effects of octopamine on locust fat body. *Gen. Comp. Endocrinol.* **48:**7–13.

Ratcliffe, N. A. (1993). Cellular defense responses of insects: Unresolved problems. In *Parasites and pathogens of insects* (N. E. Beckage, S. N. Thompson, and B. A. Federici, eds.), vol. 1, pp. 267–304. San Diego: Academic Press.

Romig, T., Luciaus, R., and Frank, W. (1980). Cerebral larvae in the second intermediate host of *Dicrocoelium dendriticum* (Rudolphi, 1819) and *Dicrocoelium hospes* Looss, 1907 (Trematodes, Dicrocoeliidae). *Z. Parasitenkd.* **63:**277–286.

Rossignol, P. A., and Rossignol, A. M. (1988). Simulations of enhanced malarial transmission and host bias induced by modified vector blood location behaviour. *Parasitology* **97:**363–372.

Rowland, M. W. and Lindsay, S. W. (1986). The circadian flight activity of *Aedes aegypti* parasitized with filarial nematode *Brugia pahangi*. *Physiol. Entomol.* **11:**325–334.

Shimizu, T., and Takeda, N. (1994). Aromatic amino acids and amine levels in the hemolymph of parasitized and unparasitized larvae of *Mythimna separata. Z. Naturforsch.* **49c:**693–695.

Simpson, S. J., and Bernays, E. A. (1983). The regulation of feeding: Locusts and blowflies are not so different from mammals. *Appetite* **4:**313–346.

Stanley-Samuelson, D. W. (1994). Prostaglandins and related eicosanoids in insects. *Adv. Insect Physiol.* **24:**115–212.

Stanley-Samuelson, D. W., Jurenka, R. A., Blomquist, G. J., and Loher, W. (1987). Sexual transfer of prostaglandin precursor in the field cricket *Telleogryllus commodus. Physiol. Entomol.* **12:**347–354.

Strambi, A., and Girardie, A. (1973). Effect de l'implanation de corpora allata actifs de *Locusta migratoria* (Orthoptère) dans des femelles *Polistes gallicus* L. (Hyménoptère) saines et parasitées par *Xenos vesparum* Rossi. *C. R. Acad. Sci.* (Paris) **276:**3319–3322.

Strambi, C., Strambi, A., and Augier, R. (1982). Protein level in the hemolymph of the wasp *Polistes gallicus* L. at the beginning of imaginal life and during overwintering. Action of strepsiterian parasite *Xenos vesparum* Rossi. *Experientia* **38:**1189–1191.

Strand, M. R., McKenzie, D. I., Grassl, V., Dover, B. A., and Aiken, J. M. (1992). Persistence and expression of *Microplitis demolitor* polydnavirus in *Pseudoplusia includens. J. Gen. Virol.* **73:**1627–1635.

Thompson, S. N. (1990). Physiological alterations during parasitism and their effects on host behaviour. In *Parasitism and host behaviour* (C. J. Bernard and J. M. Behnke, eds.), pp. 64–94. New York: Taylor and Francis.

Thompson, S. N. (1993). Redirection of host metabolism and effects on parasite nutrition. In *Parasites and pathogens of insects* (N. E. Beckage, S. N. Thompson, and B. A. Federici, eds.), vol. 1, pp. 125–144. San Diego: Academic Press.

Thompson, S. N., and Kavaliers, M. (1994). Physiological bases for parasite-induced alterations of host behaviour. *Parasitology* **109:**S119–S138.

Timmins, W. A., and Reynolds, R. E. (1992). Physiological mechanisms underlying the control of meal size in *Manduca sexta* larvae. *Physiol. Entomol.* **17:**81–89.

Tipton, K. F., and Dajas, R. (1994). *Neurotoxins in neurobiology: Their actions and applications.* New York: Harwood.

Toolson, E. C., Ashby, P. D., Howard, R. W., and Stanley-Samuelson, D. W. (1994). Eicosanoids mediate control of thermoregulatory seating in the cicada, *Tibicen dealbatus* (Insecta: Homoptera). *J. Comp. Physiol. B* **164:**278–285.

Truman, J. W., Talbot, W. S., Fahrbach, S. E., and Hogress, D. S. (1994). Ecdysone receptor expression in the CNS correlates with stage-specific responses to ecdysteroids during *Drosophila* and *Manduca* development. *Development* **120:**219–234.

Vinson, S. B. (1990). How parasitoids deal with the immune system of their host: An overview. *Arch. Insect Biochem. Physiol.* **13:**3–27.

Wigglesworth, V. B. (1972). *The principles of insect physiology.* 7th ed., London: Chapman and Hall.

Zitnan, D., Kingan, T., and Beckage, N. E. (1995). Parasitism-induced accumulation of FMRF-amide like peptides in the gut innervation and endocrine cells of *Manduca sexta. Insect Biochem. Mol. Biol.* **25:**669–678.

Zitnan, D., Kingan, T., Kramer, S. J., and Beckage, N. E. (1995). Accumulation of neuropeptides in the cerebral neurosecretory system of *Manduca sexta* larvae parasitized by the braconid wasp *Cotesia congregata. J. Comp. Neurol.* **356:**83–100.

13

Parasites, Fluctuating Asymmetry, and Sexual Selection

Michal Polak

I. Abstract

Fluctuating asymmetry is defined as the small, random deviations from perfect symmetry in a bilateral-paired trait. It arises from environmental and genetic stress occurring *during ontogeny*, and so, cannot be compensated readily in the adult form. Fluctuating asymmetry therefore can serve as a reliable indicator of individual phenotypic quality. Indeed, although there are exceptions, fluctuating asymmetry often is found to be correlated negatively with adult fitness components. In this chapter, the role of fluctuating asymmetry in sexual selection is reviewed. Of 23 animal taxa studied in this context, 65% exhibit a statistically significant positive association between morphological symmetry and estimated male mating success. Evidence suggests that female mate choice favors symmetrical males in barn swallows (*Hirundo rustica*), earwigs (*Forficula auricularia*), humans (*Homo sapiens*), scorpionflies (*Panorpa japonica)*, and zebra finches (*Taeniopygia guttata*). But there is no consensus regarding why females prefer symmetrical males, nor about the evolutionary consequences of fluctuating-asymmetry-based sexual selection. It is proposed that parasites may provide a key to understanding these issues, because: (1) parasites are ubiquitous; (2) parasitism often is correlated positively with fluctuating asymmetry, either because it elevates fluctuating asymmetry, or because high fluctuating asymmetry hosts are immunologically deficient; and (3) parasite resistance is heritable in many host taxa. Thus, female choice in favor of low fluctuating asymmetry may reflect a preference for parasite-free males, because females potentially will be rewarded with (1) reduced probability of transmitting parasites to themselves and to offspring, (2) improved paternal care, and (3) transmission of resistance genes to offspring. Evidence showing that costs of resistance against parasites can be severe physiologically is also reviewed. It is shown that when resistant individuals incur such costs during ontogeny, they may exhibit greater fluctuating asymmetry as adults relative to their parasitized counterparts. It is suggested that expression of this pattern of fluctuating asymmetry variation across host types depends on the mode of parasite transmission, and this may explain why fluctuating asymmetry is actually *positively* correlated with fitness in certain host species.

II. Introduction

Parasites as diverse as protozoa and ectoparasitic arthropods reduce male mating success in numerous host taxa (Clayton, 1991), presumably by extracting nutrients for their own growth and reproduction that would otherwise remain available to the host. The metabolic costs of parasitism may therefore damage the development of male ornamental traits used to attract receptive females (Hamilton and Zuk, 1982). This effect, for example, has been demonstrated in guppies (*Poecilia reticulata*), sticklebacks (*Gasterosteus aculeatus*), jungle fowl (*Gallus gallus*) and pheasants (*Phasianus colchius*); in these species, females favor unparasitized males as mates because of their brighter body coloration and more active courtship behavior (Houde and Torio, 1992; Milinski and Bakker, 1990; Hillgarth, 1990; Zuk et al., 1990). If parasites have sufficiently large negative effects on physiological vigor, they may also influence the outcome of male-male competitive interactions (Howard and Minchella, 1990). This effect could disrupt dominance relations among males (Freeland, 1981; Rau, 1983), and reduce parasitized males' ability to gain access to mates, territories and nuptial food items (Howard and Minchella, 1991; Mulvey and Abo, 1993; Polak and Markow, 1995).

In theory, parasites can also generate sexual selection by increasing fluctuating asymmetry (FA) in infected hosts, thereby reducing individual attractiveness and competitive ability. Fluctuating asymmetry is a feature of bilaterally symmetrical traits that reflects minor, random, deviations from perfect symmetry, and its expression depends on the magnitude and kind of stress experienced during development (Ludwig, 1932; Van Valen, 1962). Several studies have uncovered negative correlations between FA and mating success, but others have not, so that a critical evaluation of the available data is required. Moreover, there is little theoretical consensus over the origin and maintenance of mate choice for low FA, nor over the evolutionary consequences of FA-based sexual selection. Here the role of parasitism is considered because it may help to resolve these questions. My aims in this chapter are to (1) review the evidence for the role of FA in sexual selection, (2) examine the ways in which parasitism could elevate FA, and (3) emphasize the role of costs of resistance against parasites, which are seldom considered in the context of parasite-mediated sexual selection. The role of costs is considered because when costs accrue during ontogeny, they may, under conditions discussed below, lead to greater FA in resistant than in infested hosts. Thus, in order to make appropriate predictions regarding the role of an interaction between parasites and FA in sexual selection, precise knowledge of the natural history of the parasite-host association will be required.

III. Fluctuating Asymmetry and Phenotypic Quality

Central to the concept of FA is the assumption that the two components of a bilaterally symmetrical trait are the product of an identical set of genes (Van Valen, 1962). The minor fluctuations from perfect symmetry can therefore be viewed as a measure of the degree of disruption to the genetically determined, normal flow of trait development (Waddington, 1957). Disruption occurs by way of action of a force, such as an infection, teratogen, physical trauma, or mutation, that during ontogeny causes unequal growth between sides, and hence, asymmetry in the adult form (Zakharov, 1992). Of course, even after birth, traits that undergo indeterminate growth, as well as those subject to seasonal change, will likewise tend to express asymmetry under stress.

Fluctuating asymmetry is a population level phenomenon. It can be distinguished from other forms of asymmetry, namely antisymmetry and directional asymmetry, by the form of the distribution of right-minus-left values. A normal distribution of these values with a mean of zero indicates FA, whereas the variance of the distribution reflects its magnitude (Van Valen, 1962; Palmer and Strobeck, 1986). Environmental factors that elevate FA include exposure to chemical pollutants (Valentine and Soulé, 1973; Pankakoski et al., 1992; Graham et al., 1993), radiation (Møller, 1993a), and high temperatures (Parsons, 1962; Mooney et al., 1985; Clarke and McKenzie, 1992). Nutrient deprivation and audiogenic stress in pregnant mice (Sciulli et al., 1979; Siegel and Smookler, 1973), maternal alcohol intake in humans (Wilber et al., 1993), and psychological stress in pregnant pigtailed macaques, *Macaca nemestrina* (Newell-Morris et al., 1989), are associated with elevated FA in progeny.

Genetic correlates of FA are also well documented. In populations of rainbow trout (*Oncorhyncus mykiss*) and cutthroat trout (*O. clarki*), the most homozygous individuals express the greatest FA (Leary et al., 1983, 1984). Fluctuating asymmetry also increases with loss of genic coadaptation resulting from strong directional selection (Thoday, 1958; Clarke and McKenzie, 1987) and hybridization (Tanaka, 1982; Graham and Felley, 1985; Markow and Ricker, 1991). Inbreeding, which unmasks recessive alleles having negative effects on fitness, is associated with elevated FA in, for example, humans (Markow and Martin, 1993) and *Drosophila* (Mather, 1953; Biémont 1983). Some authors view these damaging recessives as having been dispersed and maintained by selection favoring disease resistance (e.g., Motulsky, 1960; Harlan, 1976). Genes influencing the expression of schizophrenia, for example, may have been selected during disease pandemics that historically invaded human populations (Motulsky, 1960; Harlan, 1976). Schizophrenia has a complex etiology, and is associated with elevated FA and reduced fitness (Slater et al., 1971; Markow and Wandler, 1986; Markow, 1992; Mellor, 1992).

Finally, chromosomal aberrations and gene mutations can also have sufficiently severe metabolic effects to disrupt development and elevate FA (Shapiro, 1983;

Clarke and McKenzie, 1987). In humans, for example, evidence suggests that changes in gene product dosage and subsequent biochemical upset may be the cause of elevated FA among Down's syndrome (trisomy 21) patients (Shapiro, 1983). Moreover, human gene defects that cause diseases such as phenylketonuria (PKU) and thalassemia (Cummings, 1988), as well as those induced by retrotransposable elements that cause, for example, haemophilia A and Duchenne muscular dystrophy (Kazazian et al., 1988; Narita et al., 1993), also are expected to elevate FA because of their severe metabolic and physiological consequences.

FA is often found to be negatively correlated with fitness components, examples of which have already been mentioned. In forest tent caterpillars (*Malacosoma disstria*) (Naugler and Leech, 1994), house flies (*Musca domestica*) (Møller, 1996), Japanese scorpionflies (*Panorpa japonica*) (Thornhill, 1992a), and male lions (*Panthera leo*) (Packer and Pusey, 1993), FA is also negatively correlated with survivorship. In human females, breast asymmetry is negatively correlated with fertility (Møller et al., 1995). Finally, in the plant *Epilobium angustifolium*, floral FA is correlated negatively with nectar production, number of visits by bumblebees, and probably pollen dispersal as well (Møller, 1995).

IV. Fluctuating Asymmetry and Sexual Selection

Researchers in sexual selection are currently engaged in the study of FA for two reasons. FA is often negatively correlated with phenotypic quality, defined as an individual's expected ability to express high fitness. Moreover, FA in adult traits cannot readily be compensated, so that FA is expected to be an honest indicator of phenotypic quality.

A. FA and Secondary Sexual Characters

Anders Pape Møller and coworkers have championed the idea that patterns of variation in FA in secondary sexual traits may help identify the mechanism by which these traits have evolved (Møller and Höglund, 1991; Møller, 1993c; Møller and Pomiankowski, 1993a, b). They have argued that trait expression will be unrelated to the genetic quality of its bearer when that trait does not exhibit a negative correlation between its degree of expression and FA. Such a trait, therefore, most likely evolved via the Fisherian process of sexual selection (Fisher, 1930; O'Donald, 1983). In contrast, traits exhibiting negative covariance between trait size and FA are viewed has having evolved via the "good genes" process, because in such cases, the nature of the relationship suggests that trait expression is condition dependent. According to the "good genes" scenario, females prefer the most exaggerated traits, and hence, drive trait evolution, because the preference is rewarded with viability genes passed to offspring. Negative correlations between trait size and FA occur in certain feather ornaments and leg spurs in birds (Møller 1990a, 1992a, 1994; Manning and Hartley, 1991;

Møller and Höglund, 1991; Sullivan et al., 1993), canines in gorillas (Manning and Chamberlain, 1994), horns in beetles (Møller, 1992a), abdominal cerci in earwigs (Radesäter and Halldórsdóttir, 1993), and petals in fireweed, *E. angustifolium* (Møller and Eriksson, 1994).

But some authors have questioned the interpretation of these negative correlations on statistical grounds (Evans and Hatchwell, 1993; Sullivan et al., 1993), whereas others have found that apparently condition-dependent traits do not always exhibit the predicted negative correlation (Balmford, Jones, and Thomas, 1993; Møller et al., 1995; Tomkins and Simmons, 1995). For example, Solberg and Sæther (1993) have reported significant positive correlations between antler size and FA in moose (*Alces alces*) despite the possibility that antler size reflects male nutritional status as well as genetic quality (Andersson, 1982; Clutton-Brock, 1982; Kitchener, 1987).

Finally, Balmford, Jones, and Thomas (1993) and Evans and Hatchwell (1993) have proposed an alternative explanation for the negative correlations between trait size and FA. They build upon the fact that aerodynamic and balance disorders that result from structural asymmetry can result in substantive energetic costs. According to their view, such costs increase disproportionately with trait size (see Fig. 1 in Balmford, Thomas, and Jones, 1993), and so, natural selection optimizes both size and symmetry simultaneously.

B. FA and Mating Success

When FA is an honest indicator of phenotypic quality, symmetry and mating success may be positively associated for several reasons:

(1) Developmentally stable, and therefore symmetrical, phenotypes may be healthier and more vigorous than asymmetrical phenotypes, an effect that could lead to an incidental association between symmetry and mating success. This effect may help explain a positive association between mating success and symmetry in traits unlikely to be perceived visually or otherwise during courstship and mating. In these cases, *symmetry per se is not the target of sexual selection.*

(2) Alternatively, degree of symmetry *per se* in secondary sexual characters (and associated structures) used in combat could confer a fighting advantage against sexual rivals (Møller, 1992a). In deer, for example, asymmetry in the structure and strength of antlers and of muscles and bones of the neck, shoulders, and legs, could influence the outcome of wrestling matches between stags when they lock antlers, twist from side to side, and try to knock each other off balance.

(3) Finally, assessment of symmetry *per se* in morphological (e.g., feathers and antlers, see Malyon and Healy, 1994) or behavioral traits (e.g., chest pounding in gorillas and chimpanzees, see Schaller, 1963) in potential mates and fighting opponents could be adaptive to the extent that symmetry is indicative, for example, of parasite load, parenting ability, genetic makeup, or physical strength and vigor (Møller and Pomiankowski, 1993a; Polak and Trivers, 1994; Watson and

Thornhill, 1994). It is noteworthy in this context of assessment that many secondary sexual traits exhibit, in males, up to ten times the observable FA relative to the same traits expressed in females, and nonsexually selected traits in males (Møller and Höglund, 1991; Møller, 1992a; cf. Balmford, Jones, and Thomas, 1993; Manning and Chamberlain, 1993; Tomkins and Simmons, 1995). Although the reasons for this pattern are unknown, it may be that the production and maintenance of costly secondary sexual traits require the allocation of limiting resources at the expense of homeostatic mechanisms (Møller, 1993b). Nonetheless, elevated FA in costly secondary sexual traits is expected to magnify differences in developmental stability between individuals, reduce costs (e.g., in terms of time investment) associated with evaluating the phenotypic quality of others, and thereby favor the evolution and maintenance of adaptive assessment of phenotypic quality via FA. On the other hand, nonadaptive explanations for mate choice favoring low FA also seem feasible (Enquist and Arak, 1994; Jennions and Oakes, 1994). For example, Johnstone (1994) has demonstrated using artificial neural networks that a preference for symmetry could originate incidentally from selection for mate recognition.

The hypothesis that variation in FA influences mating success has been evaluated in at least 23 animal taxa (Table 13-1). In 15 (65%) of these, statistically significant positive associations have been uncovered between symmetry and some estimate of mating success. In 6 (26%) species, there is no significant association. In 2 (9%) others, namely paradise whydahs and *Drosophila simulans*, sexual selection appears to favor asymmetric phenotypes. In the study of the whydah, females exhibited a preference for males whose tail feathers were experimentally rendered asymmetrical (Oakes and Barnard, 1994). However, because the manipulation also resulted in greater absolute tail length (see fig.1 in Oakes and Barnard, 1994), it is unclear whether perceived tail length, asymmetry, or both, elicited the female preference (Brookes and Pomiankowski, 1994). In the *Drosophila* study, mean asymmetry in mated males was significantly greater than in unmated males (Markow and Ricker, 1992), but numerous pairwise statistical contrasts were conducted in the *Drosophila* study without adjusting the experiment-wide error rate. This renders rejection of the null hypothesis questionable. Thus, there is no definitive evidence supporting a negative association between symmetry and sexual success.

Evidence for a positive association between symmetry and mating success was first reported by Markow (1987) in *D. melanogaster*: mated males exhibited significantly lower bristle FA relative to unmated flies. The cause of this association, however, was not elucidated. In fact, of all subsequent studies reporting a positive association between symmetry and mating success, the mechanism of sexual selection has been elucidated in six species (Table 13-1), although in earwigs and dungflies, circumstantial evidence implicates inter-male competition. Moreover, Radesäter and Halldórsdóttir (1993) have shown that male earwigs sporting symmetrical cerci (a secondary sexual character in this species) enjoy

Table 13-1. Within-Species Relationships between Morphological Symmetry and Sexual Selection

Animal	Symmetry and sexual success, character examined	Mechanism	Heritability estimate h^2 (FA) ± SE	Source
Invertebrates				
Damselfly (*Coenagrion puella*)	+, wing length	U	U	Harvey & Walsh, 1993
Damselfly (*Ischnura denticollis*)	+, backwing length	U	U	Córdoba-Aguilar, 1995
	+, forewing length	U	U	
	+, tibia length	U	U	
Decorated cricket (*Gryllodes sigillatus*)	0, file tooth number*		U	Eggert & Sakaluk, 1994
	0, forewing vein length		-0.022 ± 0.21 (1)	
Dung fly (*Scathophaga stercoraria*)	+, wing length	U	U	Liggett et al., 1993
	+, tibial length	U	U	
Drosophila melanogaster	+, thoracic bristle number	U	U	Markow, 1987;
	0, wing length		U	Markow & Sawka, 1992
D. mojavensis	0, wing length		U	Markow & Ricker, 1992
	0, thoracic bristle number		U	
D. pseudoobscura	+, wing length	U	U	Markow & Ricker, 1992
	0, aristal branch number		U	
D. simulans	–, aristal branch number	U	U	Markow & Ricker, 1992
	–, wing length	U	U	
	0, head bristle number		U	
	0, thoracic bristle number		U	
D. nigrospiracula	0, thoracic bristle number		U	Polak, 1996b
	0, thoracic bristle position		U	
Earwig (*Forficula auricularia*)	+, cercus length*	U	U	Radesäter & Halldórsdóttir, 1993
Field cricket (*Gryllus campestris*)	+, tibial length	U	U	Simmons, 1995

House fly (*Musca domestica*)	+, wing length	U	U	Møller, 1996
	+, tibial length	U	U	
Ladybird beetle (*Harmonia axyridis*)	0, hindwing length		U	Ueno, 1994
Midge (*Chironomus plumosus*)	+, wing length	U	U	McLachlan & Cant, 1995
Scorpionfly (*Panorpa japonica*)	+, forewing length	MCIC	U	Thornhill 1991, 1992a
Scorpionfly (*P. vulgaris*)	+, forewing length	IC	1.07 ± 0.44 (f + 1)	Thornhill & Sauer, 1992
	+, tarsal segment length	IC	U	
Vertebrates				
Cricket frog (*Acris crepitans*)	0, head, tibial, laryngeal*, and aural characters		U	Ryan et al., 1995
Barn swallow (*Hirundo rustica*)	+, tail feather length*	MC	0.80 ± 0.33 (f)	Møller, 1992b, 1993d, 1994
Guppy (*Poecilia reticulata*)	0, body coloration*		U	Nordell, 1995
Human (*Homo sapiens*)	+, facial morphology*	MC	U	Grammer & Thornhill, 1994;
	+, body morphology	U	U	Thornhill & Gangestad, 1994; Thornhill et al., 1995
Oribus (*Ourebia ourebia*)	+, horn shape*	U	U	Arcese, 1994
Paradise whydah (*Vidua paradisaea*)	–, tail feather length*	MC	U	Oakes & Barnard, 1994
Zebra finch (*Taeniopygia guttata*)	+, colored leg bands	MC	U	Swaddle & Cuthill, 1994a, b
	+, breast feathers*	MC	U	

*: male secondary sexual trait
I.C.: Intra-sexual competition
M.C.: Mate choice
U: Undetermined
f: Data for heritability estimate derived from field
1: Data for heritability estimate derived from laboratory

+: positive relation between symmetry and sexual success
−: negative relation
0: no relation

shorter latency times prior to copulation with virgin females, a result which suggests female choice for symmetrical males.

In Japanese scorpionflies, Thornhill (1991) demonstrated female choice for the pheromone of males with relatively low forewing FA. Subsequent manipulation of forewing FA did not alter the female preference for the pheromone, indicating that intrinsic differences among males govern the attractiveness of pheromone produced (Thornhill, 1991). Female choice in scorpionflies is also based on the quantity of food controlled by males, so it is noteworthy that symmetry and competitive ability are also positively correlated in males (Thornhill 1992a, b; Thornhill and Sauer, 1992).

In barn swallows, males whose tail feathers were experimentally rendered symmetrical mated earlier in the breeding season and enjoyed greater annual reproductive success (RS) (Møller, 1992b). Though it is possible that the asymmetric treatment influenced the outcome of male-male competitive interactions (Balmford and Thomas, 1992), the fact that frequency of fights was unaffected by the treatment weakens this hypothesis (Møller, 1992c). In fact, Møller (1992b, c) proposed that females respond directly to symmetry in male tail feathers, and discriminate against asymmetrical phenotypes. But the tail alteration may have interfered with the aerodynamic efficiency of male birds and reduced their courtship performance (Møller, 1991a; Evans et al., 1994). In a subsequent experiment, Møller (1993d) manipulated tail symmetry using white and black correction fluid, a procedure that should have minimized any disruptive effect on aerodynamic performance. Here again, apparent symmetry conferred highly significant increments in male RS (Møller, 1993d). Swaddle and Cuthill (1994a), however, have suggested that the application of white correction fluid resulted in abnormal appearance of male tail plumage (by creating a flashing lights effect during flight), which could have reduced male attractiveness independently of perceived asymmetry.

Another example of female choice for symmetry *per se* involves zebra finches. Females in aviaries preferred males adorned with symmetrically oriented colored leg bands (Swaddle and Cuthill, 1994b), and fledged more young when mated to symmetrical males (Swaddle, 1996). Female zebra finches likewise responded to symmetry in the width of dark bands on male breast feathers, which constitute a secondary sexual trait in this species (Swaddle and Cuthill, 1994a). These studies are intriguing because they demonstrate a response favoring symmetrical patterns in general, not necessarily restricted to an anatomical trait. But experiments were conducted on domesticated birds, so perhaps artificial selection for some aspect of life in captivity caused this preference for symmetry, which may not occur in wild zebra finch populations. Finally in humans, Grammer and Thornhill (1994) have shown that faces with symmetrical features are more attractive to members of the opposite sex (but see Swaddle and Cuthill, 1995), although a simultaneous preference for averageness may have confounded the interpretation of these results.

In summary, no generalizations regarding an association between FA and mating success are yet possible. We do not have sufficient data to know why this effect is so species specific. In approximately one-third of the animal species examined, there exists no association between degree of symmetry and individual mating success (Table 13-1). Only a weak majority of species supports the original prediction; namely, that symmetry and mating success will be positively associated. Among the latter species, strong evidence for a mechanism comes from experiments with barn swallows and zebra finches which have documented female choice in favor of symmetry.

But there is currently little consensus regarding the origin, maintenance, and evolutionary consequences of a female preference for low FA. Indeed, there are two very different alternative explanations for the maintenance of such a preference. The null model considers the preference a mere manifestation of a preexisting sensory bias favoring symmetry, which arose in the context, for example, of foraging, predator avoidance, or of general pattern recognition (Møller, 1993c; Enquist and Arak, 1994; Johnstone, 1994, Ryan et al., 1995). This model predicts that no tangible benefits will accrue to females from their preference for symmetry in their mate (but see Enquist and Arak, 1994, p. 172). In contrast, the theory of adaptive mate choice predicts that the preference will yield fitness increments such as improved survivorship and paternal care (direct benefits), or "good genes" passed to offspring (indirect benefits) (Kirkpatrick and Ryan, 1991; Borgia and Wilkinson, 1992; Møller, 1992b; Thornhill, 1991; Thornhill and Sauer, 1992). Discriminating between these hypotheses will require careful determination of the causes of FA variation within populations in which females exhibit a preference for symmetry, and testing for fitness gains to females mated to males that differ in their degree of symmetry.

Below I demonstrate that FA and parasitism are positively correlated in numerous taxa, indicating that a female preference for symmetry could be explained by models of adaptive mate choice, because low FA in males may be indicative of their low parasite burdens, as well as their superior immunocompetence. Moreover, when FA and parasitism are positively correlated, and given appropriate genetic variability for resistance, FA-based sexual selection could drive the evolution of resistance within host populations.

V. Parasitism and Fluctuating Asymmetry

During the course of infection, parasites compete with their host for available nutrients (e.g., Thompson, 1983). Although nutritionally specialized parasites can deplete the host of specific dietary constituents (Gordon and Webster, 1971; Rutherford and Webster, 1978), others can limit host nutrient availability by reducing host food intake, digestion, absorption, and nutrient assimilation (Thompson, 1983; Whitefield, 1979). In humans, for example, *Plasmodium* infections lead to significant depletions of amino acids, fats, and sugars, due to malabsorption as well as to competition between parasite and host specifically for circulating glucose (Karney and Tong, 1972). It is no surprise, therefore, that parasitic infections commonly impair host metabolism, growth rate and development (Sykes and Coop, 1976, 1977; Thompson, 1983; Schall et al., 1982; Goater et al., 1993). Parasite-induced nutritional deprivation therefore is predicted to destabilize host development and elevate host FA. It is noteworthy that experimentally increased nutritional stress elevates FA in both rats (Sciulli et al., 1979) and European starlings (*Sturnus vulgaris*) (Swaddle and Witter, 1994).

Alternatively, different pathological effects (e.g., cell and tissue damage) on one side of the body versus the other, could lead to FA (Watson and Thornhill, 1994). For example, congested blood vessels with *Plasmodium*-infected erythrocytes that occlude microvascular circulation (Karney and Tong, 1972) could elevate asymmetry, if one member of a bilateral trait was deprived more severely of oxygen during ontogeny than the other. Moreover, mutations and cell membrane damage in embryonic cells could affect all cellular "progeny" in a side-specific cell lineage (Stern, 1968; Ewald, 1994). The timing of such an effect, of course, would determine the proportion of cells affected, and hence, degree of asymmetry.

Therefore, parasitism may cause elevated host FA, but for different reasons. Yet a correlation between parasitism and FA can be an incidental result of other mechanisms, such as inbreeding. Distinguishing between these possibilities has important implications for the genetic and evolutionary consequences of sexual selection (see below), so I review the evidence for a causal link between parasitism and FA with this caveat in mind.

A. Causal Linkage via Maternal Effects

Livshits et al. (1988) found that infection in mothers during pregnancy explained 1.5 percent of the variation in FA of Israeli newborn infants. However, identities of the infectious agents were not reported, and their association with infant FA was marginally nonsignificant ($P = 0.06$). This study, therefore, can only be considered suggestive of maternal effects.

An experimental study examined the effects of both maternal age and ectoparasitism on offspring FA in *Drosophila nigrospiracula* (Polak, 1996b). The parasite was *Macrocheles subbadius* (Macrochelidae), an ectoparasitic mite that consumes fly hemolymph, and causes significant reductions in host fecundity, survivorship, and mating success in nature (Polak and Markow, 1995; Polak, 1996). Maternal parasite load had a strong significant effect on bristle FA in offspring, but maternal age had no effect. Moreover, the interaction between parasitism and age was not significant (Polak, 1997).

B. Causal Linkage via Effects on Immediate Host

Two correlational studies, one involving *D. nigrospiracula* and the other involving western fence lizards (*Sceloporus occidentalis*), suggest that parasitism can cause FA in some cases. *D. nigrospiracula* found infected in nature with allantonematid nematodes had higher FA than either uninfected flies or flies burdened with mites (Polak, 1993). These results were predicted because nematodes infect fly larvae and are therefore more likely to disrupt development than mites which infest adult flies. In western fence lizards, FA in belly coloration was greater in males infected with *Plasmodium mexicanum*, than in uninfected individuals (Schall, 1996). Belly coloration changes with age after hatching, so its expression could have been affected by malaria even in maturing males.

However, the correlational nature of both the above studies creates interpretational difficulties. For example, association between high FA and nematode infection in *D. nigrospiracula* could have come about simply because larvae developing in a poorer milieu that might have caused FA, were also more susceptible to infection by nematodes.

In this light, experimental studies would be more convincing. Indeed, the hypothesis that parasites cause FA has been tested experimentally in barn swallows (*H. rustica*), reindeer (*Rangifer tarandus*), and *Drosophila*, but with mixed results. Møller (1992d) studied the effect of infestation by the hematophagous fowl mite, *Ornithonyssus bursa,* on tail asymmetry in adult barn swallows. Mite loads were either increased, decreased (by fumigation) or left unmanipulated in nests of adult birds at the onset of their breeding season. Following the annual molt of males, FA in their elongated, outermost, tail feathers (a secondary sex trait) was contrasted with that in the same males before the mite treatment. Fluctuating asymmetry was found to be significantly greater in males whose nests were infested with mites than in those whose nests were fumigated or unmanipulated. However, because mite load on barn swallows was not measured directly, we cannot be certain that the manipulation actually produced high mite infestation and corresponding variation in adult FA. It is true that mite load on birds upon arrival to their nesting grounds was correlated with mite numbers in nests later in the breeding season (Møller, 1990b). But swallows are effective groomers, able to significantly reduce mite numbers from their bodies (Møller, 1991b), so the assumption that high rates of infestation in nests increases mite loads on adults remains to be verified.

In the reindeer study, Folstad et. al. (1996) compared degree of antler length FA among female reindeer whose parasite (gastrointestinal nematodes and oestrid larvae) loads were experimentally reduced using antihelminthic treatment and control females administered a placebo. Despite small sample sizes ($N = 5$ treated; $N = 8$ untreated), antler FA in treated females was significantly less than that among controls approximately six months following treatment.

In the *Drosophila* study, each of four species (*D. neotestacea*, *D. falleni*, *D. putrida*, and *D. quinaria*) was experimentally infected in the laboratory with an allantonematid nematode, *Howardula aoronymphium* (Polak and Jaenike, unpublished). During the course of infection, female nematodes penetrate the cuticle of their larval host, feed by absorbing nutrients directly through their body wall, and release juvenile nematodes into the hemocoel of adult flies shortly following their eclosion (Poinar, 1983). In the this study, FA in thoracic bristle number was compared among susceptible infected, "resistant" (flies exposed to nematodes but no nematodes present within their hemocoel), and control (unexposed) flies. For all species, bristle FA among infected hosts did not differ from that of controls (Fig. 13-1). Thus, these data do not support the hypothesis that parasites cause increased FA of hosts.

C. Noncausal Linkage

Cases of noncausal linkage between parasitism and FA are expected when developmentally unstable phenotypes are less able to avoid parasites or combat disease. For example, either inbreeding or hybridization may reduce developmental stability and independently increase susceptibility to parasitism (Sage et al., 1986; O'Brien and Evermann, 1988; Moulia et al., 1991; Clarke, 1993).

Convincing evidence for noncausal linkage of this sort comes from studies of western fence lizards (*S. occidentalis*), house flies (*M. domestica*) and humans. In northern California, U.S.A., *S. occidentalis* is frequently infected with *Plasmodium mexicanum*, cause of malaria in these lizards (Schall, 1996). Among infected males, FA in femoral pore number was significantly greater than in uninfected individuals. But *Plasmodium* is transmitted via biting arthropods to lizards after hatching (Schall, 1996). Therefore, because pore number is ontogenetically determined before hatching, high pore number FA and susceptibility to malaria appear to be independent manifestations of developmentally unstable phenotypes. Likewise, high FA is a feature of house flies susceptible to the fungus *Entomophaga muscae* that infects only adult flies: dead, infested females exhibited higher wing length FA relative to neighboring flies that died of causes other than fungal infection (Møller, 1996). When same-sex pairs of flies were experimentally exposed to fungal spores, high FA flies of both sexes succumbed to fungus more often than low FA individuals (Møller, 1996).

In humans, Down's syndrome patients exhibited significantly elevated FA in numerous characters relative to the general population (Shapiro, 1983), and they also experienced higher rates of hepatitis (Blumberg et al., 1970), leukemia, and other cellular malignancies (Shapiro, 1983). The greater susceptibility of afflicted patients extends to other disease factors including a three-fold higher sensitivity to transformation by oncogenic virus SV 40 in laboratory-cultured fibroblasts (Torado and Martin, 1967). Furthermore, the effect of chickenpox and measles virus on random chromosome aberrations was two- to tenfold greater in Down's sufferers' lymphocytes than in control cells (Higurashi et al., 1973, 1976).

VI. Parasites, Fluctuating Asymmetry, and Sexual Selection

There is considerable evidence indicating that parasitism and FA can be positively correlated, but the correlation may have different sources. Distinguishing causal versus noncausal linkage will enhance our ability to predict the consequences of FA-based sexual selection, and to identify potential benefits accrued to females that favor low-FA males. In cases of both causal and noncausal linkage, females that prefer low-FA males could reduce the probability of transmitting parasites to themselves and to their offspring. Females may also secure better paternal care for young if parasites diminish a male's ability to provide such care. Indirect benefits could accrue in the form of genetic quality from a mate, "probably influenced more or less by its entire genome." (*sensu* Andersson, 1982, p. 82; and see Trivers, 1972; West-Eberhard, 1979; Kodric-Brown and Brown, 1984). Mate choice for low FA might also reflect a preference for elevated genome-wide heterozygosity (Thornhill and Gangestad, 1993).

A distinct kind of "good genes" benefit could accrue when parasites cause elevated host FA (causal linkage). In such cases, differences in host FA could reliably reflect allelic variation at a few major genes specifically involved in avoiding parasites and resisting disease. Mate choice for low FA could therefore be rewarded in terms of genetic resistance passed to offspring, to the extent that appropriate genetic variability exists in the host population (Hamilton and Zuk, 1982). In cases of causal linkage, we might also expect FA-based sexual selection to promote the evolution of hosts better able to defend themselves against parasites currently invading the host population.

Unfortunately, the FA-parasitism and FA–sexual selection connections have only rarely been examined simultaneously in the same species. In barn swallows, females prefer males with the most symmetrical tail feathers, and mite infestation appears to increase FA in the same trait (but see above). Males with tail feathers experimentally rendered asymmetrical actually provide more paternal care than symmetrical males (Møller, 1993c), weakening the hypothesis that females favoring low-FA males would enjoy enhanced assistance. Females could, however, reduce the probability of transmitting parasites to themselves and to fledglings. Moreover, both mite load and FA have a heritable component (Møller, 1991b; and see Table 1), so females may also accrue "good genes" in terms of parasite resistance.

In *D. nigrospiracula*, nematodes and ectoparasitic mites can elevate host FA, the latter operating via maternal effects, but sexual selection studies in nature failed to show a difference in FA between mated and unmated males (Polak, 1997). Although parasites influence the degree of FA in these flies, their damaging effects on developmental stability are apparently not sufficient to impair male mating ability.

Finally, in *M. domestica* and human beings, susceptibility to parasitism and FA covary positively, apparently because each are an independent manifestation of developmental instability. In humans, this elevated FA could be due to reduced heterozygosity at loci throughout the genome including the major histocompatibility complex (MHC) (discussed by Thornhill and Gangestad 1993). In both species, high FA is associated with impaired reproductive success. Thus FA-based sexual selection in these cases could favor general phenotypic vigor, perhaps influenced by the condition of the entire genome.

A. *Heritability of FA and Parasitism*

The possibility that FA-based sexual selection could favor "good genes" is weakened by the finding that heritability estimates of developmental stability are often small and nonsignificant (Mather, 1953; Thoday, 1958; Reeve, 1960; Bailit et al., 1970; Leamy, 1984; cf. Hagen 1973; Livshits and Kobyliansky, 1989). However, Møller and Thornhill (1997) have demonstrated a significant mean heritability of developmental stability across previously reported estimates, however, Møller and Thornhill (1997) have demonstrated a significant mean heritability of development stability across previously reported estimates (cf. Markow and Clarke, 1997), most of which were derived from laboratory experiments. But laboratory heritability estimates should be interpreted cautiously when evaluating "good genes" models of FA-based sexual selection, because gene loci influencing FA of a given trait in the laboratory may be different from those influencing FA of that trait in nature (see also Falconer, 1989, p. 322, for a general discussion). Indeed, the magnitude of FA is influenced by an impressive assortment of environmental stressors (see section III), which itself implies that different loci will influence FA in different environments. Moreover, the expressivity of a given gene influencing FA is expected to differ across environments, due to potentially different stress intensities encountered. Consider, for example, the role of a parasitic species that elevates host FA. In nature, effects of host loci involved in resistance against parasitism (e.g., Wakelin, 1978, 1985; Wakelin and Blackwell, 1988) could generate significant parent-offspring regressions for FA (and lead to high heritability values) by mediating rates of parasitism similarly in parents and in their offspring. But in the laboratory, and thus in the absence of most parasites, resistance loci would play a lesser role in generating FA heritability. In summary, field-derived estimates of FA heritability (see Table 1) are predicted to be higher than those derived in the laboratory, and to be more appropriate for evaluating "good genes" models of FA-based sexual selection.

VII. Costs of Parasite Resistance and Fluctuating Asymmetry

The physiological costs of resistance against parasites may be widespread and severe. Indeed, the very presence of susceptible forms in host populations harboring virulent parasites has been viewed by some as evidence for the existence of such costs (Harlan, 1976; Burdon, 1980; Read, 1990, Reboud and Till-Bottraud, 1991; Rowland, 1991; Antonovics and Thrall, 1994). The prediction is that if costs accrue during an organism's development, they could function to elevate host FA. Below, I review evidence for this prediction and discuss its implications for sexual selection. First, however, I briefly review the literature on costs of resistance.

A. Costs of Resistance against Parasites

Costs of resistance can arise when limiting resources are diverted toward defense at the expense of other fitness-related functions. For example, Bergelson (1994) recently demonstrated that lettuce (*Lactuca sativa*) resistant against leaf root aphids and downy mildew had slowed development, reduced plant height, and depressed seed production relative to susceptible forms in the absence of the parasites. These effects were more pronounced under limited nutrient availability, implicating resource trade-offs in the expression of costs. Likewise in brown rats (*Rattus norvegicus*), warfarin resistance alters vitamin K–dependent physiological processes and leads to blood clotting disorders and reduced body weight (Martin, 1973; Smith et al., 1991). Costs may also accrue when a mounted immune response itself damages host tissue. In humans, for example, vigorous immunological activity elicited against hepatitis B virus causes liver damage (Dudley et al., 1972), and hypersensitive autoimmune responses mounted against malarial *Plasmodium* and some African trypanosomes cause necrosis of the heart and kidneys (Cohen and Lambert, 1977; Hudson, 1985). Likewise, certain host chemicals, such as hydrogen cyanide in plants that confer protection against herbivorous insects, can also be self-toxic (Simms, 1992).

Studies of *Escherichia coli* and their T4 phage parasites provide some of the strongest evidence that resistance mutations pleiotropically impose costs by altering metabolic functions. Lenski and Levin (1985) showed that bacterial growth rate fell significantly when resistance against phages first evolved, because the same mutation(s) that conferred resistance also prevented efficient uptake of nutrients. In another experiment, significant growth rate reductions were recorded in bacteria transformed with a plasmid that conferred antibiotic resistance (Bouma and Lenski, 1988). In both examples, genetic modifiers that integrated the resistance-conferring factors into the bacterial genome and restored fitness evolved within 400–500 generations.

B. Costs of Resistance and FA

The best evidence that parasite resistance might elevate FA comes from studies of resistance against the pesticide diazinon in the Australian sheep blow fly, *Lucillia cuprina*. Resistance arose in flies by allelic substitution at a single locus on chromosome IV some 10 years after initial application of the pesticide. Clarke and McKenzie (1987) demonstrated that the presence of the resistance (R) allele could disrupt developmental stability and elevate FA, but that this effect depended upon the genetic background in which the allele was expressed. Thus, FA in resistant flies, sampled from nature 14 years after the evolution of resistance, did not differ from that in susceptible laboratory flies, but FA rose dramatically when the R allele was inserted by repeated backcrossing into susceptible flies in the laboratory (Clarke and McKenzie, 1987). The modifier gene in the field population that alleviated the disruptive effects was localized to chromosome III (McKenzie and Clarke, 1988). Furthermore, the disruptive effects of the R allele also reduced fly viability (McKenzie et al., 1982), and it is interesting that the gene for this effect mapped to the same region of the genome as that responsible for reductions in FA. This finding demonstrates that the same gene, or gene complex, causes variability in both FA and fitness (Clarke and McKenzie, 1992).

New or rare alleles conferring resistance against parasites could elevate FA when driven into the host population by frequency-dependent selection (Polak and Trivers, 1994). Host FA could remain high, and even oscillate between high and low values, to the extent that coevolution occurs between the parasite and the host. This effect could operate in systems involving gene-for-gene interactions, exemplified by the classic flax–flax rust interaction (Flor, 1956), as well as when resistance is polygenic (Gallun, 1972; Wakelin, 1985). In the latter case, allelic substitution at any of the loci involved in resistance could have deleterious effects (Harlan, 1976), especially if associated with vital metabolic pathways (Zakharov, 1992). In a similar vein, when parasite-mediated selection changes the frequency of alleles at one or more of the loci that confer resistance, costs could also accrue if the selected allele(s) were genetically linked to one or more deleterious mutations. Thus, genetic hitchhiking (e.g., Kojima and Schaffer, 1967) may be yet another mechanism by which costs of resistance might accrue, and elevate host FA.

A second line of evidence that costs of resistance can elevate FA comes from the *Drosophila-Howardula* laboratory experiments described in the preceding section. In *D. falleni* and *D. neotestacea*, "resistant" individuals (flies exposed to parasites but uninfected) were significantly more asymmetric than unexposed, control flies (Fig. 13-1). An important assumption is that all host larvae were equally exposed to worms during the process of experimental infection in the laboratory, otherwise, differential parasitism could not safely be attributed to inherent differences among larvae in their ability to evade penetration by nematodes. For example, if worms were heterogeneously distributed within the infection vials, and if relatively nematode-free zones were also environmentally more stressful (e.g., limited nutrient availability), then the elevated FA observed among uninfested hosts could have resulted from environmental effects, independent of inherent costs of parasite resistance.

However, the observation that *D. falleni* and *D. neotestacea* co-occur with the worm in nature, together with the fact that worms cause high fitness variance among adult flies (Jaenike, 1992), makes it is reasonable to expect that some form of evolved immunity exists in fly populations. Larval immunity could involve thickening of the cuticle and/or active avoidance of nematodes. Behavioral avoidance occurs, for example, in *Anopheles quadrimaculatus*, whose larvae protect themselves against penetration by mermithid nematodes by moving about rapidly and continually and by "snapping" at attacking worms (Woodard and Fukuda, 1977). These putative responses probably cost energy and nutrients, which, if diverted from vital developmental processes, could elevate FA in these behaviorally resistant phenotypes. Because *D. quinaria* is not parasitized by *H.*

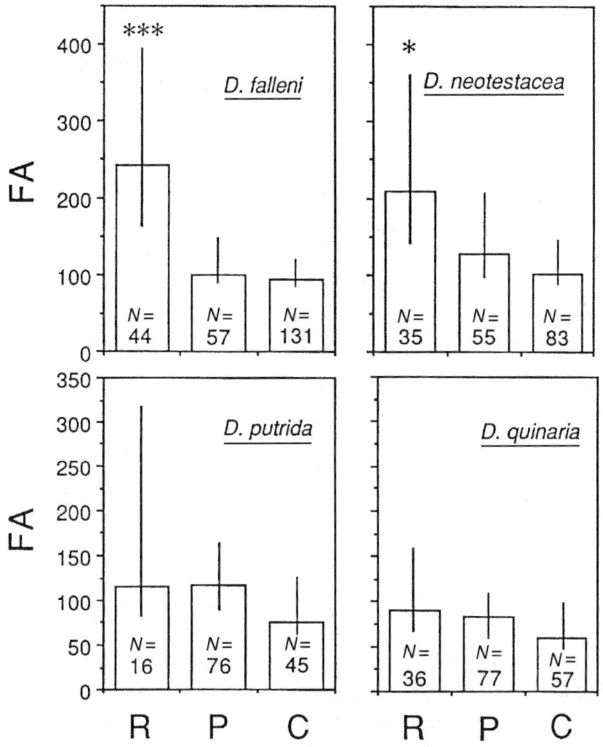

Figure 13-1. Bristle number fluctuating asymmetry (var[R–L]) in resistant (R), parasitized (P), and unexposed, control (C) flies in the genus *Drosophila*. Sample sizes (N) and 95% confidence intervals are shown. Significance was assessed using F-max tests (Zar, 1984) and is indicated * $p < 0.05$; *** $p < 0.005$. Source: M. Polak, and J. Jaenike, unpublished data.

aoronymphium in nature (*D. quinaria* breeds in skunk cabbages, J. Jaenike personal communication), and therefore, unlikely to harbor evolved specific resistance, it is not surprising that "resistant" individuals of this species were no more asymmetrical than control or parasitized flies (Fig. 13-1). However in *D. putrida*, which is a species that is naturally parasitized by this nematode, resistant individuals were also not more asymmetrical, for unknown reasons.

Results of the *Drosophila-Howardula* experiments also demonstrate that resistant hosts can exhibit higher FA relative even to their parasitized counterparts. This apparently paradoxical pattern could arise if, during character ontogeny, costs of resistance (C_R) exceed costs of parasitism (C_P) (see Fig. 13-2). Species harboring parasites selected for benign effects during host ontogeny would satisfy this condition, despite strong negative effects of the same parasite on the fitness of adult hosts. Figure 13-2 depicts elements of this trade-off between benefits

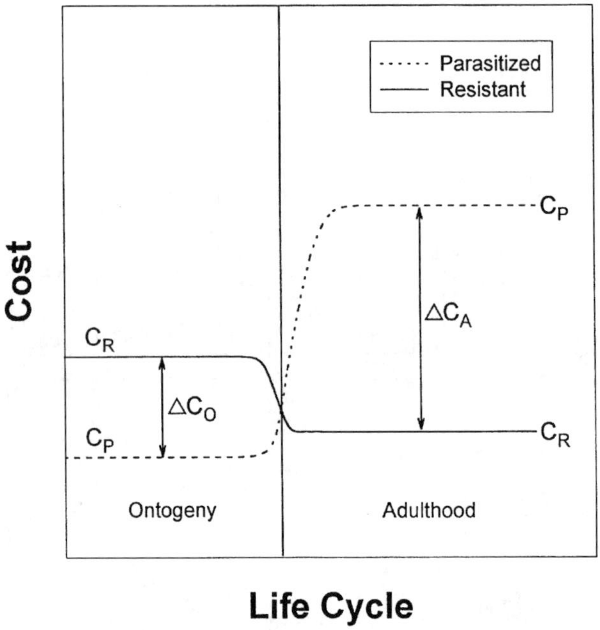

Figure 13-2. Costs of resistance (C_R) and costs of parasitism (C_P) in a hypothetical host species. C_R during ontogeny is greater than C_P, and leads to elevated FA, because the parasite is selected for benign effects during host ontogeny. In this way, the parasite avoids impairing the mobility of adult hosts, on which it depends for its own transmission. Resistance is maintained in the host population because $\Delta C_A >> \Delta C_O$.

of resistance accrued during adulthood and costs of resistance accrued during ontogeny. The difference between the higher costs of resistance and that of parasitism during ontogeny is labeled ΔC_O, whereas the difference between the higher costs of parasitism and that of resistance during adulthood is labeled ΔC_A. Thus, even though $C_R > C_P$ occurs during ontogeny, and hence elevating FA of resistant hosts, resistance is beneficial and is maintained in the host population because $\Delta C_A >> \Delta C_O$. Due to the high costs of parasitism during adulthood, low-FA adult hosts are predicted to express relatively low fitness because they will harbor more parasites due to their elevated susceptibility.

Why should parasites be selected for benign effects during host ontogeny? Ewald (1983, 1994) proposed that reduced efficiency of transmission could select for reduced virulence. This same selection may also favor parasites that minimize harm specifically during host ontogeny to safeguard mobility on the adult host on which the parasite relies for its own transmission. If the parasite damaged host developmental pathways to the extent that the host suffered malformation, behavioral phenodeviance, or any other debilitating handicap which reduces host mobility, the parasite could incur fitness losses from reduced transmission.

To illustrate, consider again the *Drosophila-Howardula* association (Jaenike, 1992). Nematode transmission depends on a mobile, uninjured adult fly: infected flies must overwinter successfully, and in the spring, disperse to fresh mushrooms, mate and oviposit (Jaenike, 1992). Nematodes then colonize these mushrooms after exiting through the anus and ovipositor of flies, and go in search of fly larvae to parasitize. Consistent with the above prediction, larval survival is unaffected by parasitism (Jaenike, 1992; J. Jaenike personal communication). If resistant larvae suffer costs that elevate their FA above that of susceptibles, these costs would probably be outweighed by the benefits of resistance during adulthood, as infection leads to massive reductions in adult fertility. Irrespective of the intensity their of occurrence within flies, nematodes reduce fertility of adult *D. falleni*, *D. neotestacea*, and *D. putrida* by ca. 50, 90, and 100%, respectively (Jaenike, 1992).

VIII. Costs of Parasite Resistance, Fluctuating Asymmetry, and Sexual Selection

If resistance is costly, and if these costs accrue during development, then resistant individuals may exhibit greater FA than susceptible hosts that have not been exposed to parasites. Symmetry will therefore be unreliable as indicator of resistance, unless all members of the host population were equally exposed. Heterogeneity in exposure may be common in nature (Crofton, 1971; Anderson and Gordon, 1982; Jaenike and Anderson, 1992). Furthermore, when parasites are selected for benign effects during host development (see above), FA in hosts with costly resistance could be greater than in unexposed as well as in parasitized individuals. In these host species, high FA could indicate resistance. This leads to the prediction that in host species burdened with 1) costly resistance and 2) parasites selected for benign effects during host ontogeny, FA in adult hosts will be correlated positively with fitness to the extent that parasites reproductively debilitate adult hosts. Interestingly, whisker spot asymmetry in lions (*Pant. leo*) is positively correlated with survivorship in females (Packer and Pusey, 1993), and feather asymmetry in European starlings (*Sturnus vulgaris*) is positively correlated with dominance (Swaddle and Witter, 1994; Witter and Swaddle, 1994).

IX. Concluding Remarks

Fluctuating asymmetry and male mating success are negatively correlated in 15 of 23 animal taxa, either because high FA per se reduces competitive ability and sexual attractiveness, or because FA is only correlated negatively with health and vigor. Because parasitism and high FA will often be associated (for a battery of different reasons), female mate choice in favor of symmetrical males may be explained by both direct and indirect benefits related to parasitism, and FA-based sexual selection in general may drive the evolution of host resistance and influence the genetic structure of host populations. The validity of these possibilities, however, remains virtually untested directly, primarily because very few studies have evaluated both the FA–mating success and the parasitism-FA connections simultaneously in one host population. More studies of this nature are therefore required, and workers studying the FA–mating success connection might examine their systems with the potential effects of parasitism in mind.

Some studies have failed to uncover any association between FA and mating success, but the short-term, cross-sectional, approach for estimating mating success employed in some of these studies may be inappropriate in the study of FA. Following known individuals over an extended period (e.g., an entire breeding season) is likely to yield more precise estimates of fitness components (Arnold and Wade, 1984), as well as help control for an interaction between FA and survivorship, which can obscure an otherwise detectable association between FA and mating success (Markow, 1995). Students of FA are also encouraged to take their measurements blind, since differences between sides of a trait will often be very subtle, and thus, potentially easily influenced by unconscious observer bias. The final, and most essential point of this chapter, is that costs of parasite resistance may altogether eliminate the expected negative relationship between FA and mating success, because the most resistant, and hence, parasite-free individuals may express the greatest FA. Detailed knowledge of the natural history of the parasite-host system, including especially parasite transmission dynamics, will help to generate appropriate predictions regarding the link between parasite burden and FA, and hence, between FA and sexual selection.

X. Acknowledgments

I wish to thank J. Alcock, J. Jaenike, R. Janssen, T. Markow, R. Thornhill and two anonymous reviewers for comments on the manuscript. I benefited from discussions of the present topics with J. Alcock, M. Forbes, T. Markow, P. Miller, S. Pitnick, T. Starmer, R. Thornhill, and R. Trivers. I also wish to thank I. Folstad, A. Møller, S. Nordell, M. Ryan, J. Schall, L. Simmons, R. Thornhill for kindly providing manuscripts or for discussing with me their unpublished results. I wish also to thank J. Alcock for providing continued encouragement while I was writing this chapter, and the DuPont Company for financial support.

XI. References

Anderson, R. M., and Gordon, D. M. (1982). Processes influencing the distribution of parasite numbers within host populations with special emphasis on parasite-induced host mortalities. *Parasitology* **85:**373–398.

Andersson, M. (1982). Sexual selection, natural selection and quality advertisement. *Biol. J. Linn. Soc.* **17:**375–393.

Antonovics, J., and Thrall, P. H. (1994). The cost of resistance and the maintenance of genetic polymorphism in host-pathogen systems. *Proc. R. Soc. Lond. B Biol. Sci.* **257:**105–110.

Arcese, P. (1994). Harem size and horn symmetry. *Anim. Behav.* **48:**1485–1488.

Arnold, S. J., and Wade, M. J. (1984). On the measurement of natural and sexual selection: Applications. *Evolution* **38:**720–734.

Bailit, H. L., Workman, P. L., Niswander, J. D., and MacLean, C. J. (1970). Dental asymmetry as an indicator of genetic and environmental conditions in human populations. *Hum. Biol.* **42:**626–638.

Balmford, A., Jones, I., and Thomas, A. L. R. (1993). On avian asymmetry: Evidence of natural selection for symmetrical tails and wings in birds. *Proc. R. Soc. Lond. B Biol. Sci.* **252:**245–251.

Balmford, A., and Thomas, A. (1992). Swallowing ornamental asymmetry. *Nature* (Lond.) **359:**487.

Balmford, A., Thomas, A. L. R., and Jones, I. (1993). Aerodynamics and the evolution of long tails in birds. *Nature (Lond.)* **361:**628–631.

Bergelson, J. (1994). The effects of genotype and the environment on costs of resistance in lettuce. *Am. Nat.* **143:**349–359.

Biémont, C. (1983). Homeostasis, enzymatic heterozygosity and inbreeding depression in natural populations of *Drosophila melanogaster. Genetica* **61:**179–189.

Blumberg, B. S., Gerstley, B. J. S., Sutnick, A. I., Millman, I., and London, W. T. (1970). Australia antigen, hepatitis virus, and Down syndrome. *Ann. N. Y. Acad. Sci.* **171:**486–499.

Borgia, G., and Wilkinson, G. (1992). Swallowing ornamental asymmetry. *Nature (Lond.)* **359:**487–488.

Bouma, J. E., and Lenski, R. E. (1988). Evolution of a bacteria/plasmid association. *Nature (Lond.)* **335:**351–352.

Brookes, M., and Pomiankowski, A. (1994). Symmetry and sexual selection: A reply. *Trends Ecol. Evol.* **9:**440.

Burdon, J. J. (1980). Variation in disease-resistance within a population of *Trifolium repens. J. Ecol.* **68:**737–744.

Clarke, G. M. (1993). The genetic basis of developmental stability. I. Relationships between stability, heterozygosity and genomic coadaptation. *Genetica* **89:**15–23.

Clarke, G. M., and McKenzie, J. A. (1987). Developmental stability of insecticide resistant phenotypes in blowfly; a result of canalizing natural selection. *Nature (Lond.)* **325:**345–346.

Clarke, G. M., and McKenzie, J. A. (1992). Fluctuating asymmetry as a quality control indicator of insect mass rearing processes. *J. Econ. Entomol.* **85:**2045–2050.

Clayton, D. H. (1991). The influence of parasites on host sexual selection. *Parasitol. Today* **7:**329–334.

Clutton-Brock, T. H. (1982). The function of antlers. *Behaviour* **79:**108–125.

Córdoba-Aguilar, A. (1995). Fluctuating asymmetry in paired and unpaired damselfly males *Ischnura denticollis* (Burmeister) (Odonata: Coenagrionidae) *J. Ethol.* **13:**129–132.

Cohen, S., and Lambert, P. H. (1977). Malaria. In *Immunology and parasitic infections,* 2nd ed. (S. Cohen and K. S. Warren, eds.), pp. 422–474. England: Blackwell Scientific.

Crofton, H. D. (1971). A quantitative approach to parasitism. *Parasitology* **62:**179–193.

Cummings, M. R. (1988). *Human heredity.* St. Paul: West Publishing.

Dudley, F. J., Fox, R. A., and Sherlock, S. (1972). Cellular immunity and hepatitis-associated, Australia antigen liver disease. *Lancet* April 1:723–726.

Eggert, A.-K., and Sakaluk, S. K. (1994). Fluctuating asymmetry and variation in the size of courtship food gifts in decorated crickets. *Am. Nat.* **144:**708–716.

Enquist, M., and Arak, A. (1994). Symmetry, beauty and evolution. *Nature* (Lond.) **372:**169–172.

Evans, M. R., and Hatchwell, B. J. (1993). New slants on ornamental asymmetry. *Proc. R. Soc. Lond. B Biol. Sci.* **251:**171–177.

Evans, M. R., Martins, T. L. F., and Haley, M. (1994). The asymmetrical cost of tail elongation in red-billed streamer-tails. *Proc. R. Soc. Lond. B Biol. Sci.* **256:**97–103.

Ewald, P. W. (1983). Host-parasite relations, vectors, and the evolution of disease severity. *Annu. Rev. Ecol. Syst.* **14:**465–485.

Ewald, P. W. (1994). *Evolution of infectious disease.* U.K.: Oxford University Press.

Falconer, D. S. (1989). *Quantitative genetics.* Essex: Longman Scientific & Technical.

Fisher, R. A. (1930). *The genetical theory of natural selection.* Oxford: Clarendon.

Flor, H. H. (1956). The complementary genetic systems in flax and flax rust. *Adv. Genet.* **8:**29–54.

Folstad, I., Arneberg, P., and Karter, A. J. (1996). Antlers and parasites. *Oecologia* **105:**556–558.

Freeland, W. J. (1981). Parasitism and behavioral dominance among male mice. *Science (Washington, D.C.)* **213:**461–462.

Gallum, R. L. (1972). Genetic interrelationships between host plants and insects. *J. Environ. Qual.* **1:**259–265.

Goater, C. P., Raymond, R. D., and Bernasconi, M. V. (1993). Effects of body size and parasite infection on the locomotory performance of juvenile toads, *Bufo bufo. Oikos* **66:**129–136.

Gordon, R., and Webster, J. M. (1971). *Mermis nigrescens:* Physiological relationship with its host, the adult desert locust *Schistocerca gregaria. Exp. Parasitol.* **29:**66–79.

Graham, J. H., and Felley, J. D. (1985). Genomic coadaptation and developmental stability within introgressed populations of *Enneacanthus gloriousus* and *E. obesus* (Pisces, Centrarchidae). *Evolution* **39:**104–114.

Graham, J. H., Roe, E. R., and West, T. B. (1993). Effects of lead and benzene on the developmental stability of *Drosophila melanogaster. Exotoxicology* **2:**185–195.

Grammer, K., and Thornhill, R. (1994). Human *(Homo sapiens)* facial attractiveness and sexual selection: The role of symmetry and averageness. *J. Comp. Psychol.* **108:**233–242.

Hagen, D. W. (1973). Inheritance of numbers of lateral plates and gill rakers in *Gasterosteus aculeatus. Heredity* **30:**303–312.

Hamilton, W. D., and Zuk, M. (1982). Heritable true fitness and bright birds: A role for parasites? *Science (Washington, D.C.)* **218:**384–387.

Harlan, J. R. (1976). Diseases as a factor in plant evolution. *Annu. Rev. Phytopathol.* **14:**31–51.

Harvey, I. F., and Walsh, K. J. (1993). Fluctuating asymmetry and lifetime mating success are correlated in males of the damselfly *Coenagrion puella* (Odonata: Coenagrionidae). *Ecol. Entomol.* **18:**198–202.

Higurashi, M., Tada, A., Miyahara, S., and Hirayama, M. (1976). Chromosome damage in Down's syndrome induced by chickenpox infection. *Pediatr. Res.* **10:**189–192.

Higurashi, M., Tamura, T., and Nakateke, T. (1973). Cytogenetic observations in cultured lymphocytes from patients with Down's syndrome and measles. *Pediat. Res.* **7:**582–587.

Hillgarth, N. (1990). Parasites and female choice in the ring-necked pheasants. *Am. Zool.* **30:**227–233.

Houde, A. E., and Torio, A. J. (1992). Effect of parasitic infection on male color pattern and female choice in guppies. *Behav. Ecol.* **3:**346–351.

Howard, R. D., and Minchella, D. J. (1990). Parasitism and mate competition. *Oikos* **58:**120–122.

Hudson, L. (1985). Autoimmune phenomena in chronic chagastic cardiopathy. *Parasitol. Today* **1:**6–9.

Jaenike, J. (1992). Mycophagous *Drosophila* and their nematode parasites. *Am. Nat.* **139:**893–906.

Jaenike, J., and Anderson, T. J. C. (1992). Dynamics of host-parasite interactions: The *Drosophila-Howardula* system. *Oikos* **64:**533–540.

Jennions, M. D., and Oakes, E. J. (1994). Symmetry and sexual selection. *Trends Ecol. Evol.* **9:**440.

Johnstone, R. A. (1994). Female preference for symmetrical males as a by-product of selection for mate recognition. *Nature (Lond.)* **372:**172–175.

Karney, W. W., and Tong, M. T. (1972). Malabsorption in *Plasmodium flaciparum* malaria. *Am. J. Trop. Med. Hyg.* **21:**1–5.

Kazazian, H. H., Jr., Wong, C., Youssoufian, H., Scott, A. F., Phillips, D. G., and Antonarakis, S. E. (1988). Haemophilia A resulting from *de novo* insertion of L1 sequences represents a novel mechanism for mutation in man. *Nature (Lond.)* **332:**164–166.

Kitchener, A. (1987). Fighting behaviour of the extinct Irish elk. *Modern Geol.* **11:**1–28.

Kirkpatrick, M., and Ryan, M. J. (1991). The evolution of mating preferences and the paradox of the lek. *Nature (Lond.)* **350:**33–38.

Kodrick-Brown, A., and Kodrick-Brown, J. H. (1984). Truth in advertising: the kinds of traits favored by sexual selection. *Am. Nat.* **124:**309–323.

Kojima, K., and Schaffer, H. E. (1967). Survival processes of linked mutant genes. *Evolution* **21:**518–531.

Leamy, L. (1984). Morphometric studies in inbred and hybrid house mice. V. Directional and fluctuating asymmetry. *Am. Nat.* **123:**579–593.

Leary, R. F., Allendorf, F. W., and Knudsen, K. L. (1983). Developmental stability and enzyme heterozygosity in rainbow trout. *Nature (Lond)* **301:**71–72.

Leary, R. F., Allendorf, F. W., and Knudsen, K. L. (1984). Superior developmental stability of heterozygotes at enzyme loci in salmonid fishes. *Am. Nat.* **124:**540–551.

Lenski, R. E., and Levin, B. R. (1985). Constraints of the coevolution of bacteria and virulent phage: A model, some experiments, and predictions for natural communities. *Am. Nat.* **125:**585–602.

Liggett, A. C., Harvey, I. F., and Manning, J. T. (1993). Fluctuating asymmetry in *Scathophaga stercoraria* L: Successful males are more symmetrical. *Anim. Behav.* **45:**1041–1043.

Livshits, G., Davidi, L., Kobyliansky, E., Ben-Amitai, D., Levi, Y., and Merlob, P. (1988). Decreased developmental stability as assessed by fluctuating asymmetry of morphometric traits in preterm infants. *Am. J. Med. Genet.* **29:**793–805.

Livshits, G., and Kobyliansky, E. (1989). Study of genetic variance in fluctuating asymmetry of anthropometrical traits. *Ann. Human Biol.* **116:**121–129.

Ludwig, W. (1932). *Das Rechts-Links Problem im Tierreich und beim Menschen.* Berlin: Springer.

Malyon, C., and Healy, S. (1994). Fluctuating asymmetry in antlers of fallow deer, *Dama dama,* indicates dominance. *Anim. Behav.* **48:**248–250.

Manning, J. T., and Chamberlain, A. T. (1993). Fluctuating asymmetry, sexual selection and canine teeth in primates. *Proc. R. Soc. Lond. B Biol. Sci.* **251:**83–87.

Manning, J. T., and Chamberlain, A. T. (1994). Fluctuating asymmetry in gorilla canines: A sensitive indicator of environmental stress. *Proc. R. Soc. Lond. B Biol. Sci.* **255:**189–193.

Manning, J. T., and Hartley, M. A. (1991). Symmetry and ornamentation are correlated in the peacock's train. *Anim. Behav.* **42:**1020–1021.

Markow, T. A. (1987). Genetic and sensory basis of sexual selection in *Drosophila.* In *Evolutionary genetics of invertebrate behavior* (M. D. Huettel, ed.), pp. 89–95. New York: Plenum.

Markow, T. A. (1992). Genetics and developmental stability: an integrative conjecture on aetiology and neurobiology of schizophrenia. *Psychol. Med.* **22:**295–305.

Markow, T. A. (1995). Evolutionary ecology of developmental instability. *Annu. Rev. Entomol.* **40:**105–120.

Markow, T. A., and Martin, J. F. (1993). Inbreeding and developmental stability in a small human population. *Ann. Human Biol.* **20:**389–394.

Markow, T. A., and Ricker, J. P. (1991). Developmental stability in hybrids between the sibling species pair, *Drosophila melanogaster* and *Drosophila simulans. Genetica* **84:**115–121.

Markow, T. A., and Ricker, J. P. (1992). Male size, developmental stability, and mating success in natural populations of three *Drosophila* species. *Heredity* **69:**122–127.

Markow, T. A., and Sawka, S. (1992). Dynamics of mating success in experimental groups of *Drosophila melanogaster* (Diptera: Drosophilidae). *J. Ins. Behav.* **5:**375–383.

Markow, T. A., and Wandler, K. (1986). Fluctuating dermatoglyphic asymmetry and the genetics of liability to schizophrenia. *Psychol. Res.* **19:**323–328.

Martin, A. D. (1973). Vitamin K requirement and anticoagulant response in warfarin resistant rats. *Biochem. Soc. Trans.* **1:**1206–1208.

Mather, K. (1953). Genetical control of stability in development. *Heredity* **7:**297–336.

McKenzie, J. A., and Clarke, G. M. (1988). Diazinon resistance, fluctuating asymmetry and fitness in the Australian sheep blowfly, *Lucilia cuprina. Genetics* **120:**213–220.

McKenzie, J. A., Whitten, M. J., and Adena, M. A. (1982). The effect of genetic background on the fitness of diazinon resistant genotypes of the Australian sheep blowfly, *Lucilia cuprina. Heredity* **49:**1–9.

McLachlan, A., and Cant, M. (1995). Small males are more symmetrical: mating success in the midge *Chironomus plumosus* L. (Diptera: Chironomidae). *Anim. Behav.* **50:**841–846.

Mellor, C. S. (1992). Dermatoglyphic evidence of fluctuating asymmetry in schizophrenia. *Br. J. Psychiat.* **160:**467–472.

Milinski, M., and Bakker, T. C. M. (1990). Female stickelbacks use male coloration in mate choice and hence avoid parasitized males. *Nature (Lond.)* **344:**330–333.

Møller, A. P. (1990a). Fluctuating asymmetry in male sexual ornaments may reliably reveal male quality. *Anim. Behav.* **40:**1185–1187.

Møller, A. P. (1990b). Effects of parasitism by a haematophagous mite on reproduction in the barn swallow. *Ecology* **71:**2345–2357.

Møller, A. P. (1991a). Sexual ornament size and the cost of fluctuating asymmetry. *Proc. R. Soc. Lond. B Biol. Sci.* **243:**59–62.

Møller, A. P. (1991b). Parasites, sexual ornaments, and mate choice in the barn swallow. In *Bird-parasite interactions* (J. E. Loye and M. Zuk, eds.), pp. 328–343. Oxford: Oxford University Press.

Møller, A. P. (1992a). Patterns of fluctuating asymmetry in weapons: Evidence for reliable signalling of quality in beetle horns and bird spurs. *Proc. R. Soc. Lond. B Biol. Sci.* **248:**199–206.

Møller, A. P. (1992b). Female swallow preference for symmetrical male sexual ornaments. *Nature (Lond.)* **357:**238–240.

Møller, A. P. (1992c). Swallowing ornamental asymmetry. *Nature (Lond.)* **359:**488.

Møller, A. P. (1992d). Parasites differentially increase the degree of fluctuating asymmetry in secondary sexual characters. *J. Evol. Biol.* **5:**691–699.

Møller, A. P. (1993a). Morphology and sexual selection in the barn swallow *Hirundo rustica* in Chernobyl, Ukraine. *Proc. R. Soc. Lond. B Biol. Sci.* **252:**51–57.

Møller, A. P. (1993b). Developmental stability, sexual selection and speciation. *J. Evol. Biol.* **6:**493–509.

Møller, A. P. (1993c). Developmental stability, sexual selection, and the evolution of secondary sexual characters. *Etología* **3:**199–208.

Møller, A. P. (1993d). Female preference for apparently symmetrical male sexual ornaments in the barn swallow *Hirundo rustica. Behav. Ecol. Sociobiol.* **32:**371–376.

Møller, A. P. (1994). Sexual selection in the barn swallow *(Hirundo rustica).* IV. Patterns of fluctuating asymmetry and selection against asymmetry. *Evolution* **48:**658–670.

Møller, A. P. (1995). Bumblebee preference for symmetrical flowers. *Proc. Natl. Acad. Sci. U.S.A.* **92:**2288–2292.

Møller, A. P. (1996). Sexual selection, viability selection, and developmental stability in the domestic fly *Musca domestica. Evolution* **50:**746–752.

Møller, A. P., and Eriksson, M. (1994). Patterns of fluctuating asymmetry in flowers: Implications for sexual selection. *J. Evol. Biol.* **7:**97–113.

Møller, A. P., and Ḧoglund, J. (1991). Patterns of fluctuating asymmetry in avian feather ornaments: Implications for sexual selection. *Proc. R. Soc. Lond. B Biol. Sci.* **245:**1–5.

Møller, A. P., and Pomiankowski, A. (1993a). Fluctuating asymmetry and sexual selection. *Genetica* **89:**267–279.

Møller, A. P., and Pomiankowski, A. (1993b). Why have animals got multiple sexual ornaments? *Behav. Ecol. Sociobiol.* **32:**167–176.

Møller, A. P., Soler, M., and Thornhill, R. (1995). Breast asymmetry, sexual selection, and human reproductive success. *Ethol. Sociobiol.* **16:**207–219, 1995.

Møller, A. P., and Thornhill, R. (1997). A meta-analysis of the heritability of developmental stability. *J. Evol. Biol.* (in press).

Mooney, M. P., Siegel, M. I., and Gest, T. R. (1985). Prenatal stress and increased fluctuating asymmetry in the parietal bones of neonatal rats. *Am. J. Phys. Anthropol.* **88:**131–134.

Motulsky, A. G. (1960). Metabolic polymorphisms and the role of infectious diseases in human evolution. *Hum. Biol.* **32:**28–62.

Moulia, C., Le Brun, N., Dallas, J., Orth, A., and Renaud, F. (1991). Wormy mice in a hybrid zone: A genetic control of susceptibility to parasite infection. *J. Evol. Biol.* **4:**679–687.

Mulvey, M., and Abo, J. M. (1993). Parasitism and mate competition: Liver flukes in white-tailed deer. *Oikos* **66:**187–192.

Narita, N., Nishio, H., Kitoh, Y., Ishikawa, Y., Minami, R., Nakamura, H., and Masafumi, M. (1993). Insertion of a 5′ truncated L1 element into the 3′ end of exon 44 of the dystrophin gene resulted in skipping of the exon during splicing in a case of Duchenne muscular dystrophy. *J. Clin. Invest.* **91:**1862–1867.

Naugler, C. T., and Leech, S. M. (1994). Fluctuating asymmetry and survival ability in the forest tent caterpillar moth *Malacosoma disstria:* Implications for pest management. *Entomol. Exp. Appl.* **70:**295–198.

Newell-Morris, L. L., Fahrenbruch, C. E., and Sackett, G. P. (1989). Prenatal psychological stress, dermatogliphic asymmetry and pregnancy outcome in the pigtailed macaque *(Macaca nemestrina). Biol. Neonate* **56:**61–75.

Nordell, S. E. (1995). Mechanisms of selection: Mate choice and predation assessment in guppies *(Poecilia reticulata).* Ph.D. Thesis, University of New Mexico.

Oakes, E. J., and Barnard, P. (1994). Fluctuating asymmetry and mate choice in paradise whydahs, *Vidua paradisaea:* an experimental approach. *Anim. Behav.* **48:**937–943.

O'Brien, S. J., and Evermann, F. F. (1988). Disease and genetic diversity in natural populations. *Trends Ecol. Evol.* **3:**254–259.

O'Donald, P. (1983). Sexual selection by female choice. In *Mate choice* (P. Bateson, ed.), pp. 53–66. New York: Cambridge.

Packer, C., and Pusey, A. E. (1993). Should a lion change its spots? *Nature (Lond.)* **363:**695.

Palmer, A. R., and Strobeck, C. (1986). Fluctuating asymmetry: Measurement, analysis, patterns. *Annu. Rev. Ecol. Syst.* **17:**391–421.

Pankakoski, E., Koivisto, I., and Hyvärinen, H. (1992). Reduced developmental stability as an indicator of heavy metal pollution in the common shrew *Sorex araneus. Acta Zool. Fenn.* **191:**127–144.

Parsons, P. A. (1962). Maternal age and developmental variability. *J. Exp. Biol.* **39:**251–260.

Poinar, G. O., Jr. (1983). *The natural history of nematodes.* Englewood Cliffs, N. J.: Prentice-Hall.

Polak, M (1993). Parasites increase fluctuating asymmetry of male *Drosophila nigrospiracula:* implications for sexual selection. *Genetica* **89:**255–265.

Polak, M. (1996). Ectoparasitic effects on host survival and reproduction: The *Drosophila-Macrocheles* association. *Ecology* **77:**1379–1389.

Polak, M. (1997). Ectoparasitism in mothers causes higher positional fluctuating asymmetry in their sons: implications for sexual selection. *Am. Nat.* (in press).

Polak, M., and Markow, T. A. (1995). Effect of ectoparasitic mites on sexual selection in a Sonoran desert fly. *Evolution* **49:**660–669.

Polak, M., and Trivers, R. (1994). The science of symmetry in biology. *Trends Ecol. Evol.* **9:**122–124.

Radesäter, T., and Halldórsdóttir, H. (1993). Fluctuating asymmetry and forceps size in earwigs, *Forficula auricularia. Anim. Behav.* **45:**626–628.

Rau, M. E. (1983). Establishment and maintenance of behavioural dominance in male mice infected with *Trichinella spiralis. Parasitology* **86:**319–322.

Read, A. F. (1990). Parasites and the evolution of host sexual behavior. In *Parasitism and host behavior* (C. J. Barnard and J. M. Behnke, eds.), pp. 117–157. U. K.: Taylor and Francis.

Reboud, X., and Till-Bottraud, I. (1991). The cost of herbicide resistance measured by a competition experiment. *Theor. Appl. Genet.* **82:**690–696.

Reeve, E. C. R. (1960). Some genetic tests on asymmetry of sternopleural chaeta number in *Drosophila. Genet. Res.* **1:**151–172.

Rowland, M. (1991). Behaviour and fitness of HCH/dieldrin resistant and susceptible females of *Anopheles gambiae* and *A. stephensi* in the absence of insecticide. *Med. Vet. Entomol.* **5:**193–206.

Rutherford, T. A., and Webster, J. M. (1978). Some effects of *Mermis nigrescens* on the haemolymph of *Schistocerca gregaria. Can. J. Zool.* **56:**339–347.

Ryan, M. J., Warkentin, K. M., McClelland, B. E., and Wilcsynski, W. (1995). Fluctuating asymmetries and advertisement call variation in the cricket frog, *Acris crepitans. Behav. Ecol.* **6:**124–131.

Sage, R. D., Heyneman, D., Lim, K.-C., and Wilson, A. C. (1986). Wormy mice in a hybrid zone. *Science (Washington, D.C.)* **324:**60–63.

Schall, J. J. (1996). Malarial parasites of lizards: diversity and ecology. *Adv. Parasitol.* **37:**255–333.

Schall, J. J., Bennet, A. F., and Putman R. W. (1982). Lizards infected with malaria: physiological and behavioral consequences. *Science (Washington, D.C.)* **217:**1057–1059.

Schaller, G. B. (1963). *The Mountain Gorilla.* Chicago: Chicago University Press.

Sciulli, P. W., Doyle, W. J., Kelley, C., Siegel, P., and Siegel, M. I. (1979). The interaction of stressors in the induction of increased levels of fluctuating asymmetry in the laboratory rat. *Am. J. Phys. Anthropol.* **50:**279–284.

Shapiro, B. L. (1983). Down syndrome—A disruption of homeostasis. *Am. J. Med. Genet.* **14:**241–269.

Siegel, M. I., and Smookler, H. H. (1973). Fluctuating dental asymmetry and audiogenic stress. *Growth* **37:**35–39.

Simms, E. L. (1992). Costs of plant resistance to herbivory. In *Plant resistance to herbivores and pathogens* (R. S. Fritz and E. L. Simms, eds.), pp. 392–425. Chicago: The University of Chicago Press.

Simmons, L. W. (1995). Correlates of male quality in the field cricket, *Gryllus campestris* L.: Age, size and symmetry determine pairing success in field populations. *Behav. Ecol.* **6:**376–381.

Slater, E., Hare, E. H., and Price, J. S. (1971). Marriage and fertility of psychiatric patients compared with national data. *Social Biol. (Suppl.)* **18:**60–73.

Smith, P., Townsend, M. G., and Smith, R. H. (1991). A cost of resistance in the brown rat? Reduced growth rate in warfarin-resistant lines. *Funct. Ecol.* **5:**441–447.

Solberg, E. J., and Sæther, B.-E. (1993). Fluctuating asymmetry in the antlers of moose *(Alces alces):* Does it signal male quality? *Proc. R. Soc. Lond. B Biol. Sci.* **254:**251–255.

Stern, C. (1986). *Genetic mosaics and other essays.* Cambridge: Harvard.

Sullivan, M. S., Robertson, P. A., and Aebischer, N. A. (1993). Fluctuating asymmetry measurement. *Nature (Lond.)* **361:**409–410.

Swaddle, J. P. (1996). Reproductive success and symmetry in zebra finches. *Anim. Behav.* **51:**203–210.

Swaddle, J. P., and Cuthill, I. C. (1994a). Female zebra finches prefer males with symmetric chest plumage. *Proc. R. Soc. Lond. B Biol. Sci.* **258:**267–271.

Swaddle, J. P., and Cuthill, I. C. (1994b). Preference for symmetric males by female zebra finches. *Nature (Lond.)* **367:**165–166.

Swaddle, J. P., and Cuthill, I. C. (1995). Asymmetry and human facial attractiveness: Symmetry may not always be beautiful. *Proc. R. Soc. Lond. B Biol. Sci.* **261:**111–116.

Swaddle, J. P., and Witter, M. S. (1994). Food, feathers and fluctuating asymmetry. *Proc. R. Soc. Lond. B Biol. Sci.* **255:**147–152.

Sykes, A. R., and Coop, R. L. (1976). Intake and utilization of food by growing lambs with parasitic damage to the small intestine caused by daily dosing with *Trichostrongylus colubriformis* larvae. *J. Agric. Sci.* **86:**507–515.

Sykes, A. R., and Coop, R. L. (1977). Intake and utilization of food by growing sheep with abomasal damage caused by daily dosing with *Ostertagia circumcincta* larvae. *J. Agric. Sci.* **88:**671–677.

Tanaka, S. (1982). Variations in nine-spine stickelbacks, *Pungitius pungitius* and *P. sinensis,* in Honsu. *Jpn. J. Ichthyol.* **29:**203–212.

Thoday, J. M. (1958). Homeostasis in a selection experiment. *Heredity* **12:**401–415.

Thompson, S. N. (1983). Biochemical and physiological effects of metazoan endoparasites on their host species. *Comp. Biochem. Physiol.* **74B:**183–211.

Thornhill, R. (1991). Female preference for the pheromone of males with low fluctuating asymmetry in the Japanese scorpionfly *(Panorpa japonica:* Mecoptera). *Behav. Ecol.* **3:**277–283.

Thornhill, R. (1992a). Fluctuating asymmetry and the mating system of the Japanese scorpionfly, *Panorpa japonica. Anim. Behav.* **44:**867–879.

Thornhill, R. (1992b). Fluctuating asymmetry, interspecific aggression and male mating tactics in two species of Japanese scorpionflies. *Behav. Ecol. Sociobiol.* **30:**357–363.

Thornhill, R., and Gangestad, S. W. (1993). Human facial beauty. *Human Nature* **4:**237–269.

Thornhill, R., and Gangestad, S. W. (1994). Human fluctuating asymmetry and sexual behavior. *Psychol. Sci.* **5:**297–302.

Thornhill, R., Gangestad, S. W., and Comer, R. (1995). Human female orgasm and mate fluctuating asymmetry. *Anim. Behav.* **50:**1601–1615.

Thornhill, R., and Sauer, P. (1992). Genetic sire effects on the fighting ability of sons and daughters and mating success of sons in a scorpionfly. *Anim. Behav.* **43:**225–264.

Tomkins, J. L., and Simmons, L. W. (1995). Patterns of fluctuating asymmetry in earwig forceps: No evidence for reliable signalling. *Proc. R. Soc. Lond. B Biol. Sci.* **259:**89–96.

Torado, G. J., and Martin, G. M. (1967). Increased susceptibility of Down syndrome fibroblasts to transformation by SV40. *Proc. Soc. Exp. Biol. Med.* **124:**1232–1236.

Trivers, R. L. (1972). Parental investment and sexual selection. In *Sexual selection and the descent of man* (B. Campbell, ed.), pp. 136–179. Chicago: Aldine.

Ueno, H. (1994). Fluctuating asymmetry in relation to two fitness components, adult longevity and male mating success in a ladybird beelte, *Harmonia axyridis* (Coleoptera: Coccinelidae). *Ecol. Entomol.* **19:**87–88.

Valentine, D. W., and Soulé, M. E. (1973). Effect of p, p′-DDT on developmental stability of pectoral fin rays in the grunion, *Leuresthes tenius. Fisheries Bull (Dublin)* **71:**920–921.

Van Valen, L. (1962). A study of fluctuating asymmetry. *Evolution* **16:**125–142.

Waddington, C. H. (1957). *The strategy of the genes.* U.K.: George Allen and Unwin.

Wakelin, D. (1978). Genetic control of susceptibility and resistance to parasitic infection. *Adv. Parasitol.* **16:**219–308.

Wakelin, D. (1985). Genetic control of immunity to helminth infections. *Parasit. Today* **1:**17–23.

Wakelin, D. M., and Blackwell, J. M. (1988). *Genetics of resistance to bacterial and parasitic infection.* U.K.: Taylor & Francis.

Watson, P. J., and Thornhill, R. (1994). Fluctuating asymmetry and sexual selection. *Trends Ecol. Evol.* **9:**21–25.

West-Eberhard, M. J. (1979). Sexual selection, social competition, and evolution. *Proc. Am. Philos. Soc.* **123:**222–234.

Whitefield, P. J. (1979). *The biology of parasitism: An introduction to the study of associating organisms.* Baltimore: University Park Press.

Wilber, E., Newell-Morris, L. L., and Pytkowicz Streissguth, A. (1993). Dermatoglyphic asymmetry in fetal alcohol syndrome. *Biol. Neonate* **64:**1–6.

Witter, M. S., and Swaddle, J. P. (1994). Fluctuating asymmetries, competition and dominance. *Proc. R. Soc. Lond. B Biol. Sci.* **256:**299–303.

Woodard, D. B., and Fukuda, T. (1977). Laboratory resistance of the mosquito *Anopheles quadrimaculatus* to the mermithid nematode *Diximermis peterseni. Mosquito News* **37:**192–195.

Zakharov, V. M. (1992). Population phenogenetics: Analysis of developmental stability in natural populations. *Acta Zool. Fenn.* **191:**7–30.

Zar, J. H. (1984). *Biostatistical analysis.* Englewood Cliffs: Prentice Hall.

Zuk, M., Thornhill, R., Ligon, J. D., and Johnson, K. (1990). Parasites and mate choice in red jungle fowl. *Am. Zool.* **30:**235–244.

14

Hormones and Sex-Specific Traits: Critical Questions

Diana K. Hews and Michael C. Moore

I. Introduction

Behavioral ecologists and evolutionary biologists have become increasingly more focused on studying the physiological processes underlying phenotype expression. Greater attention to proximate mechanisms should improve our understanding of the physiological targets of selection (e.g., Finch and Rose, 1995) and of how physiological processes affect patterns of evolution (e.g., Endler and McLellan, 1988; Ryan, 1991). Workers interested in sexual selection have increasingly turned to questions about the hormonal mechanisms regulating expression of sex-specific secondary traits. Many organisms exhibit traits that differ between the sexes, and such traits often are thought to contribute to differential reproductive success. At the proximate level, these traits are examples of secondary sexual characters—sex-specific traits that develop after primary sexual differentiation and whose expression is dependent on the actions of sex steroid hormones (Van Tienhoven, 1983).

In this chapter we selectively review and critique the working model that behavioral ecologists and evolutionary biologists often use when considering how steroid hormones mediate the expression of secondary sexual characters. A commonly held view is that there is substantial empirical support for a relatively tight relationship between variation in androgen levels and resulting variation in androgen effects. For example, Ligon and colleagues (1990) make this assumption when suggesting a relationship between trait expression and variation in body condition resulting from disease in red jungle fowl (see also, for example, Wittenberger, 1981, pp. 151–153; Folstad and Karter, 1992). Our critique focuses on the nature of hormone–trait relationships.

We concentrate on two themes of central importance for understanding hormone-trait relationships that also have implications for the evaluation of evolutionary arguments about immune function and sexual selection (Hamilton and

Zuk, 1982; Folstad and Karter, 1992) presented below. First, we focus on the nature of hormone dependency for secondary sexual traits, the interaction between hormone-dependent and hormone-independent control of the phenotype, and the complexity of endocrine system regulation. In the second part we focus on the notion of evolutionary plasticity in hormone–trait relationships. For simplicity we discuss primarily, though not exclusively, sex differences in behavior, though the same arguments apply to any phenotypic trait.

We integrate our presentation with discussion of the immunocompetence handicap hypothesis (Folstad and Karter, 1992). This hypothesis considers relationships among the immune system, androgens, and secondary sexual traits. Hamilton and Zuk (1982) proposed that elaborate secondary sexual traits evolve in males because such traits signal to females the presence of genes that confer greater resistance to parasites. Males with lower parasite loads can display with greater vigor or exhibit greater trait elaboration, thereby signaling their underlying genetic quality. At the same time, expression of elaborate traits can be considered a handicap (Zahavi, 1975, 1977; Maynard Smith, 1985) because their expression could reduce a male's chance of survival.

The immunocompetence handicap hypothesis (IHH) of Folstad and Karter (1992) provides a mechanism for the feedback system necessary for the Hamilton–Zuk hypothesis. In brief, the IHH proposes that tradeoffs exist between immune system activation and level of immunocompetence and the elaboration of secondary sexual traits, and that these tradeoffs are mediated by plasma levels of testosterone. The IHH is based on integrating separate bodies of research indicating that: (1) elevated testosterone levels are necessary for elaboration of male secondary sexual traits, and (2) elevated testosterone levels reduce immunocompetence. The IHH posits that parasite-infected males must reduce testosterone levels to increase immunocompetence and thereby suffer the unavoidable cost of reduced elaboration of testosterone-dependent secondary sexual traits. The causal argument of the IHH for the necessary feedback between trait elaboration and parasite resistance is a valuable approach for refining (or redefining) the Hamilton–Zuk hypothesis. However, some of the underlying assumptions of the IHH deserve more careful consideration, which are highlighted below.

II. Deterministic Models of Hormone-Behavior Relationships

Historical reasons may partially explain why many behavioral ecologists and evolutionary biologists hold an overly deterministic view of hormone action, especially with respect to hormone–behavior relationships. At the root of this may be a too-literal interpretation of the term *hormone dependence.* In this section we discuss three aspects of hormone dependence, including (1) two models of hormone action (dose-dependent versus threshold response), (2) how hormone-dependent and hormone-independent factors mediate trait expression, and (3) the complexity of the endocrine system and its regulation.

A. Hormone Dependence: Dose Dependency versus Threshold Response

Many sex-specific behavior patterns and morphological traits are androgen-dependent, and not expressed in the absence of androgen (e.g., van Tienhoven, 1983; Becker et al., 1992; and citations in Folstad and Karter, 1992). Typically researchers establish hormone dependency by determining how trait expression is affected by removal of the endogenous source of the hormone, and by subsequent exogenous replacement of the hormone (for more detailed discussion see Moore and Lindzey, 1992; Nelson, 1995). The typical response for hormone-dependent traits is a dramatic decline or complete abolition of trait expression upon removal of the major source of the hormone, typically achieved via surgical or pharmacological castration.

Testosterone is a key androgen mediating many sex differences in vertebrates. Testosterone could affect the elaboration of male secondary sexual traits in one of two ways. In a dose-dependent model (Fig. 14-1a), elaboration of the trait is increased for each incremental hormone level. In a threshold model (Fig. 14-1b), hormones act as simple on–off switches: full expression of the trait is turned on when levels exceed some minimum threshold, and enhancement of expression is not affected by a rise above this minimum.

An understanding of the endocrinological methods and the models of hormone action that underlie claims of hormone dependence is critically important when constructing and evaluating evolutionary hypotheses. For example, most theories relating to sexual selection, such as the IHH, must consider phenotypic variation

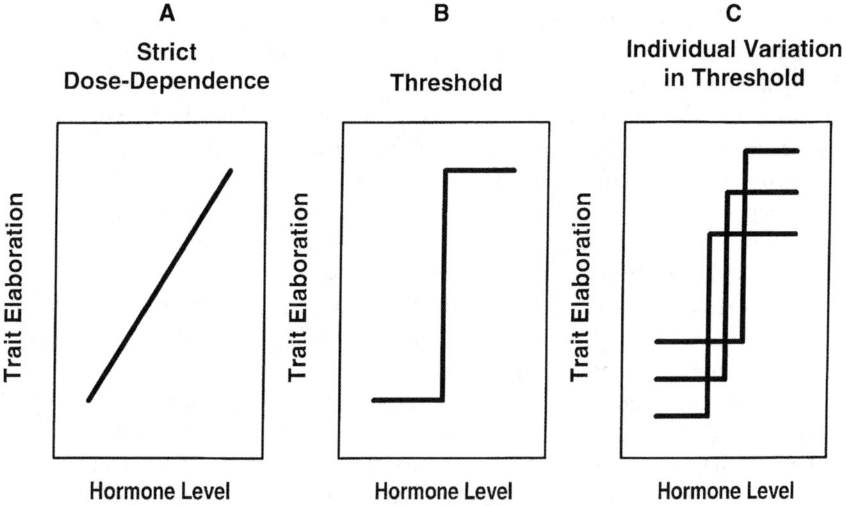

Figure 14-1. Models of relationships between hormone levels and trait elaboration: A) dose-dependent, B) threshold response, and C) individual variation in threshold of response. See text for further discussion.

among reproductively competent individuals of a sex (typically males). However, the vast majority of studies documenting androgen dependence, and all of those cited by Folstad and Karter (1992) in support of the IHH, do not compare reproductive males that have various elevated, physiological levels of testosterone. Rather they examine the effects of the presence and absence of testosterone by comparing, for example, breeding and non-breeding males, or castrated versus intact animals.

This is critical to the IHH. A central assumption of this hypothesis is that there is a tight dose-dependent relationship within individuals between relatively small variations in testosterone levels and the degree of trait expression (Folstad and Karter, 1992). This dose-dependency is essential to the IHH in that (1) it produces variations in signal expression in reproductively competent males, and (2) the tight coupling of signal elaboration and androgen levels would maintain the honesty of the signal, since androgen levels affect immunocompetence.

There are few experimental demonstrations of a relationship between increasing physiological levels of testosterone and increasing trait elaboration; even correlational data supporting such a relationship are rare. One of the few such correlations for sexual behavior is the relationship between singing behavior and testosterone levels in breeding songbird males (Sossinka, 1980). And in reproductive red jungle fowl males, plasma testosterone levels correlate with comb length, a trait upon which females sometimes base their mate choice (Zuk et al., 1995). In barn swallows, testosterone levels correlate with tail length, also a trait that females assess in choosing mates (Saino and Møller, 1995) However, in the latter study, testosterone levels were first statistically adjusted for effects of season and breeding stage, and it is unclear if the residual variance was sufficiently greater than the error arising from intra- and inter-assay variation in measuring the hormone levels.

Most data for secondary sexual traits in vertebrates either better support the threshold model of hormone action or are not sufficient to discriminate between the two models of hormone action. To illustrate the general problem of studying dose dependency, consider the studies cited by Folstad and Karter (1992). Although different hormone doses were administered, these studies have methodological and/or interpretational problems. First, some were not replicated or had sample sizes of less than five individuals (Rohwer and Rohwer, 1978; Watson and Parr, 1981; Post et al., 1987). Second, the dosages varied a thousand-fold and thus possibly generated pharmacologic artifacts at the higher doses. Further, other studies that demonstrated a linear relationship between hormone levels and response used a logarithmic scale for the dose axis, making an inherently nonlinear relationship linear (Lofts and Murton, 1973; Norton and Wira, 1977). Data viewed on untransformed axes are much more consistent with the threshold response hypothesis. Consequently, these data cannot be used to distinguish whether the variation in androgen levels in reproductively competent males acts either in a dose- or threshold-dependent manner.

Some studies indicate that elevations in circulating androgens above levels normally found in breeding males progressively increase the level of trait expression. For example, the "challenge hypothesis" study by Wingfield and coworkers (1990), a review of mating systems and patterns of testosterone secretion, is cited as support for the concept that display and aggression behaviors track testosterone profiles (Folstad and Karter, 1992). Although the challenge hypothesis proposes different thresholds for different traits, it does not require strict dose dependency. Furthermore, empirical data relating to the challenge hypothesis have focused on the hormonal consequences of social interactions, not the behavioral effects of the documented hormonal changes. Data directly testing for dose-dependent effects of socially stimulated hormone changes are rare (e.g., Moore, 1984, 1987) and do not distinguish between true dose dependency versus the existence of multiple thresholds (see below).

A difficult problem in discriminating between threshold and dose-dependent models is that the existence of direct dose dependency within an individual may be very difficult to demonstrate rigorously. A common approach to this problem is to correlate variation in expression of male traits with variation in hormone levels. Data indicating a positive relationship is taken as support for dose dependency. However, such results can also be produced if individuals vary in their threshold (Fig. 14-1c). A more direct approach is to manipulate hormone levels, but these studies also suffer from the difficulties of discriminating true dose dependency from individual variation in threshold. There are few manipulative studies directly testing dose dependency, and they provide mixed support. For example, Davidson (1972) found that castrated adult male rats exhibited dose dependency in expression of copulatory behavior when administered doses between 0 percent and 10 percent of those found in normal reproductive males, but when given testosterone at levels between 10 percent and 100 percent of physiological levels, exhibited no relationship between trait expression and hormone level. Thus, androgens would have to be decreased by more than 90 percent in a reproductive male for an effect to be observed. Ferkin and colleagues (1994) found that in meadow voles increasing doses of testosterone given to castrated males resulted in production of pheromones that were increasingly attractive (as measured by time investigated by a conspecific). However, of the five doses given to castrated males (no testosterone implant, 5 mm, 10 mm, 15 mm, or 20 mm implants) only one (10 mm) yielded plasma hormone levels that fell within the range measured in breeding season males.

Thus, a thorough consideration of the studies cited in support of the IHH and of other research reveals that most data (1) do not provide conclusive demonstration of dose dependency, and (2) may be more consistent, or at least equally consistent, with a threshold effect of testosterone. Considering these problems and the results of other studies supporting simple threshold effects of steroid hormones (e.g., Crews et al., 1991), we conclude that there is not strong direct evidence for the type of dose-dependent action of testosterone required by the

IHH. Further support is needed for the link between increasing androgen levels, specifically dose dependency, in reproductively competent males and increasing expression of androgen-dependent traits. It also would be valuable to examine the IHH model itself to determine if it may be modified to accommodate threshold expression.

Despite the importance to the IHH of demonstrating direct dose dependency, a direct attack on this problem may not be productive because of the difficulties in designing an experiment to discriminate between dose dependency and individual variation in threshold. We therefore suggest a more feasible experiment that, while it does not test the dose-dependence assumption, does test the fundamental assumption of a link between parasite loads, testosterone levels, and trait elaboration. The IHH predicts that males with high parasite loads and reduced trait expression (and presumably low testosterone levels) should respond to increased testosterone levels with an increase in trait elaboration. If such males showed an increase in trait elaboration in response to testosterone supplements (at the high end of the physiological range), it would support this critical feature of the IHH. Failure of testosterone supplements to increase trait elaboration would suggest that reduced trait elaboration is due to mechanisms not involving testosterone levels.

Finally, Poulin and Vickery (1993) present similar arguments about the androgen–immune system link of the IHH. They state that the IHH, as proposed, implies a linear relationship between hormone levels and immune function. They suggest there is little information supporting this view based on observations of *in vivo* systems.

B. Hormone-Dependent and -Independent Factors

Most laboratory studies of hormone action on secondary sexual characters intentionally minimize the influence of other factors in order to maximize the probability of detecting a hormonal effect (Feder, 1984; Moore, 1987). Failure to recognize this has led to an overly deterministic view of hormone action (Moore, 1984; Wingfield and Ramenofsky, 1985; Moore and Marler, 1988; Wingfield et al., 1990) and has advanced the misconception that a direct correspondence between small variations in androgen levels and individual variations in secondary sexual characters is well-supported. This view is misconceived for at least two other reasons aside from the concerns about analysis of dose dependency outlined above.

First, many nonhormonal factors affect trait expression, even for traits that are clearly hormone-dependent. Biologists are well aware of the multiplicity of factors affecting trait expression, but some mistakenly hold the view that hormones represent the "final common pathway" (Fig. 14-2a) through which these multiple cues are integrated. A more correct view is that hormones are one of many control pathways that (directly and indirectly) affect trait expression (Fig. 14-2b). Second, in contrast to the actions of androgens under rigorously controlled

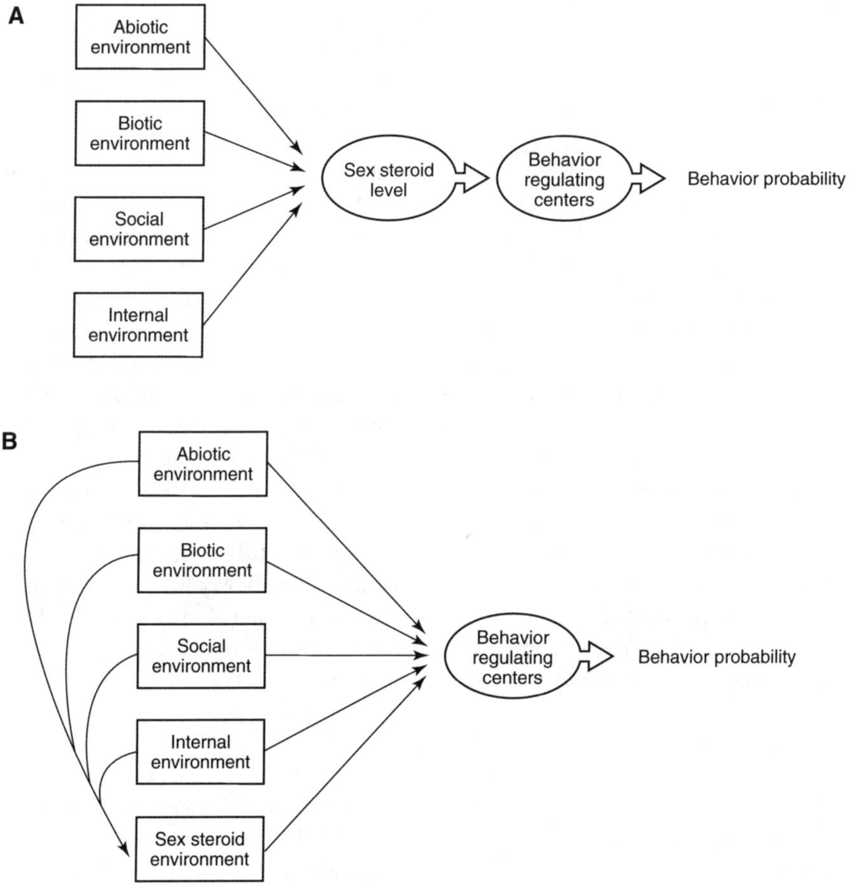

Figure 14-2. Conceptual models of relationships between factors with direct and indirect effects on trait expression: A) final common pathway, and B) multiple pathways. See text for further discussion.

and simplified laboratory settings, androgens in the natural environment act synergistically with many other cues regulating individual differences in behavior and morphology (e.g., Wingfield and Farner, 1980; Wingfield and Moore, 1987; Moore and Marler, 1988; Becker et al., 1992).

The two following studies are examples of the importance of multiple cues. In rats levels of aggression decline, both for resident and nonresident males, following castration (Christie and Barfield, 1979). This illustrates that aggression is hormone-dependent. However, dramatic differences in aggression levels between residents and nonresidents persist even following castration. Thus some factor other than circulating plasma hormone levels, perhaps social experience or motivation, must contribute to differences in aggression. Similarly, in mountain spiny lizards, castration and testosterone-implantation experiments on free-living animals reveal that territorial aggression in males is mediated by both testosterone-dependent and testosterone-independent factors (reviewed in Moore and Lindzey, 1992).

Given that multiple factors regulate trait expression, is such an androgen-dependent trait a good predictor of immune functioning and of genetic quality (i.e., an indicator of genes that confer resistance to infection)? A substantial proportion of individual variation in trait expression may not be the result of variation among individuals in testosterone titers, but the result of variation in other controlling factors. For example, in barn swallows, only 12% of variation in tail length (a potentially androgen-dependent secondary sexual trait that females assess in choosing a mate) is explained by naturally occurring variation in androgen levels (statistically adjusted for effects of season and stage of breeding cycle) in reproductive males (Saino and Møller, 1994). Thus, not surprisingly, other factors such as genetics and nutritional condition must also be mediating variation in tail length. Further, even if the androgen–immune system link in this species is proven in future studies, the amount of variation attributable to variation in testosterone levels remains relatively small and tail length would not be a particularly good predictor of immune function as a basis for female mate choice. Thus the complexity of the system renders it less informative to females, even if females are using this trait as an indicator of male quality.

C. Complexity of the Endocrine System

Regulation of the endocrine system and the interactions among endocrine system components are complex (Brown, 1994; Nelson, 1995). Many hormones exert negative feedback at several levels in the brain and affect levels of other hormones. For example, increased circulating levels of testosterone can result in increased or decreased levels of other hormones. Further, a single hormone acts on multiple target systems, each of which can also have direct and indirect effects on the immune system. Grossman (1989) provides an excellent overview of this hormonal complexity as it relates to the immune system, reviewing how gonadal steroids, adrenal steroids, growth hormone, and prolactin from the pituitary, thymic hormones, and substances generated by activated lymphocytes all interact to affect developing lymphocytes and to stimulate T- and B-cell functioning. An especially relevant example is the relationship between testosterone and prolactin levels. Immunosuppressive effects of testosterone may vary depending on levels of prolactin, which is immunostimulatory. If elevated testosterone levels are correlated with elevated prolactin levels, then the immunosuppressive effects of the androgen have a greatly reduced impact on immune functioning. Prolactin is often considered a "female" hormone, but males of a number of mammalian and passerine bird species exhibit significant elevated prolactin levels that function during the breeding season (Goldsmith, 1983; Bronson, 1989). Finally, some evidence confirms that testosterone is not universally immunosuppressive. For example, Cohn (1979) reviews numerous studies demonstrating that androgens have no effect or are immunostimulatory, depending on what component of immune function is being assessed for a response.

Møller (1995) also argues that such complexity needs further examination when evaluating the IHH, and notes that other endocrine factors such as adrenal steroids (e.g., corticosterone, cortisol) may be particularly important (see Chapter 7). Corticosterone, for example, is known to influence the immune response, and levels rise during some parasitic infections. In some host species, this phenomenon may be directly manipulated by the parasite. Møller argues that the most fundamental solution for addressing this complexity will be to take a broad perspective and measure the actual fitness consequences of several biochemical factors affecting immune function, not just testosterone, which will obviously be a difficult task.

III. Evolutionary Plasticity of Hormone-Trait Relationships

Developmental processes are often viewed as constraints to evolution (Gould and Lewontin, 1979). In this view these processes are considered to lack evolutionary plasticity, and instead are considered relatively conserved. Along these lines, the IHH posits an obligate, evolutionarily inflexible coupling between trait expression and androgen levels. Thus the IHH assumes that this relationship is evolutionarily stable and resistant to mutation. In evolutionary game theory (e.g., Maynard Smith, 1985), the system is thus viewed as stable and immune to invasion by a "cheater" strategy (a mutant) that would uncouple the relationship between trait elaboration and androgen levels. However, we argue that any such system is indeed vulnerable to invasion by "cheaters," and thus is evolutionarily unstable and would not likely persist. Our argument is based on the extensive evidence of evolutionary plasticity in hormone–trait relationships in a wide array of vertebrates (see below). Therefore, existing hormone–trait relationships should not be construed as evolutionarily inflexible. This potential for evolutionary uncoupling indicates that the IHH may be less general than first conceived.

Why do we argue that the IHH scenario would not be stable, but susceptible to "invasion" by a "cheater" mutant? The IHH proposes that testosterone is the causal link between signal expression and parasite loads, maintaining the "honesty" of the signal. Only males with genes conferring resistance to parasite infection—resistance that resides outside normal immune system functioning—can afford to exhibit elevated testosterone levels and suffer the subsequent reduction in immunocompetence of both the cellular and humoral immune system components. Males lacking such genes cannot afford this cost. However, if a mutation arose (representing a "cheater" strategy in game theory parlance) that conferred the ability to greatly elaborate a trait with relatively low androgen levels, functionally "decoupling" trait expression from circulating androgen levels, this mutation could be favored by selection. Any male that could exhibit a high degree of trait elaboration but have relatively low androgen levels would have an advantage over other males in the population, if this mutation allowed them to express the showy trait and increase their reproductive success and not suffer reduced immune function. Thus, such a mutation could successfully increase in frequency and quickly invade a population.

The primary reason such a scenario is possible is that the hormonal control system is now recognized to be evolutionarily labile (Crews and Moore, 1986). Such lability suggests that the evolutionary uncoupling of androgen levels and trait expression, via a mutation or a series of mutations, is also possible. A now classic example of such an evolutionary uncoupling of the androgen dependency of trait expression is reproductive behavior in vertebrates, which was formerly considered to be obligately dependent on elevated steroid levels. In some species, courtship and breeding behavior do not require activation by gonadal steroids, and instead are activated by environmental cues such as temperature or light cycles (see reviews in Moore and Lindzey, 1992; Crews, 1992; Nelson, 1995). This uncoupling reveals the evolutionary malleability of physiologically controlled mechanisms (Crews and Moore, 1986; Moore and Marler, 1988).

Another classic example of evolutionary uncoupling particularly relevant to the IHH is control of aggressive behavior in female Wilson's phalaropes, a polyandrous bird species with sex role reversal. Females exhibit high levels of aggressive behavior and bright plumage, though such traits are typically expressed only in males in many vertebrates. It is unclear how the showy plumage characters of females are controlled in this species, and there are few examples of androgen-dependent plumage characters in birds. Indeed, in many oscine passerines (songbirds), elaborate male plumage results from the absence of high estrogen levels; in females, with higher estrogen levels, development of male plumage is inhibited. Owens and Short (1995) discuss these data with reference to application of IHH and cite species with secondary sexual traits (mostly showy plumage characteristics), whose expression is not coupled to androgen levels but are under primarily genetic control.

In female phalaropes we can speculate about possible evolutionary changes in the hormonal mediation of aggressive behavior, the other trait elaborated in females. This behavior is not accompanied by high circulating androgen levels (Fig. 14-3), which are in the range typical for other female birds (Fivizzani et al., 1986). Thus, female phalaropes show aggression despite low plasma androgen levels. Selection presumably has altered tissue-specific aspects of hormonal control, instead of altering circulating levels of testosterone. Recent data indicate that females have unusually high levels of androstenedione (Fivizzani et al., 1994), a precursor of testosterone. This suggests that some target tissues in females might have high levels of the enzyme 17β-HSD, which converts this precursor steroid to the biologically active steroid testosterone. Other particularly relevant examples of evolutionary plasticity in the occurrence and activity of

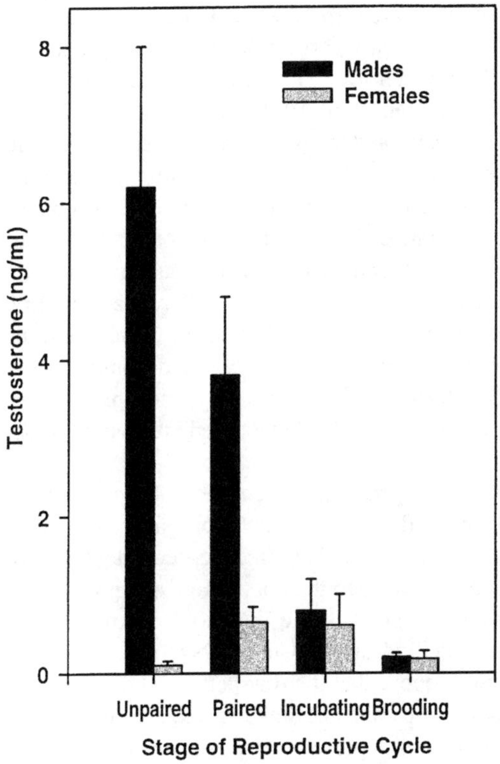

Figure 14-3. Levels of testosterone as a function of reproductive stage in male and female phalaropes, a shorebird species with sex role–reversal. Data redrawn from Fivizzani et al. (1986).

enzymes important in converting androgens to biologically active metabolites include 5-α reductase activity in female phalaropes (Schlinger et al., 1989) and the occurrence of hen-feathering as a consequence of unusually high levels of aromatase activity, the enzyme that converts testosterone to estradiol, in male Sebright chickens (George et al., 1981). Sexual selection could similarly favor male cheaters with characteristics allowing decreased reliance of trait elaboration on a certain high level of circulating androgen by changing the level of response of a specific target tissue to a hormone.

In general, changing target-tissue characteristics would have fewer consequences for the entire organism than would altering circulating levels of a steroid hormone that has multiple effects. Indeed, we predict that circulating levels would be altered evolutionarily only when selection favors altering the timing of the entire reproductive cycle. In such a case, an entire suite of reproductive characters would be affected. For example, many North American lizards breed in the late spring, but some species are fall-breeding. In the former species, seasonally elevated hormone levels in late spring and early summer contribute to seasonal expression of reproduction-linked traits and behaviors. In the latter species, this seasonal elevation is shifted to the fall months (see Moore and Lindzey, 1992).

Other types of evolutionary uncoupling are also possible. For example, even if dose-dependent trait expression occurred in any population, a mutation producing individual variation in sensitivity to a given level of circulating androgens may appear (e.g., Bartke, 1974; Cohn and Hamilton, 1976; Post et al., 1987) and invade the population. Changes in receptor density or receptor specificity are other mechanisms underlying an evolutionary change in sensitivity, as would be the increase in enzymes needed to metabolize a hormone precursor to its biologically active form discussed above, or changes in hormone metabolism and inactivation. Such evolutionary changes in the endocrine system could functionally remove the dose dependency of signal expression, assuming there was variation among individuals. This variation in sensitivity to testosterone, which could be tissue-specific, would produce variation in signal expression decoupled from parasite loads, and thus would be "dishonest." Theoretically, the link between testosterone and immune functioning also could be "uncoupled" if favored by selection.

The preceding discussion raises an interesting speculation. If indeed most secondary sexual traits do not exhibit dose-dependent expression, it may be because these traits were formerly "honest" signals with dose-dependent expression, but a cheater with threshold control invaded the population, or a progressive sequence of evolutionary changes occurred, and the threshold control thereby became fixed in the population. Wedekind and Folstad (1994) discuss similar arguments regarding "cheaters," but propose that the link between hormones and immune function (a negative relationship) and the link between hormones and trait elaboration (a positive relationship) represent adaptive links maintained because immune function and trait elaboration represent competing demands. They suggest that energy requirements or a limiting factor needed both for trait expression and immune function could be the basis for this competition. Whether immune functioning and trait elaboration are processes that compete for a common limited resource remains to be established.

IV. Conclusions

We offer insights into key aspects of hormone-trait relationships, and about the nature of the underlying endocrinological framework for these relationships. Most importantly, we have discussed why biologists should be cautious in using a deterministic model of hormone action. There is ample evidence indicating that if a hormone is removed, intensity of trait expression declines dramatically. Such hormone-dependent traits, however, have not been shown experimentally to exhibit graded expression in response to physiologically relevant gradations in hormone levels, and few such correlations have even been found. Further, hormones are only one of many interacting factors influencing trait expression. Finally, many examples of evolutionary plasticity in hormone-trait relationships have now been documented, suggesting that selection could alter the nature of these relationships at numerous points in the hormonal control pathways.

The IHH has been extremely valuable in focusing attention on several critical issues. Current support for the link between physiological variation in plasma androgen levels and graded variation in trait expression is not as strong as originally suggested. We propose an experimental paradigm to assess the links between androgen levels, degree of trait expression, and immunocompetence. Such experiments would provide valuable direct tests of the IHH, in addition to more general information about the interplay between testosterone control of secondary sexual characteristics and sexual selection. Whether or not the IHH can be modified to accommodate threshold androgen effects instead of simply reflecting graded dose dependency should also be explored.

V. Acknowledgments

We thank R. Knapp, M. Polak, and S. Woodley for discussions and comments on earlier versions of this manuscript. This work was supported by an Individual National Research Service Award (NIMH 5F32 MH10074 PYB) to D.K.H, and by NIMH MH 48654 PYB to M.C.M.

VI. References

Bartke, A. (1974). Increased sensitivity of seminal vesicles to testosterone in a mouse strain with low plasma testosterone levels. *J. Endocrinol.* **60:**145–148.

Becker, J. B., Breedlove, S. M., and Crews, D., eds. (1992). *Behavioral endocrinology.* Cambridge, Mass.: MIT Press.

Bronson, F. H. (1989). *Mammalian reproductive biology.* Chicago: University of Chicago Press.

Brown, R. E. (1994). *An introduction to neuroendocrinology.* Cambridge: Cambridge University Press.

Christie, M. H., and Barfield, R. J. (1979). Effects of castration and home cage residency on aggressive behavior in the rat. *Horm. Behav.* **13:**85–91.

Cohn, D. A., and Hamilton, J. B. (1976). Sensitivity to androgen and the immune response: Immunoglobulin levels in two strains of mice, one with high and one with low-target organ responses to androgen. *J. Reticuloendo. Soc.* **20:**1–10.

Crews, D. (1992). Diversity of hormone–behavior relations in reproductive behavior. In *Behavioral endocrinology* (J. B. Becker, S. M. Breedlove and D. Crews, eds.), pp. 143–186. Cambridge: MIT Press.

Crews, D., Bull, J. J., and Wibbels, T. (1991). Estrogen and sex reversal in turtles: A dose-dependent phenomenon. *Gen. Comp. Endocrinol.* **81:**357–364.

Crews, D., and Moore, M. C. (1986). Evolution of mechanisms controlling behavior. *Science* (Wash., D.C.) **231:**121–125.

Davidson, J. M. (1972). Hormones and behavior. In *Hormones and reproductive behavior* (S. Levine, ed.), pp. 63–103. New York: Academic Press.

Endler, J. A., and McLellan, T. (1988). The processes of evolution: towards a newer synthesis. *Ann. Rev. Ecol. Syst.* **19:**395–421.

Feder, H. H. (1984). Hormones and sexual behavior. *Annu. Rev. Psychol.* **34:**165–200.

Ferkin, M. H. Sorokin, E. S. Renfroe, M. W., and Johnston, R. E. (1994). Attractiveness of male odors to females varies directly with plasma testosterone concentration in meadow voles. *Physiol. Behav.* **55:**347–353.

Finch, C. E., and Rose, M. R. (1995). Hormones and the physiological architecture of life history evolution. *Q. Rev. Biol.* **70:**1–52.

Fivizzani, A. J., Colwell, M. A., and Oring, L. W. (1986). Plasma steroid hormone levels in free-living Wilson's phalaropes, *Phalaropus tricolor. Gen. Comp. Endocrinol.* **62:**137–144.

Fivizzani, A. J., Delehanty, D., Oring, L., Wrege, R., and Emlen, S. (1994). Elevated female androstenedione levels in two sex role–reversed avian species. *Amer. Zool.* **34:**22A.

Folstad, I., and Karter, A. J. (1992). Parasites, bright males, and the immunocompetence handicap. *Amer. Natur.* **139:**603–622.

George, F. W., Noble, J. F., and Wilson, J. D. (1981). Female feathering in Sebright cocks is due to conversion of testosterone to estradiol in the skin. *Science* **213:**557–559.

Goldsmith, A. R. (1983). Prolactin in avian reproductive cycles. In *Hormones and behavior in higher vertebrates* (J. Balthazart, E. Pröve and R. Giles, eds.) pp. 375–387. Berlin: Springer-Verlag.

Gould, S. J., and Lewontin, R. C. (1979). The spandrels of San Marco and the Panglossian paradigm: A critique of the adaptationist program. *Proc. Roy. Soc. Lond. B* **205:**581–598.

Grossman, C. 1989. Possible underlying mechanisms of sexual dimorphism in the immune response, fact and hypothesis. *J. Steroid Biochem.* **34:**241–251.

Hamilton, W. D., and Zuk, M. (1982). Heritable true fitness and bright birds: A role for parasites? *Science* (Wash., D.C.) **218:**384–387.

Ligon, J. D., Thornhill, R., Zuk, M., and Johnson, K. (1990). Male–male competition, ornamentation, and the role of testosterone in sexual selection in the red jungle fowl. *Anim. Behav.* **40:**367–373.

Lofts, B., and Murton, R. K. (1973). Reproduction in birds. In *Avian biology* (D. S. Farner and J. R. King, eds.), Vol. 3, pp. 1–107. New York: Academic.

Maynard Smith, J. (1985). Mini review: Sexual selection, handicaps, and true fitness. *J. Theor. Biol.* **115:**1–8.

Møller, A. P. (1995). Hormones, handicaps, and bright birds. *Trends Ecol. Evol.* **10:**121–122.

Moore, M. C. (1984). Changes in territorial defense produced by changes in circulating levels of testosterone: A possible hormonal basis for mate-guarding behavior in white-crowned sparrows. *Behaviour* **88:**215–226.

———. (1987). Castration affects territorial and sexual behavior of free-living male lizards, *Sceloporus jarrovi. Anim. Behav.* **35:**1193–1199.

Moore, M. C., and Lindzey, J. (1992). The physiological basis of sexual behavior in male reptiles. In *Hormones, brain, and behavior* (C. Gans and D. Crews, eds.), Biology of The Reptilia, Vol. 18, Physiology E, pp. 70–113. Chicago: University of Chicago Press.

Moore, M. C., and Marler, C. A. (1988). Hormones, behavior, and the environment: an evolutionary perspective. In *Processing environmental information in vertebrates* (M. H. Stetson, ed.), Proceedings in Life Sciences, pp. 71–83. New York: Springer-Verlag.

Nelson, R. J. (1995). *An introduction to behavioral endocrinology.* Sunderland: Sinauer Assoc.

Norton, J. M., and Wira, C. R. (1977). Dose-related effects of the sex hormones and cortisol on the growth of the bursa of fabricius in chick embryos. *J. Steroid Biochem.* **8:**985–987.

Owens, I. P. F., and Short, R. V. (1995). Hormonal basis of sexual dimorphism in birds: Implications for new theories of sexual selection. *Trends Ecol. Evol.* **10:**44–47.

Post, T. B., Christensen, H. R., and Seifert, G. W. (1987). Reproductive performance and productive traits of beef bulls selected for different levels of testosterone response to GnRH. *Theriogenology* **27**:317–328.

Poulin, R., and Vickery, W. L. (1994). Parasite distribution and virulence: Implications for parasite-mediated sexual selection. *Behav. Ecol. Sociobiol.* **33:**429–436.

Rohwer, S., and Rohwer, F. C. (1978). Status signaling in Harris sparrows: Experimental deceptions achieved. *Anim. Behav.* **26:**1012–1022.

Ryan, M. J. (1991). Sexual selection, sensory systems, and sensory exploitation. *Oxf. Surv. Evol. Biol.* **7:**157–195.

Saino, N. and Møller, A. P. (1994). Secondary sexual characters, parasites, and testosterone in the barn swallow, *Hirundo rustica. Anim. Behav.* **48:**1325–1333.

Schlinger, B. A., Fivizzani, A. J., and Callard, G. V. (1989). Aromatase, 5α- and 5β-reductase in brain, pituitary, and skin of the sex role–reversed Wilson's phalarope. *J. Endocrinol.* **122:**573–581.

Van Tienhoven, A. (1983). *Reproductive physiology of vertebrates.* Ithaca: Cornell Press.

Watson, A., and Parr, R. (1981). Hormone implants affecting territory size and aggressive and sexual behaviour in red grouse. *Ornis Scandinavica* 12:55–61.

Wedekind, C., and Folstad, T. (1994). Adaptive or nonadaptive immunosuppression by sex hormones? *Amer. Natur.* **143:**936–938.

Wingfield, J. C., and Farner, D. S. (1980). Control of seasonal reproduction in temperate zone birds. *Progr. Reprod. Biol.* **5:**62–101.

Wingfield, J. C., Hegner, R. E., Dufty, A. M., Jr., and Ball, G. F. (1990). The "challenge hypothesis": Theoretical implications for patterns of testosterone secretion, mating systems, and breeding strategies. *Amer. Natur.* **136:**829–846.

Wingfield, J. C., and Moore, M. C. (1987). Hormonal, social, and environmental factors in the reproductive biology of free-living male birds. In *Psychobiology of reproductive behavior: An evolutionary perspective* (D. Crews, ed.), pp. 149–175. Englewood Cliffs: Prentice-Hall.

Wingfield, J. C., and Ramenofsky, M. (1985). Testosterone and aggressive behavior during the reproductive cycle of male birds. In *Neurobiology* (R. Gilles and J. Balthazart, eds.), pp. 92–104. Berlin: Springer-Verlag.

Wittenberger, J. F. (1981). *Animal social behavior.* Boston: Duxbury Press.

Zahavi, A. (1975). Mate selection—a selection for a handicap. *J. Theor. Biol.* **53:**205–214.

———. (1977). The cost of honesty. *J. Theor. Biol.* **67:**603–605.

15

Host Behavior Modification: An Evolutionary Perspective

Armand M. Kuris

I. Introduction

Behavioral modification of hosts by infectious agents can potentially increase the fitness of the parasite. Its ecological importance has been recognized (Dobson, 1988; Keymer and Read, 1991; Lafferty and Morris, 1996; Chapter 16) and an evolutionary perspective on these behaviors is developing (Smith Trail, 1980; Moore, 1983, 1984; Moore and Gotelli, 1990; Lafferty, 1992; Poulin, 1994). Unraveling the physiological mechanisms of host behavior modification is also an active research area (Helluy and Holmes, 1990; Hurd, 1990; Thompson and Kavaliers, 1994).

Here I describe the factors that predict the nature and degree of behavior modification induced by parasites and set this analysis in the context of an evolutionary classification of exploitative symbiotic interactions that recognizes types of parasites with distinctive host-parasite responses and ecologies (Kuris, 1974). Some ecological correlates of these symbiotic interactions are noted, and the predicted behavior modifications associated with each type of interactions are discussed. I then consider complex life cycles and the behavior modifications that are predicted to be associated with them. Complex life cycles are of particular interest because host behavior modifications associated with trophic transmission of parasites are frequently reported and are often quite striking. Finally, I will present a synthesis based on the premise that parasite life histories explain host behavior modification, draw attention to some new or little recognized patterns of host behavior modification, and offer predictions to test the adaptive hypotheses that may explain those patterns.

First it is necessary to discuss other sorts of behavioral modifications associated with parasitic infections, as these types of behaviors have sometimes been invoked as counterexamples to the adaptationist program that is discussed here. For example, some host behaviors associated with parasitism are not behavior modifications induced by parasites. These include host defensive behaviors and behaviors that increase risk of parasitization.

Host defensive behaviors may be associated with higher rates of parasitization because such behaviors are more likely to be elicited when risk of infection is high. Grooming is an important, relatively obvious, anti-infection behavior (reviewed in Chapter 11). A good invertebrate example is the use of specialized grooming appendages (fourth pereiopods) by porcelain crabs to avoid infection by sacculinid barnacles, which are parasitic castrators (Ritchie and Høeg, 1981). For these hosts, it is likely that grooming efficiency is so high that susceptibility to infection is limited to the postmolt period when the cuticle is soft and host mobility is reduced. Experimental removal of the grooming appendages has established that infection can occur at other stages of the molting cycle. Another dramatic example may be the activities of minima workers carried on the leaf fragments transported by larger worker leaf-cutter ants. These tiny ants apparently ward off the attack of larger workers by adult phorid flies whose larvae develop as parasitoids in the ants (Eibl-Eibesfeldt and Eibl-Eibesfeldt, 1968). Quantification of grooming efficacy is an area that might add substantial insight into host-parasite population dynamics and factors governing dispersion of parasites among hosts.

Some host behaviors promote contact with parasites and are thus associated with higher rates of parasitization, but are clearly not pathological sequelae. For humans, certain occupations obviously present greater risks. Herdsmen have a higher prevalence of *Echinococcus granulosus* compared with other occupations (Katz et al., 1989) and fishermen have a higher prevalence of schistosomiasis (Farooq et al., 1966). More subtly, among vertebrates, social relationships involving dominance/subordinance hierarchies are generally hormonally mediated by steroids. Because high steroid levels are often immunosuppressive, the individuals with the highest steroid titers are at greatest risk of parasitic infection (Davis and Read, 1958), or at least, of more marked pathology from comparable parasite burdens if the immune response diminishes pathology. Such hosts will often be dominant or subdominant individuals. A good example may be the baboon social groups studied by Hausfater & Watson (1976), in which the highest nematode egg counts were from dominant individuals. Even more difficult to analyze is the association between pica and infection rates of *Toxocara* spp. (Glickman et al., 1979) in developmentally disabled and other institutionalized populations. Clearly, oral contact with items other than food will increase uptake rates of helminth eggs. But, since *Toxocara* spp. exhibit visceral larval migrans, often with central nervous system involvement, infection with this parasite can cause behavior modification that might further increase rates of parasitization (Glickman et al., 1981).

Parasites may induce pathologies, including behavioral changes, that yield no advantage to the resident parasites. For example, loss of social status of the host might occur but this would increase neither parasite dissemination nor survival.

It is likely that, with high parasite intensities, infected hosts may suffer severe consequences that do not provide a gain to the parasites, or may be of obvious detriment to the parasites. An excellent example is acute schistosomiasis, in which recent heavy infections, acquired over a short time period, can prove rapidly fatal. Parasite longevity and the period of egg production can be greatly shortened. Cheever (1968) reported such a fatal case in a boy from whom 1,600 worm pairs were recovered by perfusion at autopsy only 3½ months after he experienced dermatitis, presumably caused by schistosomes; the mean longevity of human schistosomes is about three years (Hairston, 1973).

Due to space constraints, this review will not further consider behavior associated with parasitism unless it potentially increases parasite fitness. "Potentially" is emphasized because only a few studies have directly assessed effects of host behavior on parasite fitness.

II. An Evolutionary Classification of Symbiotic Interactions

The term "symbiotic" will be used here in its original sense (Frank, 1877; De Bary, 1879), as an all encompassing term to include all types of intimate interactions, be they mutualistic, commensalistic or exploitative (parasitic). The classification is termed evolutionary because it is explicitly determined by the effect of an *individual* symbiont on its *individual* host. As previously noted (Kuris 1974), individual symbionts would include the asexual progeny of an individual that establishes the infection (e.g., larval trematode parthenitae produced from a single miracidium). In complex multihost life cycles, the nature of the interaction is quite likely to change as the symbiont (usually a consumer) engages successive hosts.

Exploitative interactions that are symbiotic provide the greatest challenge to an ecologically useful classification. This can be resolved by displaying the different types of exploitative interspecific interactions as a 2 × 3 table based on the number of hosts attacked and killed by an individual consumer (Table 15-1). Here, the symbiotic types of interactions are defined in an operationally exclusive

Table 15-1. A Classification of Exploitative Symbiotic Interactions Based on the Number of Hosts Successfully Attacked by an Individual Consumer and the Number of Hosts Killed by that Consumer

Number of hosts killed by individual consumer	Number of hosts attacked by individual consumer: 1	Number of hosts attacked by individual consumer: > 1
0	parasite	micropredator
1	parasitoid parasitic castrator	—*
> 1	—*	predator

* Blanks indicate logically impossible or biologically improbable (no examples known) categories.

manner that also distinguishes them from other (nonintimate) exploitative interactions (predation). Ambiguous interactions are few; but sometimes we lack sufficient information to assess the nature of the interaction. The five types of exploitative symbiotic interactions in Table 15-1 may represent the set of adaptive peaks for consumers. If so, intermediate types of interactions would be rare. In any case, this analysis is developed for interactions that fully satisfy categories used in the evolutionary classification. The most striking feature of this approach is the discrimination of parasitic castrators from typical parasites (Kuris, 1973; Baudoin, 1975; Obrebski, 1975; Blower and Roughgarden, 1987) and the recognition of their relation to parasitoids (Kuris, 1974; 1990; Lafferty and Kuris, 1994; 1996). Finally, note that in the preceding sections, "parasite" was used in a loose, vernacular sense including all four types of exploitative interactions. In what follows, the vernacular term "parasites" refers to any or all types of symbiotic consumers, while "typical parasites" always is used to refer explicitly to parasites exclusive of parasitoids, parasitic castrators, and micropredators as defined in Table 15-1.

The evolutionary basis for a parasite adaptation must depend on its effect on the fitness of individual parasites. However, because a host is a microhabitat for a population of parasites, host behavior modification and other pathological sequelae are parasite population effects. The value of the evolutionary classification used here is that important ecological distinctions emerge when host-parasite associations are categorized on the basis of the effect of an individual symbiont on its host (Kuris, 1974). Some important ecological correlates concern the relationship between intensity and pathology, the relative size of the parasite and the host, and the dispersion pattern of the parasites among the hosts. The nature of the exploitative symbiotic interaction may also effectively predict the function and magnitude of host behavior modification.

Typical parasites are those symbiont consumers that infect a single host but do not kill that host. They usually cause little (often unmeasurable) pathology in that host. For typical parasites, although the impact of an individual parasite is small, pathology is intensity dependent, parasite body size is small (usually less than one percent of host weight) and parasites are highly aggregated among the host population. Good examples are adult schistosomes, monogenes, and pinworms.

Individual micropredators, like typical parasites, cause little damage to their hosts, and they move from host to host like predators. Like typical parasites, pathology is intensity dependent, and their body size is small relative to that of their host. They also tend to be highly aggregated among their hosts. Examples include mosquitoes, ticks, and caligoid copepods.

Parasitoids feed on but a single host and always kill that host. Thus, pathology is not intensity dependent. Their body size is large relative to that of their host (usually 1–50 percent of host weight) and they are less aggregated among the host population; some dispersion patterns are more uniform than random. Examples include braconid hymenopterans, some hyperiid amphipods, and horsehair worms.

Parasitic castrators, like parasitoids, feed on but a single host. Although the host does not die as a result of a single infection, reproduction is fully suppressed. Thus, in terms of the fitness of individual hosts, the outcome is equivalent to that of the parasitoid. As with parasitoids, pathology is not intensity dependent, parasitic castrators are comparably large relative to host body size and dispersion patterns are relatively uniform. It is essential to note that pathology affecting reproductive output of the host is the only sort of pathology that cannot simultaneously diminish the viability of the host (nor, *ipso facto*, that of the included parasites). Examples include rhizocephalans, larval trematodes in first intermediate hosts, and orthonectids.

Parasitoids and parasitic castrators are akin to predators in terms of their ecological impact on the individual host and perhaps the host population. However, the physiology and anatomy are like typical parasites because the degree of intimacy is certainly comparable.

However, important differences delineate parasitoids and parasitic castrators. Most notably, hosts infected with parasitoids die, reducing intraspecific competition among survivors; whereas, for parasitic castrators, infected hosts remain alive to compete against reproductively competent hosts (Lafferty, 1993). Although the mechanisms of competition are intraspecific, it is, in a sense, a form of interspecific competition; because, as aptly noted by O'Brien and Van Wyk (1985), the parasitically castrated host displays a host phenotype with a parasitic genotype.

III. Behavior Modifications Associated with Types of Symbiotic Exploitative Interactions

A. Parasitic Castrators

Parasitic castrators usually infect immature hosts and parasitize the host for its remaining life span. This period of opportunity, coupled with the loss of counterselection by the host for defenses that mitigate pathology, provides the environmental circumstances for selection to effect behavioral changes in the host that are markedly adaptive for the parasitic castrator. The relatively large body size of the parasitic castrator also alleviates an important constraint experienced by typical parasites: the relative mass and energetics needed to produce behavior-inducing substances in sufficient quantity to alter host behavior.

The fitness of parasitic castrators will generally increase in proportion to increases in host longevity because they will produce many more broods than they would in short-lived hosts. While host death is a major mortality factor for all parasites, it is virtually the only mortality factor for the parasitic castrator (but see Kuris et al., 1980). Thus, the prediction is that hosts infected with parasitic castrators will engage in behaviors to reduce their mortality relative to comparable uninfected individuals. They are potentially able to do this because their hosts need not engage in risky behaviors to facilitate host reproductive success. Such behavior modifications might include predator avoidance, reduced exposure to predators due to lack of mating, reduced altruism, etc.

Parasitic castrators divert energy that would normally go to host reproduction to fuel their own ultimate reproductive output (see Walkey and Meakin, 1970; Chapter 9). Thus, I predict that altering host behavior to increase energy acquisition will be more frequent in parasitically castrated hosts. Expected behavior modifications would include increased food consumption, reduced energy expenditures, and perhaps lessened activity levels (other than feeding), should this be adaptive for the parasitic castrator. A possible example here might be the movement of male green crabs, *Carcinus maenas*, castrated by a rhizocephalan barnacle, *Sacculina carcini*, to the mouths of bays and estuaries along with the uninfected ovigerous female crabs (Rasmussen, 1959).

As with the castration effect, behavior modification by parasitic castrators is generally not intensity dependent (Bethel and Holmes, 1973; Dence, 1958). This is in contrast to behavioral effects of typical parasites (see below).

For castrators, as for all symbionts, behavior modifications that increase transmission to the next host will be selectively favored because successful transmission is the least probable event in most symbiont life cycles (most parasitoids are the principal exception to this generalization). When transmission is immediate (through trophic transmission) or when the transmissive stage is ephemeral, selection should favor behavior modifications which expose hosts to appropriate predators or result in movement to sites favorable for transmission. A spectacular example of behavior modification to increase likelihood of transmission is the "eternal seeker" behavior of bumble bee queens parasitically castrated by the nematode, *Sphaerularia bombi* (Poinar and van der Laan, 1972). Infected queens repeatedly initiate nest sites in prime habitat for nesting. While doing so, these "eternal seekers" release larval infective nematodes. As these nesting sites have a high probability of being used by uninfected queens that arrive later, the transmission rate is greatly increased. Altered vertical migration patterns of marine snails infected with some species of larval trematodes exemplify the induction of altered behavior to enhance parasite dissemination in appropriate habitats (Rothschild, 1940; Lambert and Farley, 1968).

B. Parasitoids

Parasitoids, compared with parasitic castrators, are less likely to alter host behavior so as to increase host longevity. They are going to kill their hosts, and the more rapidly they develop within the host, the less likely they will be to experience mortality along with their host. However, risky host behavior (such as that associated with mating) should also be avoided by hosts infected with parasitoids and many suppress the metamorphosis of their host (Chapter 1). In a sense, selection on parasitoids should avoid "wrongful death" due to causes that compete with parasitoid-induced host mortality (Fritz, 1982). For example, avoidance of exposure to hyperparasitoid attack can increase longevity (Brodeur and McNeil, 1992). Stamp (1981) has experimentally shown that behavioral changes improve survival of parasitized hosts and MacLellan (1958) has quantified the magnitude of these changes in nature.

Parasitoid reproduction occurs after the larval parasitoids have completed development in the host and emerged as free-living adults. Hence, behavior modification leading directly to increased parasitoid reproduction will not likely occur. However, increased feeding of the host may both improve parasitoid longevity and increase its adult body size and likely reproductive output (Schmid-Hempel and Schmid-Hempel, 1991).

As with parasitic castrators, parasitoids might alter host behavior to increase transmission. However, parasitoid transmissive stages are often the relatively vagile adults. Thus, selection to modify the behavior of the host to improve transmission will be weak because host location is an activity of the free-living adult and is often very efficient (Kuris and Norton, 1985). Some parasitoids alter host behavior to deliver adult stages to appropriate habitats. Increased thirst and the resultant "water drive", reportedly induced by nematomorphs in their insect hosts, is a well-known example of such behavior (Poinar, 1991).

As with parasitic castrators, parasitoids are relatively large compared to their hosts and potentially have the energetic resources and anatomical and physiological machinery to alter host behavior. For both parasitoids and parasitic castrators, this physiological capability is maximized when the symbiont is closely related to the host taxon (agastoparasitism, Ronquist, 1994). The most spectacular examples are the numerous hymenopteran and dipteran parasitoids which attack other insects, the rhizocephalan barnacle and epicaridean isopod parasitic castrators developing in other crustaceans, and the strepsipteran parasitic castrators of other insects.

C. Typical Parasites

Consistent with their short life spans relative to parasitic castrators and parasitoids, there is scant evidence that typical parasites alter host behavior so as to increase parasite longevity or reproductive output. Rather, behavior modifications are often manifested as parasite-increased susceptibility to predation (PISP). Considerable recent work suggests that PISP may be a very widespread and important host-parasite interaction when trophic transmission is the most efficient route to the next appropriate host in a complex life cycle (Holmes and Bethel, 1972; Moore, 1984; Moore and Gotelli, 1990; Milinski, 1990).

The ability of typical parasites to induce host behavior modification is constrained by their relatively small size (with associated anatomical and physiological simplicity) and their often ephemeral nature. Consequently, when behavior modification is evident, it is often associated with the specific location of parasites within key host organs (e.g., eyes, brain), and the magnitude of the modification is usually intensity dependent (Crowden and Broom, 1980; Lafferty and Morris, 1996). The larval trematode metacercaria (*Dicrocoelium dendriticum*) induces parasitized ants to adhere to exposed vegetation in the cool of the evening when other ants have retreated to their nest. Thus exposed, they are more likely to be ingested by grazing sheep final hosts; if uneaten, they resume foraging in the morning as the day warms up (Carney, 1969). The presence of metacercariae of *Euhaplorchis californiensis* in the brain case of the killifish, *Fundulus parvipinnis* demonstrates the importance of behavior modification for predation efficacy. As described by Lafferty (Chapter 16; Lafferty and Morris, 1996) field experiments indicate that parasitized fish are 10–30 times more susceptible to piscivorous shorebirds. Both behavior modification of these fish and their increased susceptibility to bird predators are intensity dependent. If specific sites are not affected, then parasite-induced behavior modification may be most often manifested as a general malaise (e.g., reduced activity level).

Beyond direct access to behavior modification, there are, of course, other important features of brains and nerves as sites for parasites. Most notably, the reduced ability of many cellular and humoral components of the immune system to cross the blood-brain barrier enables these locations to serve as potential refuges for parasites.

Parasites that alter sensory capabilities (e.g., *Diplostomum* metacercariae in the eyes of fishes, Crowden and Broom, 1980), increase risk to a variety of predators, as do those that reduce stamina such as parasites of the musculature (e.g., *Trichinella* spp.), heart (e.g., echinostome metacercariae in freshwater pulmonates, Kuris and Warren, 1980), and lungs (e.g., hydatid cysts, Rau and Caron, 1979).

Thus, host behavior modification serves to reduce mortality among parasitic castrators, reduces risk of death to causes other than parasitoid emergence (avoids wrongful death), and increases the likelihood of host death to facilitate trophic transmission of typical parasites (targeted death). Table 15-2 summarizes the benefits that host behavior modifications may provide to parasites with simple

Table 15-2. Predicted Effects of Host Behavior Modifications on Life History Characteristics of Different Types of Intimate Exploitative Symbionts

Type of interaction	Effect of behavior modification		Probability of transmission
	Longevity	Reproduction	
parasitic castrator	increase (protect from death)	increase	increase
parasitoid	increase (protect from wrongful death)	—	increase
typical parasite	—	—	increase (promote targeted death)

(one-host) life cycles. The most spectacular examples are often associated with parasites residing in the brain.

D. Micropredators

Micropredators, having relatively brief contacts with their hosts, have relatively little opportunity to affect host behavior. However, many have independently evolved use of analgesics which are injected with saliva to desensitize the host and to reduce the incidence of dislodgment behavior. Examples range from mosquitos to cymothoid isopods and leeches.

IV. Host Behavior Modifications Associated with Complex Life Cycles

In this section, life cycle complexity refers to multiple host life cycles and the discussion focuses on those involving trophic transmission in which a host acquires a parasite by predation upon the preceding host in the life cycle. The distribution of trophic transmission among parasites is described, noting its relationship to whether the hosts are the definitive or intermediate hosts, or vectors.

Symbionts in definitive hosts generally function as typical parasites (adult digenes, acanthocephalans, etc.), although some parasitic castrators are known (e.g., entoniscid isopods). In intermediate hosts, both parasitic castrators and typical parasites are common types of interactions. The parasitoid interaction is uncommon in multiple-host life cycles. However, hydatid cysts are quite large, pathogenic and often located in sites such as the brain and lungs. They may so ensure the death of the host to carnivory that they are, indeed, effectively parasitoids. This would be particularly so for *Echinococcus multilocularis*, in which budding cysts from a single infection continue to spread through host tissues.

Among helminths, trophic transmission is a widespread mode of transfer from one host to the next in a multiple-host life cycle. Table 15-3 summarizes the frequency of trophic transmission among helminth phyla and classes. As is evident, trophic transmission is widespread. In almost all taxa, it is either an obligatory aspect of all life cycles, or is at least often present. The few exceptions are the monogenes, gyrocotylid Cestodaria, and rhabditoid, oxyuroid, and filarial nematodes. In the filarial worms, parasites are not transferred by predation but by micropredators serving as vectors that feed on blood or tissue fluids of the vertebrate definitive hosts.

The likelihood that a host will acquire a parasite via ingestion of an infected prey is closely associated with the life cycle stage of the parasite. In general, definitive hosts are large, relatively vagile, and are rarely ingested by intermediate hosts. Following sexual reproduction of the parasites, the parasite progeny are typically packaged as small disseminules (eggs or larvae) which are acquired either when the first intermediate host ingests them, or, when they actively penetrate those hosts. Hence, definitive hosts generally serve to disperse the

Table 15-3. Distribution of Trophic Transmission among Helminths

Taxon	None	Once per life cycle	Twice per life cycle
Monogenea	+++	–	–
Aspidobothridea	+++	+	–
Digenea			
Strigeata	+++	++	+
Echinostomata	+++	+++	
Plagiorchiata	–	+++	
Opisthorchiata		+++	+
Cestoda			
Amphilinidea	–	+++	–
Gyrocotylidea	+++	–	–
Trypanorhynchida	–	–	+++
Tetraphyllida	–	–	+++
Pseudophyllida	–	–	+++
Cyclophyllida	+	+++	+
Acanthocephala	–	+++	+
Nematoda			
Trichurida	+++	++	–
Dioctophymida		+++	++
Rhabditida	+++	–	–
Strongylida	+++	++	+
Ascarida	+++	++	+
Oxyurida	+++	–	–
Spirurida		+++	++
Camallanata	–	+++	–
Filariata	+++	–	–

Legend: +++ represents widespread and common occurrence of trophic transmission in the taxon; ++ represents frequent, widespread occurrence or characteristic of one or a few groups within the taxon; + represents occasional report or characteristic of one small group in the taxon; – not reported; blanks indicate that insufficient information was available to characterize the taxon.

parasite, whereas parasite eggs and larvae infect more geographically localized intermediate hosts.

In contrast to passive transmission to intermediate hosts, ingestion of parasitized intermediate hosts is the general mechanism by which second intermediate or definitive hosts acquire infections. The principal exceptions are among the Digenea in which a free-living dispersal stage (cercaria) exits the molluscan first intermediate host to actively penetrate, and be ingested (or encyst prior to ingestion), by the second intermediate or definitive hosts. Paratenic hosts (transport hosts, in which no further development of the parasite occurs) always acquire parasitic infections by ingestion of parasitized prey items. Thus, life cycles involving two or more hosts generally use trophic transmission for at least one transfer between hosts. Trophic transmission is always a transfer mechanism when a parasite uses more than two hosts in a life cycle.

Two exceptions to the rule that definitive hosts do not serve as prey to complete parasite life cycles deserve comment. (1) *Trichinella* spp. are always transmitted by predation of the definitive host (although the short-lived adult worms will often no longer be present at transmission). This interesting and unusual life cycle will be discussed below. (2) Blood-sucking micropredators can serve as vectors when they ingest microfilariae.

Behavior modification of insect vectors by parasites has been a subject of intense study over the past 20 years. This work includes study of vectors of filarial diseases, protozoans and other microbial pathogens (Jenni et al., 1980; Ribeiro et al., 1985; Rossignol et al., 1985; Moore, 1993). These elegant studies have considered both the mechanisms and adaptive significance of behavior modifications. Examples of vector behavior modification include increased biting frequency (Killick-Kendrick and Molyneux, 1990), improved feeding efficiency (Rossignol et al., 1985) and decreased flight (Townson, 1970). This subject has been well-reviewed and will not be further treated here.

For parasites with complex life cycles, parasite-increased susceptibility to predation (PISP) poses some perplexing challenges to evolutionary theory and is such a widespread phenomenon that it may also be ecologically very important. Lafferty (1992) outlined the paradox concerning eating parasitized prey. Why do the predators bite on prey made more available via PISP? If the parasites will be damaging to the predator, then one would expect selection for avoidance of the easily caught, parasitized prey. The resolution lies in a cost-benefit analysis, comparing the benefit of food items easily acquired through PISP with the cost of parasitism. As Lafferty (1992) demonstrates, the benefits can often be large, especially if behavior modification is strong; while costs are often low, sometimes negligible, because many trophically transmitted parasites are avirulent in the predator. In the definitive host, they are typically small, ephemeral, or both when PISP is evident (e.g., *Echinococcus*, Fig. 15-1.).

When parasites in definitive hosts grow large and are long-lived (e.g., many taeniids, liver flukes) PISP either cannot operate (liver fluke metacercaria encyst on vegetation) or may be weak in terms of the pathogenicity of an individual parasite in the intermediate host (as seems likely for a taeniid cysticercus). It is worth noting that even if PISP is markedly intensity dependent (such as might be expected for taeniid cysticerci and *Hymenolepis* cysticercoids), the intensity-dependent pathogenicity in the definitive host would be ameliorated by very strong crowding effects (Roberts, 1961).

The cost-benefit analysis of PISP generates several predictions about the distribution of PISP in parasite life cycles. These only apply if predators have the ability to distinguish between parasitized and unparasitized prey. As noted by Lafferty (1992), PISP should not evolve when a typical parasite has high individual virulence or a steep intensity-pathology curve in the predator. The most

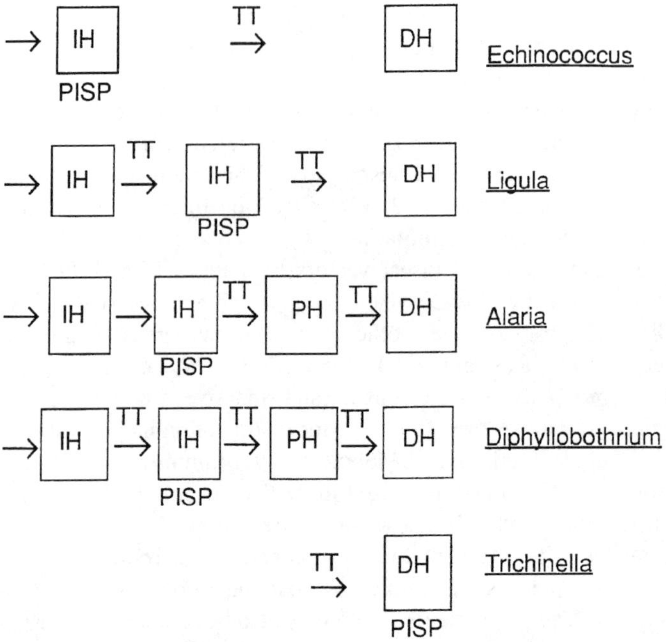

Figure 15-1. Examples of trophic transmission and location of selected helminth life cycles; IH–intermediate host, DH–definitive host, PH–paratenic host, TT–trophic transmission, PISP–parasite-increased susceptibility to predation.

interesting apparent exception (*Trichinella*) will be discussed below. Consistent with this logic, PISP is not a means of transmission for parasitic castrators or parasitoids to the next host. Indced, it appears that trophic transmission is a decidedly uncommon means to infect definitive hosts subject to these interactions. For most parasitoids or parasitic castrators, the infection process requires an active search for, and penetration of, the host. In some cases, hosts are infected through ingestion of passive infective disseminules.

The Lafferty model suggests that strong behavior modification to induce PISP can even overcome costly parasites. However, a perusal of the available case studies of host behavior modification suggests that costs to the predator may generally be quite low (approaching nil). Adult acanthocephalans, however, generally cause moderate pathology (lesions at the site of proboscis attachment), and for *Trichinella* spp., pathology is clearly high enough to reduce fitness. A careful review of host behavior modification in the context of pathogenicity in the predator host would be a valuable contribution.

The line of reasoning developed by Lafferty (1992) provides a powerful explanation for the evolution of PISP when trophic transmission occurs once in the life cycle. A careful evaluation of more complex life cycles in which trophic transmission may occur more than once (e.g., *Ligula*, *Alaria*, *Diphyllobothrium*, Fig. 15-1) suggests some further generalizations. In a life cycle involving trophic transmission during two host transfers, if PISP increases transmission to the first predator, then the parasite will generally have low virulence in that host. If virulence is low in that host, then, by definition, the parasite cannot induce PISP to transfer the parasite through carnivory to the next host (i.e., the definitive host). Similarly, if, as in *Ligula* (Fig. 15-1), the parasite transfers, via PISP, from the second intermediate host (fish) to the final host (bird); then strong behavior modification to increase susceptibility to predation from the first intermediate host (copepod) to the fish will be less likely to evolve. If PISP were also to occur at the first trophic transmission, then selection on the second intermediate host (the fish) to avoid the behaviorally modified copepod would increase because the cost of the parasite to the fish would be high (greatly increased risk of predation). Thus, the generalization emerges that PISP is likely to evolve but once in a complex life cycle or, at least, that but one trophic transfer will be strongly influenced by PISP. A recent study provides perspective for this view. The tapeworm, *Schistocephalus solidus*, affects reproduction of its second intermediate host (Arme and Owen, 1967) and markedly alters its behavior to increase susceptibility to piscivorous bird predators that serve as final hosts (Giles, 1987; Milinski, 1990; Tierney et al., 1993). Urdal et al. (1995) demonstrated that the first intermediate host, a copepod, exhibited some behavior modification when heavily infected. However, this did not lead to a significant increase in susceptibility to predation by fish second intermediate hosts, but a weak effect may have been undetected. Although the authors emphasize the cautionary point that behavior modification does not necessarily imply PISP, a strong effect would not have been expected in this case because PISP, with its associated necessary decrease in host fitness, has been so well demonstrated for the predatory host which follows. These pseudophyllidean tapeworm systems (*Ligula*, *Schistocephalus*, *Diphyllobothrium*) may provide an interesting opportunity to test Lafferty's (1992) conclusion that strong behavior modification can induce PISP, even when the cost of the parasite to the predator is high.

The involvement of paratenic hosts in trophic transmission adds additional complexity to selection for behavior modifications leading to PISP. For *Diphyllobothrium* spp. paratenic predatory fish hosts are typically incorporated into the life cycle (Freeman & Thompson 1969). For species such as the human fish tapeworm, *D. latum*, this probably increases the probability of ingestion by fish predators such as humans and bears (which selectively feed on large fishes). Here, PISP would be adaptive for the tapeworm, and avoidance measures adaptive for large fishes. In contrast, avian species, such as *D. ditremum*, will be less likely to be successfully transmitted once the plerocercoid passes from the smaller fishes (susceptible to piscivorous birds) to the larger fishes (too large to be captured by birds). In this case, behavior modifications would either be selected to target bird predators, or selection would not favor behavior modifications that would increase risk of fish predation (unless risk of bird predation was also increased).

Trichinella spp. provide a challenge to this general body of theory (Fig. 15-1) because they are important exceptions to the pattern of low pathogenicity of parasites in predator hosts that acquire parasites via PISP. Adult worms are short-lived (three weeks) in the intestinal mucosa. They produce about 1,000 larvae which migrate to active skeletal musculature, encyst and remain infective for up to 31 years (Katz et al., 1989). Behavior modification that likely leads to PISP has been well-documented (Rau, 1983). Yet the host which obtains food via increased susceptibility to predation caused by *Trichinella* will also fall prey to another predator to complete the life cycle. Since *Trichinella* larvae are long-lived and rather pathogenic (Rau, 1983), hosts should be under strong selection to avoid delivery of prey by *Trichinella*.

It is not obvious why *Trichinella* is an exception to the pattern of low virulence of trophically transmitted parasites in definitive hosts. However, several distinctive features of *Trichinella* infection dynamics may provide an explanation. (1) *Trichinella* spp. have low host specificity (Pozio et al., 1992). Hence, the risk of acquiring an infection is spread across several species, weakening selective pressure on each. (2) It may be relatively difficult for predators to detect *Trichinella* because it causes general debilitation of the prey. (3) Scavenging is an important mode of transmission. Ewald (1995) has argued that persistent transmissive stages will select for high virulence because the quicker the host dies, the more rapidly the parasite has an opportunity for transmission. Retention of infective larvae in carrion amounts to the sort of sit and wait transmissive strategy that could select for high virulence. If scavenging is more important than carnivory of living prey in the epidemiology of *Trichinella*, then behavior modification leading to PISP is, in a sense, a side-effect of the scavenging pathway. Scavenging on carrion appears important in the transmission of at least *T. nelsoni* in Africa (Campbell, 1988). Although transmission via scavenging does not appear to have been quantitatively assessed, the high prevalence in hyenas supports the role of carrion in transmission. Even humans may serve as carrion hosts when funerary practices expose corpses to scavenging carnivores such as among the Turkana in Kenya (MacPherson, 1983). Further, transmission of an infectious agent through carnivory must concentrate these parasites in the highest trophic levels, in a manner analogous to the buildup of pesticides through a food chain. The logical outcome of a one-host life cycle with transmission through carnivory is either to lead to a dead end (as for *Trichinella* in humans in the domestic cycle), or perhaps to rely on scavenging on dead top predators to regenerate the food chain in sylvatic cycles. If top predators are dead ends, strong selection would be exerted against trophic transmission to these predators. However, where relative susceptibility has been studied, these top predators often are highly susceptible hosts permitting high reproductive rates for *Trichinella* spp. (Pozio et al., 1992; Campbell, 1988). Further, studies of behavior of such predators show that they often selectively feed on sick or injured prey (Kruuk, 1972; Temple, 1987), presumably because healthy (less parasitized) prey are so very difficult to capture.

Parasitized hosts may be increasingly likely to fall prey to the next host in the life cycle if parasites cause either an increased generalized susceptibility to predation, or a more targeted susceptibility with a high likelihood of consumption by a suitable predator. A generalized increase in susceptibility to infection will be selectively advantageous only under some circumstances. If debilitation is so great that parasitized intermediate hosts fall prey to predators in which parasite development cannot occur, or frequently die due to environmental stressors without being consumed by the appropriate final host, then selection will favor decreased virulence in the intermediate host (Sogandares-Bernal et al., 1979). Thus, generalized debilitation causing increased susceptibility to predation would appear to be selectively advantageous under a limited set of conditions. These include that the intermediate host is a preferred, or at least a common, food item for the predator, or that predation by the appropriate predator is an important mortality cause for the prey species. Also, if host specificity in potential predators is broad, generalized susceptibility to predation will most likely be adaptive for the parasite. Upon consumption of the infected prey, this would open a wide window for successful development of the parasite. A good example may be *Echinococcus* tapeworms, in which the hydatid cysts greatly increase susceptibility to predation whether the intermediate host is a vole or a moose, and specificity to the final host predators is low (i.e., canids and some other carnivores) (Roberts and Janovy, 1995, p 338). Susceptibility to predators is so generalized that even humans hunting moose with rifles are more likely to kill parasitized moose early in the hunting season (Rau and Caron, 1979). Adult *Schistocephalus solidus* tapeworms have been reported from 40 species of piscivorous birds (Cooper, 1918). Paratenic hosts may be particularly prone to feeding on prey showing generalized debilitation elicited by PISP since they tend to be important in life cycles where host specificity in the predators is low.

To target a parasitized prey for consumption by a specific predator will require relatively specific interventions in host physiology, most notably with respect to host hormonal or neurological mechanisms. Parasitic castrators have the greatest *a priori* potential to achieve this specific control of host fate because they are large and have already substantially intervened in host physiological regulation. By definition, parasitically castrated hosts cannot counter expression of the parasitic castrator phenotype (including behavioral attributes). The best-studied examples here are larval plerocercoid tapeworms in freshwater fishes that increase susceptibility of these second intermediate hosts to bird predators (Van Dobben, 1952; Wilson, 1971). Altered host behavior can shift the risk of predation towards those predators that will be suitable for parasite maturation. This has been best shown for *Schistocephalus solidus*. Lester (1971) found that parasitized stickleback were much more likely to inhabit shallow water. Here, their sluggish responses exposed them almost exclusively to bird predators (where *S. solidus* can mature) rather than to the abundant trout predators, which are unsuitable for their development, in the deeper waters of the lake. This habitat preference of the infected fish is key to the targeted death of the infective fish as Sweeting (1976) has shown that sluggishness and suppressed escape responses would otherwise make *Ligula*-infected fish more susceptible to fish predation.

Typical parasites may be most able to achieve relatively specific PISP if they are located in organs such as the brain or sense organs, and if lesions in those organs produce specific deficits. Examples include *Euhaplorchis californiensis* in killifish (Lafferty and Morris, 1996), and *Dicrocoelium dendriticum* in ants (Carney, 1969).

As has been well recognized, the mode of predation will influence the type of behavior modification leading to PISP (Holmes and Bethel, 1972; Combes, 1991). I predict that the mode of predation will also strongly influence the importance of PISP in infection dynamics. Specifically: (1) selection for host behavior modification will be more readily achieved if the predator relies strongly on its senses to detect prey. Such a predator should be more able to detect a weak behavior modification signal. Raptors and snakes are notable predators that rely on sensitive prey detection mechanisms. The role of PISP should be evaluated for these predators. (2) Behavior modification resulting in PISP should also be highly advantageous for predators relying on strength to overcome prey. Carnivores already provide some of the most striking examples described above. Perhaps even *Tyrannosaurus rex* may have benefited from PISP? (3) Behavioral alterations yielding PISP will be very advantageous to predators with long prey-handling times. Thus, PISP may be an important factor altering the shape of the Type II functional response of Holling (1959).

Dobson (1988), using host-parasite population models, concluded that behavior modifications to increase parasite transmission would be most strongly selected when the definitive host was rare, ephemeral or highly mobile. Selection should promote PISP if some or all of the following are also attributes of the predatory host. These additional features reduce the cost of the parasite to the predator.

(1) The parasite is avirulent in the definitive host. Virulence will be minimized if the parasite has actually acquired the bulk of its nutrition from another host in the life cycle. Since predators are almost always larger than their prey, this implies that the parasite will be relatively smaller in the predator than in the intermediate host. It will also be more likely to be ephemeral and to be precociously mature. Excellent examples include many species of heterophyid trematodes. Adults are often minute (less than 1 mm in length), live only a few weeks, and mature within a few days (Schell, 1985), whereas they grow considerably in the molluscan host in which they may live for years. For example, marked snails infected with *Euhaplorchis californiensis* have been recaptured more than one year later (Sousa, 1983), and these infections occupy up to one half the tissue weight of the snail (Kuris, 1990). Yet, adults in the sea gull are only 0.5 mm long, mature in less than a week, and live only 2–3 weeks (Martin, 1950). The tapeworms *Schistocephalus solidus*, *Ligula intestinalis* and *Echinococcus*

spp. also satisfy these conditions. For example, *S. solidus* grows to its full weight as a plerocercoid in the fish. In the avian host, eggs are produced in 36 hours and shed for but 7–11 days before the worms die and are evacuated (Smyth, 1946; Vik, 1954; Hopkins and Smyth, 1951; McCaig and Hopkins, 1963). These tapeworms are so ephemeral that parasitized birds are rarely collected (Cooper, 1918).

(2) A strong crowding effect will minimize the cost of parasite acquisition to the predator. This will be so whether the crowding effect is due to parasite intraspecific competition or host-mediated factors. When individual parasite biomass is directly inversely proportional to parasite intensity, there is no added cost to the predator upon consuming further parasitized prey. Very strong crowding effects are reported for some trophically transmitted parasites. The rat tapeworm, *Hymenolepis diminuta*, is an example in which mean worm size decreases in high intensity infections such that the total mass of the parasite population scarcely exceeds that of a worm in a single infection (Roberts, 1961).

(3) Concomitant immunity provides cost reduction to the predator similar to the crowding effect. Pathology of typical parasites is intensity dependent. Thus, by reducing the probability of establishment of high parasite intensities, concomitant immunity greatly reduces the cost of ingestion of parasites beyond the primary inoculum.

Interestingly, avirulence and the crowding effect are generally seen as primarily host advantages. An individual parasite gains nothing *per se* from these attributes. However, concomitant immunity is clearly advantageous to the established parasites as well as to their host. For *H. diminuta*, it is actually unclear which feature affords the greatest cost reduction to the rat. An individual tapeworm does not seem to reduce the fitness of the infected rat. Its energetic drain has been quantified and is quite small (Bailey, 1975). In addition to a strong crowding effect, immunity against *H. diminuta* is also concomitant.

Some population characteristics of trophically transmitted parasites in prey intermediate hosts should also favor selection for host behavior modification (see also Poulin, 1994). Highly aggregated parasites have three features that could promote preferential ingestion by predators.

(1) By definition, a highly aggregated parasite population means that the chance of a predator acquiring parasites in a predatory event is relatively low. Avoidance behavior will therefore be more difficult to learn or to be readily selected because heavily infected prey will rarely be encountered.

(2) For typical parasites, behavior modification will be proportional to parasite intensity in the prey host (see Lafferty and Morris, 1996, for example). So, hosts with high parasite intensities will be more readily consumed the greater the host behavior modification leading to PISP. Hence, selection for PISP, and for tendencies of parasites to aggregate in prey hosts, will be mutually reinforcing.

(3) Similarly, if parasites are rare in prey hosts, then opportunities for the predator to learn avoidance are minimized. Obviously, selection cannot promote rarity. Rather, rarity is a consequence of recruitment difficulties earlier in the

life cycle. However, a rare parasite with behavior modification traits will be at a selective advantage because transmission success is obviously difficult to achieve for a rare parasite. There are intriguing hints that PISP may be so intense that even rare parasites have a remarkably high probability of reaching the host predator. In Patagonian lakes, an adult acanthocephalan, *Polymorphus patagonicus* is abundant in predatory fishes. However, samples of lake zooplankton have so far failed to recover any infected amphipod (*Hyalella patagonica*) intermediate hosts. Yet stomach contents of the fishes repeatedly yield recently ingested amphipods infected with cystacanths (L. Semenas, C. Ubeda and S. Ortubay, personal communication). It seems that the fishes may sample infected amphipods more effectively than do plankton nets.

V. Conclusions

The nature of the symbiotic interaction is a major determinant of the extent, type and function of behavior modification. Parasitic castrators and parasitoids have more physiological opportunity, can more completely overcome host resistance, and produce more marked behavioral deviations in their hosts than do typical parasites or micropredators.

Trophic transmission is widespread and may often involve prey host behavior modifications that provide parasite-increased susceptibility to consumption by predatory hosts. Intermediate hosts are more likely to experience such behavior modifications. The relationship between intensity and pathology will determine the cost to the host predator as it consumes parasitized prey. Costs will be low when there is a strong crowding effect or if concomitant immunity decreases the likelihood of high worm burdens. Rarity or high aggregation of parasites in the prey host will reducc selective pressure to avoid infected prey. It seems likely that parasite-increased susceptibility to predation will usually affect but one host in those complex life cycles with trophic transmission two or more host transfers.

VI. Acknowledgments

The insightful comments of Per-Arne Amundsen, Todd Huspeni, Kevin Lafferty, Robert Warner, and especially Janice Moore are greatly appreciated. Development of these ideas was also aided by discussions with Hilary Hurd, Brad Vinson, Nancy Beckage, Marijke de Jong-Brink, and Kimo Morris.

VII. References

Arme, C. and Owen, R. W. (1967). Infections of the three-spined stickleback *Gasterosteus aculeatus* L., with the plerocercoid larvae of *Schistocephalus solidus* with special reference to pathological effects. *Parasitology* **57**:301–314.

Bailey, G. N. A. (1975). Energetics of a host-parasite system: a preliminary report. *Int. J. Parasitol.* **5:**609–613.

Bethel, W. M., and Holmes, J. C. (1973). Altered evasive behavior and responses to light in amphipods harboring acanthocephalan cystacanths. *J. Parasitol.* **59:**945–956.

Blower, S., and Roughgarden, J. (1987). Population dynamics and parasitic castration: test of a model. *Am. Nat.* **134:**848–858.

Brodeur, J., and McNeil, J. N. (1992). Host behaviour modification by the endoparasitoid *Aphidius nigripes*: A strategy to reduce hyperparasitism. *Ecol. Entomol.* **17:**97–104.

Campbell, W. C. (1988). Trichinosis revisited—another look at modes of transmission. *Parasitol. Today* **4:**83–86.

Carney, W. P. (1969). Behavioral and morphological changes in carpenter ants harboring dicrocoelid metacercaria. *Am. Midl. Nat.* **82:**605–611.

Cheever, A. W. (1968). A quantitative post-mortem study of *Schistosoma mansoni* in man. *Am. J. Trop. Med. Hyg.* **17:**38–64.

Combes, C. (1991). Ethological aspects of parasite transmission. *Am. Nat.* **138:**867–880.

Cooper, A. R. (1918). North American pseudophyllidean cestodes from fishes. *Ill. Biol. Monogr.* **4:**1–243.

Crowden, A. E., and Broom, D. M. (1980). Effects of the eyefluke, *Diplostomum spathecum* on the behaviour of dace (*Leuciscus leuciscus*). *Anim. Behav.* **28:**287–294.

Davis, David E., and Clark P. Read. (1958). Effect of Behavior on Development of Resistance in Trichinosis. *Proc. Soc. Exptl. Biol. and Med.* **99**(1):269–272.

De Bary, A. (1879). Die Erscheinung der Symbiose. In "Vortraug auf der Versammlung der Naturforscher und Artze zu Cassel" pp. 1–30. Strassburg: Verlag von Karl L. Trubner.

Dence, W. A. (1958). Studies on *Ligula* infected common shiners (*Notropis cornutus* Agassiz) in the Adirondocks. *J. Parasitol.* **44**:334–338.

Dobson, A. P. (1988). The population biology of parasite-induced changes in host behavior. *Q. Rev. Biol.* **63:**139–165.

Eibl-Eibesfeldt, I., and Eibl-Eibesfeldt, E. (1968). The workers' bodyguard. *Animal's Magazine* **11:**16–17.

Ewald, P. W. (1995). The evolution of virulence: a unifying link between parasitology and ecology. *J. Parasitol.* **81:**659–669.

Farooq, M., Nielson, J., Samaan, S. A., Mallah, M. B., and Allam, A. A. (1966). The epidemiology of *Schistosoma haematobium* and *S. mansoni* infections in the Egypt–49 Project area. 2. Prevalence of bilharziasis in relation to personal attributes and habits. *Bull. W.H.O.* **35:**293–318.

Frank, A. B. (1877). Uber die biologischen Verhältnisse des Thallus eineger Krustenflechten. *Beitr. Biol. Pflanz.* **2:**123–200.

Freeman, R. S., and Thompson, B. H. (1969). Observations on transmission of *Diphyllobothrium* sp. (Cestoda) to lake trout in Algonguin Park, Canada. *J. Fish. Res. Board Canada* **26:**871–878.

Fritz, R. S. (1982). Selection for host modification by insect parasitoids. *Evolution* **36:**283–288.

Giles, N. (1987). Predation risk and reduced foraging activity in fish: experiments with parasitized and nonparasitized three-spined sticklebacks, *Gasterosteus aculeatus*. *J. Fish Biol.* **31:**37–44.

Glickman, L. T., Cypess, R. H., Crumrine, P. K., and Gitlin, D. A. (1979). *Toxocara* infection and epilepsy in children. *J. Pediatr.* **94**:75–79.

Glickman, L. T., Chaudry, I. V., Constantino, J., Clack, F. B., Cypress, R. H., and Winslow, L. (1981). Pica patterns, toxocariasis and elevated blood lead in children. *Am. J. Trop. Med. Hyg.* **30:**77–80.

Hairston, N. G. (1973). The dynamics of transmission. In *Epidemiology and Control of Schistosomiasis (Bilharziasis)* (N. Ansari, ed.), pp. 250–336. Basel: S. Karger.

Hausfater, G., and Watson, D. F. (1976). Social and reproductive correlates of parasite ova emission by baboons. *Nature* (Lond.) **262:**688–689.

Helluy, S., and Holmes, J. C. (1990). Serotonin, octopamine, and the clinging behavior induced by the parasite *Polymorphus paradoxus* (Acanthocephala) in *Gammarus lacustris* (Crustacea). *Can. J. Zool.* **68:**1214–1220.

Holling, C. S. (1959). The components of predation as revealed by a study of small mammal predation of the European pine sawfly. *Can. Entomol.* **91:**293–320.

Holmes, J. C., and Bethel, W. M. (1972). Modification of intermediate host behavior by parasites. In *Behavioral Aspects of Parasite Transmission* (E. U. Canning and C. A. Wright, eds.), pp. 128–149. New York: Academic Press.

Hopkins, C. A., and Smyth, J. D. (1951). Notes on the morphology and life history of *Schistocephalus solidus* (Cestoda: Diphyllobothriidae). *Parasitology* **41:**283–291.

Hurd, H. (1990). Physiological and behavioural interactions between parasites and invertebrate hosts. *Adv. Parasitol.* **29:**271–318.

Jenni, L., Molyneux, D. H., Livesy, J. L., and Galun, R. (1980). Feeding behaviour of tsetse flies infected with salivarian trypanosomes. *Nature* (Lond.) **283:**383–385.

Katz, M., Despommier, D. D., and Gwadz, R. W. (1989). Parasitic Diseases, 2nd Ed., New York: Springer-Verlag.

Keymer, A. E. and Read, A. F. (1991). Behavioural ecology: the impact of parasitism. In *Parasitism: Coexistence or Conflict* (A. Aeschlimann and C. A. Toft, eds.), pp. 36–61. Oxford University Press.

Killick-Kendrick, R., and Molyneux, D. H. (1990). Interrupted feeding of vectors. *Parasitology Today* **6:**188–189.

Kruuk, H. (1972). The spotted hyaena. Chicago: University Chicago Press.

Kuris, A. M. (1973). Biological control: implications of the analogy between the trophic interactions of insect pest-parasitoid and snail-trematode systems. *Exp. Parasitol.* **33:**365–379.

Kuris, A. M. (1974). Trophic interactions: similarity of parasitic castrators to parasitoids. *Q. Rev. Biol.* **49**:129–148.

Kuris, A. M. (1990). Guild structure of larval trematodes in molluscan hosts: prevalence, dominance and significance of competition. In *Parasite Communities: Patterns and*

Processes (G. W. Esch, A. O. Bush, and J. M. Aho, eds.) pp. 69–100. London: Chapman & Hall.

Kuris, A. M., and Norton, S. F. (1985). Evolutionary importance of overspecialization: insect parasitoids as an example. *Am. Nat.* **126:**387–391.

Kuris, A. M., and Warren, J. (1980). Echinostome cercarial penetration and metacercarial encystment as mortality factors for a second intermediate host, *Biomphalaria glabrata. J. Parasitol.* **66:**630–635.

Kuris, A. M., Poinar, G. O., and Hess, R. (1980). Mortality of the internal isopodan parasitic castrator, *Portunion conformis* (Epicaridea, Entoniscidae), in the shore crab, *Hemigrapsus oregonensis* with a description of the host response. *Parasitology* **80:**211–232.

Lafferty, K. D. (1992). Foraging on prey that are modified by parasites. *Am. Nat.* **140:**854–867.

Lafferty, K. D. (1993). Effects of parasitic castration on growth, reproduction and population dynamics of *Cerithidea californica. Mar. Ecol. Prog. Ser.* **96:**229–237.

Lafferty, K. D., and Kuris, A. M. (1994). Potential for biological control of alien marine species. In *Proceedings of the Conference & Workshop on Nonindigenous Estuarine and Marine Organisms.* (D. Cottingham, ed.) pp. 97–102. US Department of Commerce, NOAA Office of the Chief Scientist.

Lafferty, K. D., and Kuris, A. M. (1996). Biological control of marine pests. *Ecology,* **77:**1989–2000.

Lafferty, K. D., and Morris, A. K. (1996). Altered behavior of parasitized killifish greatly increases susceptibility to predation by bird final hosts. *Ecology,* **77:**1390–1397.

Lambert, T. C., and Farley, J. (1968). The effect of parasitism by the trematode *Cryptocotyle lingua* (Creplin) on zonation and winter migration of the common periwinkle, *Littorina littorea* (L.). *Can. J. Zool.,* **46:**1139–1147.

Lester, R. J. G. (1971). The influence of *Schistocephalus* plerocercoids on the respiration of *Gasterosteus* and a possible resulting effect on the behavior of the fish. *Can. J. Zool.* **49:**361–366.

Martin, W. E. (1950). *Euhaplorchis californiensis* n.g., n. sp., Heterophyidae, Trematoda, with notes on its life cycle. *Trans. Am. Micros. Soc.* **69:**194–209.

MacLellan, C. R. (1958). Role of woodpeckers in control of the codling moth in Nova Scotia. *Can. Entomol.* **90:**18–22.

MacPherson, C. L. (1983). An active intermediate host role for man in the life cycle of *Echinococcus granulosus* in Turkana, Kenya. *Am. J. Trop. Med. Hyg.* **32:**397–404.

McCaig, M. L. O., and Hopkins, C. A. (1963). Studies on *Schistocephalus solidus.* II. Establishment and longevity in the definitive host. *Exp. Parasitol.* **13:**273–283.

Milinski, M. (1990). Parasites and host decision-making. In *Parasitism and host behaviour* (J. Barnard and J. M. Behnke, eds.), pp. 95–116. London: Taylor & Francis.

Moore, J. (1984). Altered behavioural responses in intermediate hosts: An acanthocephalan parasite strategy. *Am. Nat.* **123:**572–577.

Moore, J. (1993). Parasites and the behavior of biting flies. *J. Parasitol.* **79:**1–16.

Moore, J., and Gotelli, N. J. (1990). A phylogenetic perspective on the evolution of altered host behaviours. In *Parasitism and host behaviour* (C. J. Barnard and J. M. Behnke, eds.), pp. 193–233. London: Taylor & Francis.

Obrebski, S. (1975). Parasite castration strategy and evolution of castration of hosts by parasites. *Science* (Washington, D.C.) **188:**1314–1316.

O'Brien, J., and van Wyk, P. (1985). Effects of crustacean parasitic castrators (epicaridean isopods and rhizocephalan barnacles) on growth of crustacean hosts. In *Factors in Adult Growth. Crustacean Issues 3i* (A. M. Wenner, ed.), pp. 191–218. Rotterdam: A. A. Balkema.

Poinar, G. O. (1991). Nematoda and nematomorpha. In *Ecology and classification of North American freshwater invertebrates* (J. H. Thorp and A. P. Covich, eds.), pp. 249–283. San Diego: Academic Press.

Poinar, G. O., and van der Laan, P. A. (1972). Morphology and life history of *Sphaerularia bombi. Nematologica* **18:**239–252.

Poulin, R. (1994). The evolution of parasite manipulation of host behaviour: A theoretical analysis. *Parasitology* **109:**S109-S118.

Pozio, E., La Rosa, G., Rossi, P., and Murrell, K. D. (1992). Biological characterization of *Trichinella* isolates from various host species and geographic regions. *J. Parasitol.* **78:**647–53.

Rasmussen, E. (1959). Behavior of sacculinized shore crabs (*Carcinus maenas* Pennant). *Nature* (Lond.) **183:**479–480.

Rau, M. E. (1983). Establishment and maintenance of behavioural dominance in *male* mice infected with *Trichinella spiralis. Parasitology* **86:**319–322.

Rau, M. E., and Caron, F. R. (1979). Parasite-induced susceptibility of moose to hunting. *Can. J. Zool.* **57:**2466–2478.

Ribeiro, J. M. C., Rossignol, P. A., and Speilman, A. (1985). *Aedes aegypti*: Model for blood-finding strategy and prediction of parasite manipulation. *Exp. Parasitol.* **60:**118–132.

Ritchie, L. E., and Høeg, J. T. (1981). The life history of *Lernaeodiscus porcellanae* (Cirripedia: Rhizocephala) and co-evolution with its porcellanid host. *J. Crustacean Biol.* **1:**334–347.

Roberts, L. S. (1961). The influence of population density on patterns and physiology of growth in *Hymenolepis diminuta* (Cestoda: Cyclophyllidea) in the definitive host. *Exp. Parasitol.* **11:**332–371.

Roberts, L. S., and Janovy, J. (1995). Foundations of parasitology, 5th ed. Dubuque, Iowa: William C. Brown.

Ronquist, F. (1994). Evolution of parasitism among closely related species: Phylogenetic relationships and the origin of inquilinism in gall wasps (Hymenoptera, Cynipidae). *Evolution* **48:**241–266.

Rossignol, P. A., Ribeiro, J. M. C., Jungery, M., Turell, M. J., Spielman, A., and Bailey, L. C. (1985). Enhanced mosquito blood-finding success on parasitemic hosts: evidence for vector-parasite mutualism. *Proc. Nat. Acad. Sci. USA* **82:**7725–7727.

Rothschild, M. (1940). *Cercaria pricei*, a new trematode, with remarks on the specific characters of the "Prima" group of Xiphidiocercariae. *J. Wash. Acad. Sci.* **30:**437–448.

Schell, S. C. (1985). Handbook of trematodes of North America north of Mexico. Moscow, Idaho: Univ. Press of Idaho.

Schmid-Hempel, R., and Schmid-Hempel, P. (1991). Endoparasitic flies, pollen-collection by bumblebees and a potential host-parasitic conflict. *Oecologia* **87:**227–232.

Smith Trail, D. R. (1980). Behavioral interactions between parasites and hosts: Host suicide and the evolution of complex life cycles. *Am. Nat.* 116:77–91.

Smyth, J. D. (1946). Studies on tapeworm physiology. I. The cultivation of *Schistocephalus solidus in vitro. J. Exp. Biol.* **23:**47–70.

Sogandares-Bernal, F., Hietala, H. J., and Gunst, R. F. (1979). Metacercariae of *Ornithodiplostomum ptychocheilus* (Faust, 1917) Dubois, 1936 encysted in the brains and viscera of red-sided shiners from the Clark-Fork and Bitteroot rivers of Montana: An analysis of the infected hosts. *J. Parasitol.* **65:**616–623.

Sousa, W. P. (1983). Host life history and the effect of parasitic castration on growth: A field study of *Cerithidea californica* Haldeman (Gastropoda: Prosobranchia) and its trematode parasites. *J. Exp. Mar. Biol. Ecol.* **73:**273–296.

Stamp, N. E. (1981). Behavior of parasitized aposematic caterpillars: Advantageous to the parasitoid or host? *Am. Nat.* **118:**715–725.

Sweeting, R. A. (1976). Studies on *Ligula intestinalis* (L.) effects on a roach population in a gravel pit. *J. Fish Biol.* **9:**515–522.

Temple, S. A. (1987). Do predators always capture substandard individuals disproportionately from prey populations? *Ecology* **68:**669–674.

Thompson, S. N., and Kavaliers, M. (1994). Physiological bases for parasite-induced alterations of host behaviour. *Parasitology* **109:**S119-S138.

Tierney, J. F., Huntingford, F. A., and Crompton, D. W. T. (1993). The relationship between infectivity of *Schistocephalus solidus* (Cestoda) and anti-predator behaviour of its intermediate host, the three-spined stickleback, *Gasterosteus aculeatus. Anim. Behav.* **46:**603–605.

Townson, H. (1970). The effect of infection with *Brugia pahangi* on the flight of *Aedes aegypti. Ann. Trop. Med. Parasitol.* **64:**411–420.

Urdal, K., Tierney, J. F., and Jakobsen, P. J. (1995). The tapeworm *Schistocephalus solidus* alters the activity and response, but not the predation susceptibility of infected copepods. *J. Parasitol.* **81:**330–333.

Van Dobben, W. H. (1952). The food of the cormorant in the Netherlands. *Ardea* **40:**1–63.

Vik, R. (1954). Investigations on the pseudophyllidean cestodes of fish, birds & mammals in the Anoya water system in Trøndelag. Part I. *Cyathocephalus truncatus* and *Schistocephalus solidus. Nytt Mag. Zool.* (*Oslo*) **2:**5–51.

Walkey, M. and Meakin, R.H. (1970). An attempt to balance the energy budget of a host parasite system. *J. Fish Biol.*, **2:**361–372.

Wilson, R. S. (1971). The decline of a roach *Rutilus rutilus* population in Chew Valley Lake. *J. Fish Biol.* **3:**129–137.

16

The Ecology of Parasites in a Salt Marsh Ecosystem

Kevin D. Lafferty

I. Introduction

Community ecologists aim to understand how groups of species interact with their environment and with each other (e.g., Diamond, 1975; Connor and Simberloff, 1979; Connell, 1980, 1983; Strong et al., 1984; Schoener, 1983; Sih et al., 1985; Sale, 1991). Species interactions involving predation and competition receive the bulk of our attention. What is missing is an understanding of the role that parasites play in the communities of their hosts (Sousa, 1991). Parasites are a pervasive and potentially important component of most communities. In addition, parasites with complex life cycles depend on how their different hosts interact and overlap in space and time. In turn, they may affect interactions among various host species. As described in many chapters in this volume, the effects that parasites have on their hosts may be extraordinary when parasites manipulate host hormones or behavior.

Here, I will review how parasites affect various hosts in salt marsh ecosystems found along the west coast of California (U.S.) and Baja California (Mexico). The parasites I will discuss are digenetic trematodes. As trematodes journey through their life cycle, they come across very different types of hosts. Each host presents unique challenges and, consequently, trematodes affect each host differently. A strength of this system is that interesting questions can be addressed at every stage of the life cycle. I will not discuss the mechanisms by which parasites alter their hosts. Instead, I will concentrate on why they alter their hosts and how alteration affects host individuals and populations.

The wetland systems that I will be discussing are fascinating habitats. Because their mouths usually stay open to the sea, they are subject to regular tidal influence. Aquatic beds (dominated by the pickleweed *Salicornia virginica*, and, in some areas, cord grass, *Spartina foliosa*) cover much of their area (Ferren et al., 1997). Epifaunal invertebrates are common and active during warm months on the mud flats, salt pans, and channel banks. These include the horn (or mud) snail, *Cerithidea californica*, and the crabs, *Hemigrapsus oregonensis*, *Pachygrapsus crassipes* and *Uca crenulata* (Page and Lafferty, 1997). Along with various clams (*Tagelus californianus*, *Protothaca staminea*, and *Macoma spp.*), living in the fine intertidal sediments are polychaete worms and fly larvae (Page and Lafferty, 1997). The most abundant fishes in the tidal channels are typically topsmelt (*Atherinops affinis*), arrow gobies (*Clevelandia ios*), longjaw mudsuckers (*Gillichthys mirabilis*), California killifish (*Fundulus parvipinnis*) and staghorn sculpin (*Leptocottus armatus*) (Lafferty, in preparation). Many species of West Coast waterfowl, including waders, gulls, herons, egrets, terns and various diving birds are common, particularly in larger systems (Lafferty and Morris 1996).

This chapter will follow a trematode through its life cycle and note the different effects that it has on each of its three hosts. The most common trematode in salt marsh systems is typically *Euhaplorchis californiensis*. The first intermediate host, a snail, *Cerithidea californica*, becomes infected when it consumes trematode eggs as it forages on epibenthic diatoms (other trematode species infect snails with free-swimming miracidia). After a period of asexual reproduction within the snail as rediae, the trematode produces free-swimming cercarial stages that leave to infect a second intermediate host and encyst as metacercariae. For *E. californiensis*, the second intermediate host is the California killifish *Fundulus parvipinnis* (Martin 1950). The trematode completes its life cycle when a bird eats an infected killifish (See Fig. 16-1 for a description of the life cycle). In the bird, the small worms (1.5 mm) live in the gastrointestinal tract, mate, and produce eggs that pass into the marsh with the bird's feces. Each of the 17 trematode species in the salt marsh has a different life cycle (Martin, 1972). A couple differ only in that they infect different organs in the killifish (Martin, 1972). Others use clams, crabs, and other fishes as second intermediate hosts. Blood flukes skip the second intermediate host and penetrate definitive host birds with cercariae.

II. The Snail

Once an individual trematode successfully infects a snail, it reproduces asexually to produce clones of sausage-shaped rediae. In the process, the trematode consumes the gonad of the snail and may displace a portion of the digestive gland. By consuming snail tissue and interfering with the snail's hormones, many trematodes castrate male and female snails. Why castrate? Parasitic castration may be a superb strategy for parasites of hosts that live long and put much energy into reproduction (Kuris, 1974; O'Brien and Van Wyk, 1985). In this way, they can extract the maximum amount of energy from the host without reducing the host's longevity. This is particularly true when death of the host means death of the parasite.

As the first intermediate host, *C. californica* acts as a hub through which the

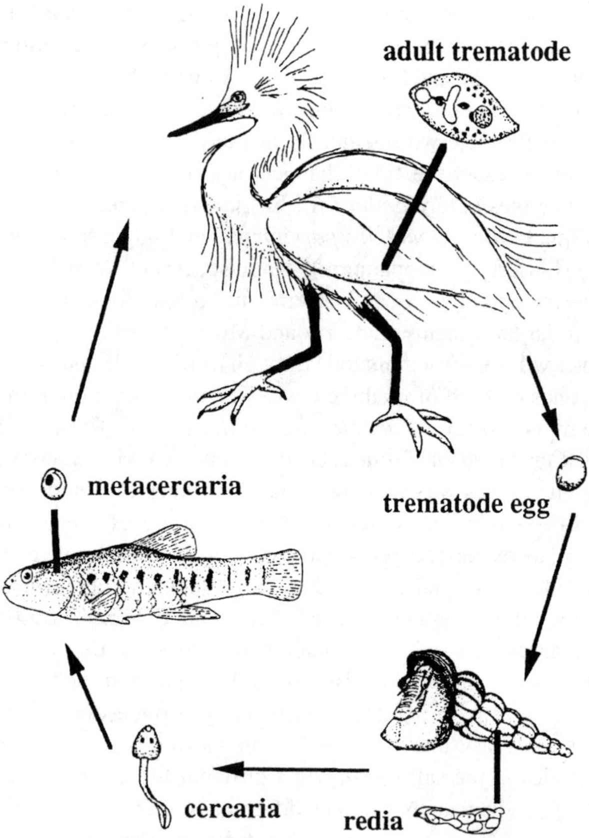

Figure 16-1. Life cycle of the trematode, *Euhaplorchis californiensis.* Hosts include the horn snail, the Pacific killifish and piscivorous birds such as the snowy egret.

different trematode species will pass. In many salt marshes, trematodes infect the majority of the larger snails (Sousa, 1983; Kuris, 1990; Lafferty, 1993b). Because an infection from a single individual trematode eventually uses the entire gonad of the snail, there is potential for competition between two trematode species that may infect the same host (Kuris, 1990; Sousa, 1992).

As an aside, trematode communities are excellent model systems to address questions about how competition can structure communities (Kuris and Lafferty, 1994). One can witness one trematode species attacking another by dissecting snails infected with more than one species or by placing different trematodes together *in vitro.* Aggregation by birds leads to several trematode species being concentrated in one area which leads to an increase in the frequency of interspecific interactions; subsequent competition among the species usually results in single-species infections (Lafferty et al., 1994).

How should parasitism affect the snail population? Castration is similar to predation because an infected individual has zero future fitness (Kuris, 1973). This should decrease the reproductive output of the host population. If an individual dies by predation, it no longer remains in the population and the resources it once used are now available to other members of the population (Lafferty, 1993a). A castrated host, on the other hand, remains in the population, and consumes resources (Hughes, 1986). Since the castrated snail produces only trematode offspring, it is essentially a parasite genotype that has a host phenotype (O' Brien and Van Wyk, 1985), a veritable wolf in sheep's clothing. If infected snails graze on the same epibenthic diatoms as uninfected snails, this sets up the odd situation where parasites compete with uninfected hosts if resources are limited. If, as happens with snails and trematodes, the parasite disperses broadly and the host does not, the host population will be disproportionately affected by the interaction (Lafferty 1991, Gaines and Lafferty, 1995).

To investigate the effects of parasitic castration, I conducted a field experiment in which I manipulated snail density and trematode prevalence independently (Lafferty, 1993a). In treatments where infected snails were replaced with uninfected snails, more than twice as many eggs were produced by the snail population (Fig. 16-2). This confirms the assumption that castration substantially reduces reproductive output at the population level. If there are to be indirect effects of castration, it is important to know whether resources are limited. In support of this, there was an increase in the amount of chlorophyll extracted from the

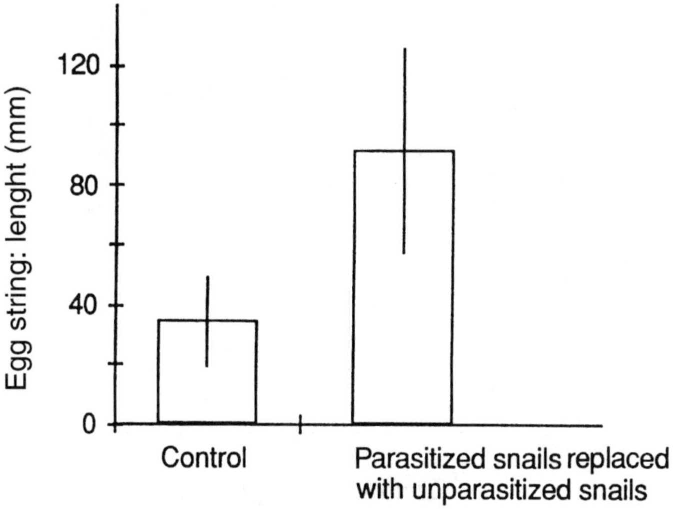

Figure 16-2. The effect of castration on the egg production of the snail population. The vertical axis represents the combined length of egg strings laid by caged populations of snails (July and August). Error bars represent standard errors.

sediment in cages where I removed all snails (Fig. 16-3). This does not necessarily indicate competition between infected and uninfected snails because, in other systems, infected and uninfected snails prefer different foods (Boland and Fried, 1984; Curtis, 1985). However, in cages where I removed parasitized snails, unparasitized snails grew faster, indicating that competition between trematodes and uninfected hosts occurs in a manner analogous to interspecific competition (Fig. 16-4). Lowered egg production might be compensated by increased juvenile survivorship, begging the question, what is the overall effect of parasitic castration on host population density? In areas with high parasitism, snail density is lower than expected, suggesting that trematodes reduce the density of the snail population (Fig. 16-5).

Some populations have a higher prevalence of a parasitism than others. Life history theory predicts that hosts should mature earlier in areas with high risks of infection (Ruiz, 1991; Keymer and Read, 1992). This is a way to ensure some chance of reproduction. For example, freshwater snails shifted energy to egg production after they were penetrated by miracidia, apparently as a last ditch effort to reproduce (Minchella, 1985,; Thornhill et al., 1986). I estimated size of maturation in subpopulations within a marsh and in snails from 18 different marshes (Fig. 16-6, Lafferty, 1993b). After accounting for environmental variation in maturation size, I assessed the association between the variation in matura-

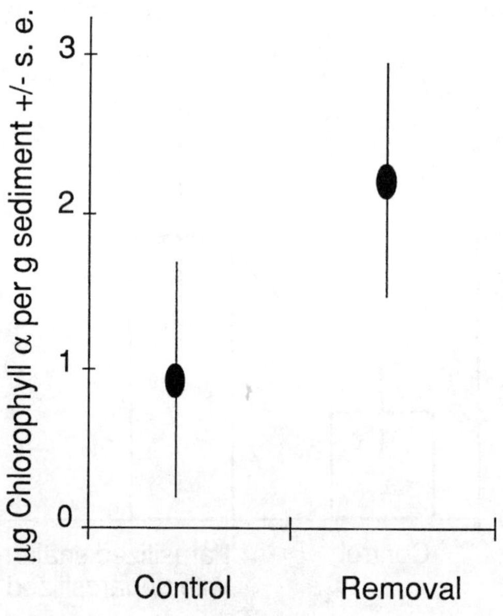

Figure 16-3. The effect of snails on epibenthic diatoms. The vertical axis represents chlorophyll levels from seven snail exclosures (snails removed) and seven snail enclosures (control).

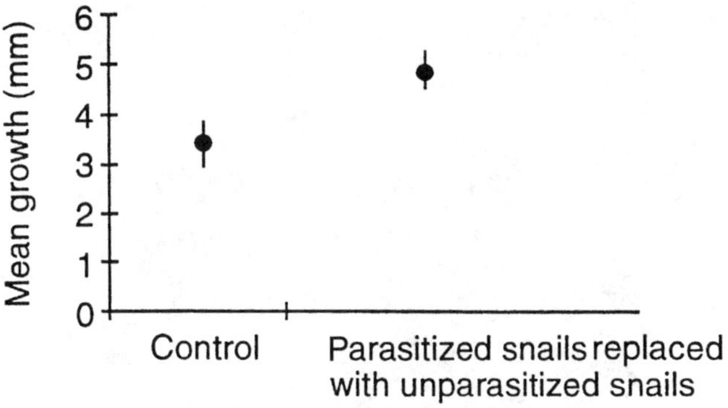

Figure 16-4. The effect of competition between infected and uninfected snails on the growth rate of uninfected snails. The vertical axis represents the change in size over three months for caged 15–20 mm snails.

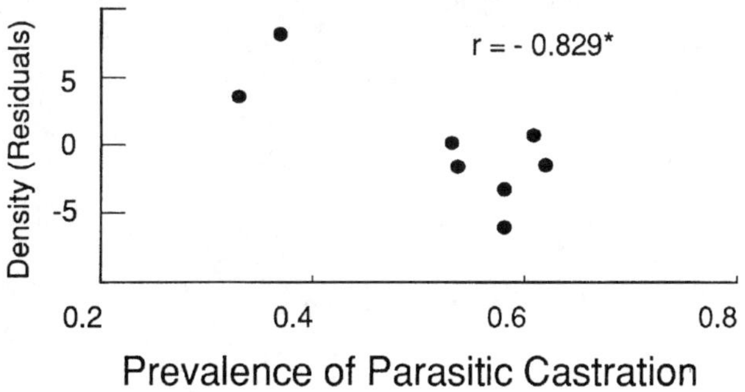

Figure 16-5. The association between snail density (residuals after accounting for mean snail size and growing conditions) and parasitic castration among eight subpopulations within Carpinteria Salt Marsh Reserve. * indicates significance at the $p < 0.05$ level.

tion size and the prevalence of parasitism. Consistent with an adaptation to parasitism, in areas with high risks of parasitism, snails matured at smaller sizes. If maturation size is a result of selection from parasites, it is due either to genetic differences among populations, or to a plastic response to the risk of parasitism. For example, uninfected snails might be able to sense infective stages in the water or determine the relative density of infected snails.

In an attempt to distinguish between a genetic and plastic response, I collected very young snails for a reciprocal transplant. Half were from a population where parasites were absent and half were from my study site, where parasites were common (Lafferty, 1993b). Consistent with the prediction of the genetic difference hypothesis, snails from the high risk area matured at smaller sizes in both habitats. There was also an environmental effect, however. Snails from both habitats matured at smaller sizes in the high risk habitat. These results are

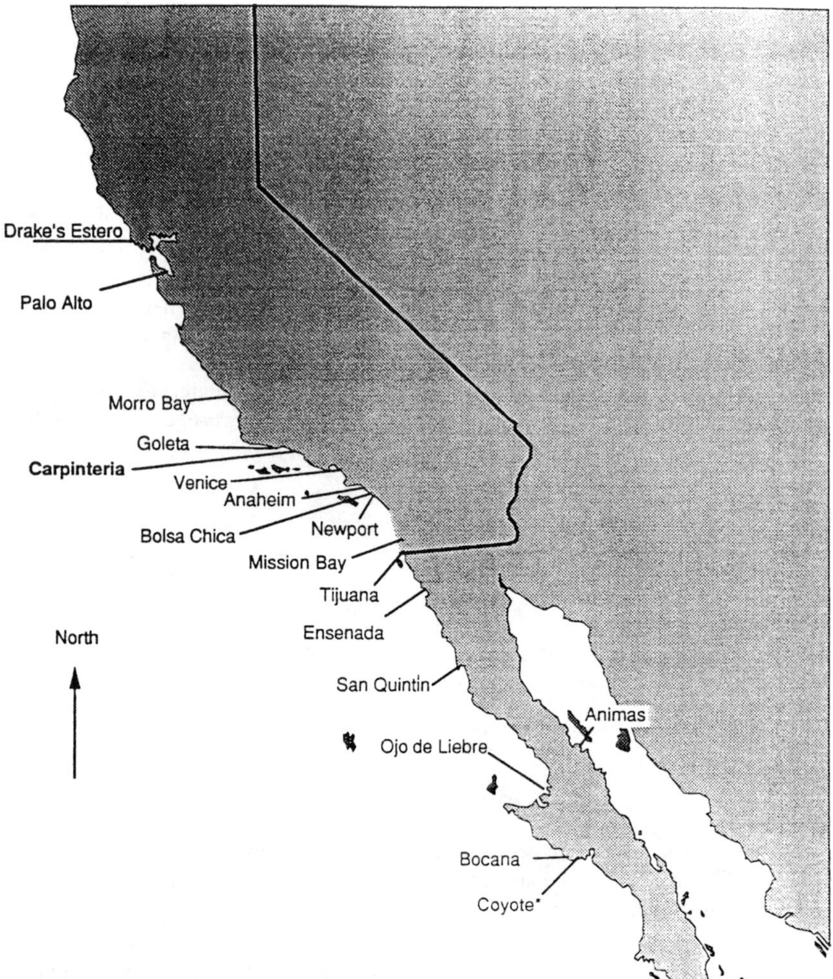

Figure 16-6. Sites surveyed along the west coast for the association between maturation size and the prevalence of parasitic castration.

somewhat inconclusive, and multigenerational studies would be required to assess the adaptive mechanism. Unfortunately, such studies cannot be done with this species yet.

In summary, trematodes, by castrating their hosts, are unlike predators in that they do not directly reduce snail density, although some infected snails may suffer higher mortality than uninfected snails (Sousa and Gleason, 1989; Lafferty,

1993a). However, parasites indirectly affect the snail population in two ways. First, trematodes substantially reduce the reproductive output of the snail population. Second, infected snails remain on the mud flat and compete for limited food with uninfected snails. Because *C. californica* has a larval stage with limited dispersal abilities, castration reduces recruitment of juveniles to the snail population and, subsequently, lowers snail density. In addition, castration appears to be an important factor that selects for early maturation of hosts.

A wide range of parasites causes parasitic castration in an equally diverse group of hosts (Table 16-1). These parasitic castrators are also likely to be able to reduce the density of their hosts and may select for adaptive life history strategies. Some anecdotal evidence suggests that parasitic, castrating barnacles can reduce the density of their host crabs (Kuris and Lafferty, 1992). In cases

Table 16-1. The Various Types of Hosts That Are Castrated by Parasites

Hosts	Parasitic castrators
Plants	Fungi
	Nematodes
	Mites
	Weevils
Jellyfish	Anemones
	Amphipods
Flatworms	Flatworms
Tapeworms	Sporozoans
Nemerteans	Orthonectids
Snails	Trematodes
Bivalves	Trematodes
Annelids	Sporozoans
Barnacles	Barnacles
	Isopods
Shrimps	Isopods
Crabs	Protozoans
	Nemerteans
	Barnacles
	Isopods
Insects	Protozoans
	Nematodes
	Strepsipterans
Brachiopods	Trematodes
Starfish	Orthonectids
	Barnacles
Sea urchins	Barnacles
Sea cucumbers	Snails
	Fishes
Sharks	Barnacles
Fishes	Protozoans
Rodents	Botflies

where the crustacean hosts support economically important fisheries, nontraditional management strategies may reduce the impact of the parasite on the fishery yield (Kuris and Lafferty, 1992).

III. The Fish

Infected snails regularly shed cercarial stages into the water. Depending on the species of trematode parasitizing it, a snail can shed hundreds to thousands of cercariae in a single day. In areas where infected snails are common, trematode cercariae could make up the bulk of the zooplankton population in the salt marsh (T. Stevens, unpublished data). Although some animals are probably able to exploit cercariae as a food source, others end up being penetrated and infected.

Euhaplorchis californiensis cercariae infect their second intermediate host, the California killifish (*Fundulus parvipinnis*), by penetrating the gills of the fish. The cercaria drops its tail, migrates within the host tissues, and forms a metacercarial cyst. Inside each cyst is a miniature adult worm that will be released when the enzymes in a bird's gut digest away the cyst wall. The strategy of the trematode in the fish differs markedly from what it is in the snail. Here, transmission only occurs when a bird eats an infected fish. Anything that the metacercaria can do to facilitate this will increase its expected fitness. *E. californiensis* specifically infects the surface of the killifish's brain. Hundreds to thousands of cysts per fish braincase are typical levels of infection. This mode of infection is suggestive that the parasite might be attempting to modify the fish's behavior to increase transmission. An alternative hypothesis is that parasites invade the central nervous system in an effort to avoid the host's immune system (Szidat, 1969; *see also* Chapter 15).

To assess the hypothesis that behavior modification increases parasite transmission, we brought infected fish into the laboratory and identified odd behaviors that might make the fish more apparent to a foraging bird (Lafferty and Morris 1996). Although fish usually schooled normally, fish occasionally swam abruptly to the surface, jerked suddenly, shimmied, contorted, and scratched themselves on the bottom of the aquaria (Fig. 16-7). We captured infected fish from the Carpinteria Salt Marsh Reserve. For comparison, we collected uninfected fish nearby at Coal Oil Point Reserve, where snails were not present. Fish from the infected population had four times as many odd behaviors. This is consistent with the behavioral modification hypothesis, but could also be due to other unknown differences between the populations. Unfortunately, we were unable to experimentally infect fish. An analysis of the infected fish indicated that the frequency of odd behaviors increased with the number of parasites in the brain. This is also consistent with the behavior modification hypothesis but could also be explained by odd behavior increasing susceptibility to infection (Moore and Gotelli, 1990). In combination, however, these results strongly suggest that the

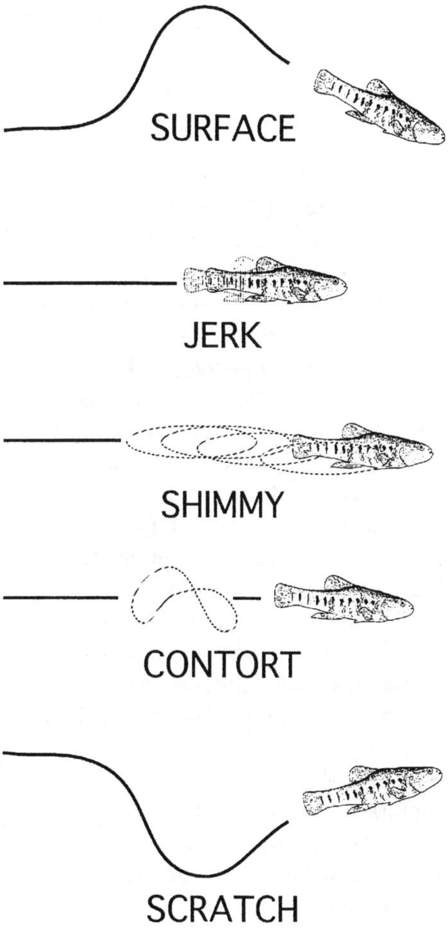

Figure 16-7. Killifish behaviors characterized as conspicuous.

trematode alters fish behavior in ways that should make the fish more conspicuous to predation by birds (Lafferty and Morris, 1996).

Behavior modification might greatly increase mortality rates for killifish and decrease their abundance in areas where snails are common. High rates of differential mortality of parasitized fish also might reduce the prevalence of parasitism in the fish population. In other words, predators may eat parasitized fish soon after the fish become infected. This creates the unusual situation where a parasite might greatly affect the host, yet seem unimportant, because parasitized hosts are relatively rare. Although mathematical models support this supposition (Lafferty, 1992), it does not appear to hold for infected killifish. At Carpinteria Salt Marsh

Reserve, *E. californiensis* is not rare; it infects all killifish which exceed a minimum threshold size.

IV. The Bird

What does the alteration of fish behavior mean for the bird? There are many examples of associations between odd host behavior and parasitism (see reviews in Holmes and Bethel, 1972; Dobson, 1988; Curio, 1988; Moore and Gotelli, 1990; Lafferty, 1992; Poulin, 1994; other chapters in this volume). Although there is good evidence for enhanced predation on infected intermediate hosts in the wild, predator exclosure experiments have not been reported.

To see whether infection by *E. californiensis* made killifish more susceptible to predation by piscivorous birds, we set up fish pens in a lagoon (Lafferty and Morris, 1996). One pen was open on the top, allowing bird access. We covered the other pen with bird netting; this arena acted as a control for nonpredation mortality and escape. Infected and uninfected fishes were matched for size, mixed together and put into each pen. Herons and egrets foraged in and around the pens. After twenty days, we collected fish from the two pens and dissected them to quantify the intensity of infection in the remaining fish. By comparing what remained in the control and experimental pens, we were able to estimate what types of fish birds ate (Fig. 16-8). The results were striking. Birds were 30 times more likely to eat infected fish than uninfected fish. In addition, heavily infected fish suffered much higher rates of predation than lightly infected fish. This suggests that behavioral modification is a parasite strategy that increases transmission in nature (Lafferty and Morris, 1996).

Birds are intelligent foragers and presumably recognize when a fish is behaving abnormally. Given this, should a bird avoid parasitized fish? In doing so, it would benefit from reduced parasitism but would suffer the cost of reduced food intake. This remains a theoretical question. Mathematical models suggest that birds that can and do avoid parasitized prey gain relatively little energy compared with those that do not practice avoidance (Lafferty, 1992). Of course, the energy gain for birds that do not avoid parasitized prey depends on how costly the parasite is. Even for costly parasites, the bird might do worse than if the parasite was not in the system, but still may do better than if it avoided parasitized prey. The only exception appears to be when the parasite is very costly and does not make the prey much easier to catch. A particularly interesting suggestion of the model is that if the parasite only generates a moderate cost, the bird can actually do better than if the parasite was not in the system. One might view this trematode as a pizza man; it delivers a normally hard to get meal and, in return, extracts a small tip for service. In general, parasites that are trophically transmitted are relatively benign in their definitive hosts. Although we do not know the costs of *E. californiensis* for the bird, it is a small and short-lived worm that probably causes very little pathology on a per worm basis.

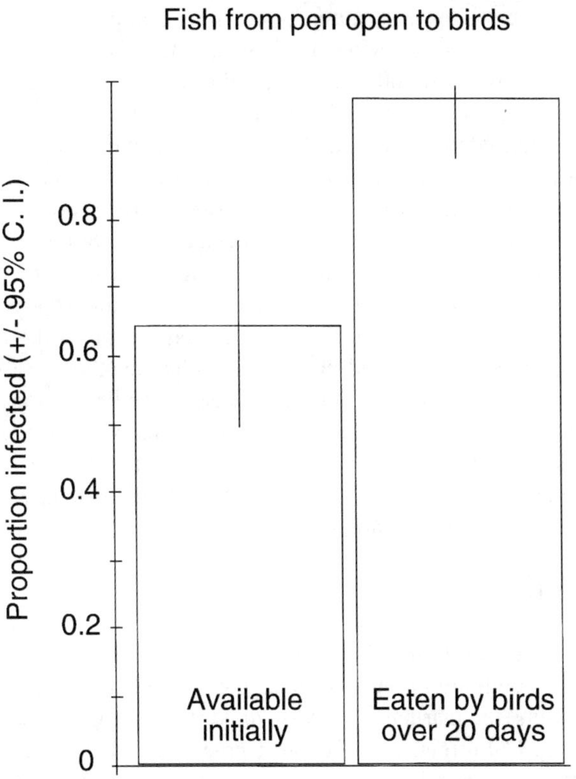

Figure 16-8. The effect of bird predation on parasitized fish, indicating that infected fish were more likely to be eaten than uninfected fish.

Easier foraging could result in increased densities of birds in these areas (Lafferty, 1992). This hypothesis might be testable by comparing bird densities between areas with and without snails, if one carefully considered the many other factors associated with bird density. Many types of parasites besides trematodes, including tapeworms, nematodes, and acanthocephalans, have complex life cycles that depend on trophic transmission. Such parasites are most likely to have important consequences for predator-prey interactions when prey are normally difficult to catch. Wolves, for example, may be dependent on the debilitating effects that tapeworm larvae (*Echinococcus granulosus*) have on moose. Without the tapeworm, which lodges in the liver, lungs and brain of moose, wolves would probably find moose too dangerous to attack. An interesting piece of evidence supporting this idea is that moose hunters are more likely to bag infected moose just after the season opens than later in the season, suggesting that infected moose are easier game (Rau and Caron, 1979).

In summary, trematodes affect their hosts differently than one might expect.

We tend to view parasites as stealing (or certainly competing with) a portion of their host's energy or tissues. Most host-parasite models assume that this effect simply raises the host's mortality rate. *E. californiensis* has strikingly different strategies depending on what host it is in. By castrating the first intermediate host snail, the trematode enjoys a long life expectancy during which it can produce multitudes of cercariae. This is the host from which the trematode takes the most energy. It reduces snail density by lowering the reproductive output of the snail population and competing with uninfected snails for food. In the killifish, the trematode does not grow and is not likely to take much, if any, nourishment. Its effect on fish behavior appears to facilitate transmission to piscivorous birds. It is not clear how much the adult trematode feeds while in the bird, but it is probably not significant. It would not appear to be adaptive for the parasite to strongly affect the bird's health. This is because a healthy bird, as it flies from marsh to marsh, acts as a dispersal agent for the trematode in the same way that ocean currents act to disperse the planktonic larvae of marine species. In a sense, the trematode employs a strategy similar to plants that encompass their seeds in juicy fruits that frugivores eat and disperse.

V. Ecosystem Health

We have subjected Californian wetlands to substantial pressures. Their presence in a coastal zone with high property values has subjected them to filling for real estate or dredging for marinas. The vast majority of original estuarine habitat has been lost in California. Flood control practices, urban encroachment, and water pollution degrade what remains. Efforts to understand and protect these vanishing habitats are ongoing in the face of continued losses.

Because their life cycles are so complex, trematodes depend on a healthy functioning ecosystem. Furthermore, in a diverse community of trematodes, each species has its own required set of hosts. Changes that alter the community of hosts, particularly birds, should result in a change in the diversity and abundance of trematodes in snails. Because the diversity and abundance of trematodes in snails are very easy to quantify, they may be useful for monitoring environmental degradation in wetland systems (Kuris and Lafferty, 1994). There is some anecdotal evidence for the utility of parasites as biomonitoring tools. For example, at Bolsa Chica, a highly disturbed segment of the wetland had a 1% prevalence of infection relative to a 25 percent prevalence in the most adjacent intact area (Lafferty, 1993b). Among other wetlands, there is substantial variation in trematode prevalence that might be associated with wetland condition. One could assess this approach by comparing aspects of wetland functioning (for example, censuses of birds, fishes, and invertebrates) with trematode prevalence and diversity. Such a comparative approach is limited to areas where the host snail is present. In addition, changes in trematode prevalence and diversity over several years could indicate wetland degradation or recovery (Cort et al., 1960).

VI. Summary

Trematode parasites of salt marsh animals provide a fine example of how parasites affect the ecology of their hosts. Many trematodes castrate their first intermediate host, snails. This leads to a substantial reduction in the reproductive output of the host snail population as well as competition for food between infected and uninfected snails. In combination, these factors reduce snail population density. Trematodes modify the behavior of their second intermediate host, fish, making them strikingly more susceptible to predation by birds that are the trematode's final host. Birds may benefit because trematodes make food easier for them to capture. Because they depend on complex ecosystems, trematodes may serve as indicators of ecosystem health.

VII. Acknowledgments

I thank Armand Kuris, Janice Moore, and three anonymous reviewers for helpful critical comments. Many people contributed to the studies that form the bulk of this paper. In particular, I thank, Don Canestro, Carolee Levick, Maria Martinez, Kimo Morris, David Sammond, and Tom Vincent. Lanny Lafferty returned the final draft of the manuscript to me following its disappearance at a Christmas party so that I could make the submission deadline. The University of California's Natural Reserve System provided access to sites.

VIII. References

Boland, L. M., and Fried, B. (1984). Chemoattraction of normal and *Echinostoma revolutum*-infected *Helisoma trivolvis* to romaine lettuce (*Lactuca sativa longefolia*). *J. Parasitol.* **70:**436–439.

Connell, J. H. (1980). Diversity and the coevolution of competitors, or the ghost of competition past. *Oikos* **35:**131–138.

Connell, J. H. (1983). On the prevalence and relative importance of interspecific competition: evidence from field experiments. *Am. Nat.* **122:**661–696.

Conner, E. F., and Simberloff, D. (1979). The assembly of species communities: chance or competition? *Ecology* **60:**1132–1140.

Cort, W. W., Hussey, K. L. and Ameel, D. J. (1960). Seasonal fluctuations in larval trematode infections in *Stagnicola emarginata angulata* from Phragmites Flats on Douglas Lake. *Proc. Helminthol. Soc. Wash.* **27:**11–13.

Curio, E. (1988). Behavior and parasitism. In H. Mehlhorn, ed., *Parasitology in focus, facts & trends,* pp. 149–160. Berlin: Springer Verlag.

Curtis (1985). The influence of sex and trematode parasites on carrion response of the estuarine snail *Ilynassa obsoleta Biol. Bull.* **169:**377–309.

Diamond, J. M. (1975). Assembly of species communities. In M. L. Cody, and J. M.

Diamond, eds. *Ecology and evolution of communities,* pp. 342–444. Cambridge, Mass.: Harvard University Press.

Dobson, A. P. (1988). The population biology of parasite-induced changes in host behavior. *Q. Rev. Biol.* **63:**139–165.

Ferren, W. J., Fiedler, P. L., Leidy, R. A., Lafferty, K. D., and Mertes, L. A. K. (1997). Wetlands of California, part III: Key to and catalogue of wetlands of the Central and Southern California coast and coastal watersheds. Madroño (in press).

Gaines, S. D., and Lafferty, K. D. (1995). Modeling the dynamics of marine species: The importance of incorporating larval dispersal. In *Ecology of marine invertebrate larvae* (L. McEdward ed.), pp. 389–412. Boca Raton: CRC Press.

Holmes, J. C., and Bethel, W. M. (1972). Modification of intermediate host behaviour by parasites. *Zool. J. Linn. Soc.* 51, Suppl. **1:**123–149.

Hughes, R. N. (1986). *A functional biology of marine gastropods.* Baltimore, Md.: The Johns Hopkins University Press.

Keymer, A. E., and Read, A. F. (1992). Behavioral ecology: The impact of parasitism. In *Parasite-host associations. Coexistence or conflict?* (C.A. Toft, A. Schlimann, and L. Bolis, eds.), pp. 37–61. New York: Oxford University Press.

Kuris, A. M. (1973). Biological control: Implications of the analogy between the trophic interactions of insect pest-parasitoid and snail-trematode systems. *Exp. Parasitol.* **33:**365–379.

Kuris, A. M. (1974). Trophic interactions: Similarity of parasitic castrators to parasitoids. *Q. Rev. Biol.* **49:**129–148.

Kuris, A. M. (1990). Guild structure of larval trematodes in molluscan hosts: Prevalence, dominance and significance of competition. In *Parasite communities: Patterns and processes* (G. W. Esch, A. O. Bush, and J. M. Aho, eds.), pp. 69–100. London: Chapman and Hall.

Kuris, A. M., and Lafferty, K. D. (1992). Modelling crustacean fisheries: Effects of parasites on management strategies. *Can. J. Fish. Aquat. Sci.* **49:**327–336.

Kuris, A. M., and Lafferty, K. D. (1994). Community structure: Larval trematodes in snail hosts. *Ann. Rev. Ecol. Syst.* **25:**189–217.

Lafferty (1991). Effects of parasitic castration on the salt marsh snail, *Cerithidea californica.* Ph.D. Dissertation, Department of Biology, University of California, Santa Barbara.

Lafferty, K. D. (1992). Foraging on prey that are modified by parasites. *Am. Nat.* **140:**854–867.

Lafferty, K. D. (1993a). Effects of parasitic castration on growth, reproduction and population dynamics of *Cerithidea californica. Mar. Ecol. Prog. Ser.* **96:**229–237.

Lafferty, K. D. (1993b). The marine snail, *Cerithidea californica,* matures at smaller sizes where parasitism is high. *Oikos* **68:**3–11.

Lafferty, K. D., and Morris, A. K. (1996). Altered behavior of parasitized killifish greatly increases susceptibility to predation by bird final hosts. *Ecology* **77:**1390–1397.

Lafferty, K. D., Sammond, D. T., and Kuris, A. M. (1994). Analysis of larval trematode community structure: Separating the roles of competition and spatial heterogeneity. *Ecology* **75:**2275–2285.

Martin, W. E. (1950). *Euhaplorchis californiensis* n.g., n.sp., Heterophyidae, Trematoda, with notes on its life cycle. *Trans. Am. Microsc. Soc.* **69:**194–209.

Martin, W. E. (1972). An annotated key to the cercariae that develop in the snail *Cerithidea californica. Bull. South Calif. Acad. Sci.* **71:**39–43.

Minchella, D. J. (1985). Host life-history variation in response to parasitism. *Parasitology* **90:**205–216.

Moore, J., and Gotelli, N. J. (1990). A phylogenetic perspective on the evolution of altered host behaviours: A critical look at the manipulation hypothesis. In *Parasitism and host behaviour* (C. J. Barnard, and J. M. Behnke, eds.), pp. 193–233. London: Taylor and Francis.

O'Brien, J., and Van Wyk, P. (1985). Effects of crustacean parasitic castrators (epicaridean isopods and rhizocephalan barnacles) on growth of crustacean hosts. In *Factors in adult growth* (A. M. Wenner, ed.), pp. 191–218. Boston: A. A. Balkema.

Page, H. M., and Lafferty, K. D. (1997). Estuarine and marine invertebrates. In *zoological resources of carpinteria salt marsh* (H. M. Page, and W. Ferren eds.) (in press).

Poulin, R. (1994). Meta-analysis of parasite-induced behavioural changes. *Anim. Behav.* **48:**137–146.

Rau, M. E., and Caron, F. R. (1979). Parasite induced susceptibility of moose to hunting. *Can. J. Zool.* **57:**2466–8.

Ruiz, G. M. (1991). Consequences of parasitism to marine invertebrates-host evolution. *Am. Zool.* **31:**831–839.

Sale, P. F. (1991). Reef fish communities: Open non-equilibrial systems. In *The ecology of fishes on coral reefs* (P. F. Sale, ed.), pp. 564–598. San Diego: Academic Press.

Schoener, T. W. (1983). Field experiments in interspecific competition. *Am. Nat.* **122:**240–285.

Sih, A., Crowley, P., McPeek, M., Petranka, J., and Strohmeier, K. (1985). Predation, competition and prey communities: a review of field experiments. *Ann. Rev. Ecol. Syst.* **16:**269–311.

Sousa, W. P. (1983). Host life history and the effect of parasitic castration on growth: a field study of *Cerithidea californica* Haldeman (Gastropoda: Prosobranchia) and its trematode parasites. *J. Exp. Mar. Biol. Ecol.* **73:**273–296.

Sousa, W. P. (1991). Can models of soft-sediment communities be complete without parasites? *Am. Zool.* **31:**821–830.

Sousa, W. P (1992). Interspecific interactions among larval trematode parasites of freshwater and marine snails. *Am. Zool.* **32:**583–592.

Sousa, W. P., and M. Gleason. (1989). Does parasitic infection compromise host survival under extreme environmental conditions? The case for *Cerithidia californica* (Gastropoda: Prosobranchia). *Oecologia* **80:**456–464.

Strong, D. R., Simberloff, D., Abele, L. G., and Thistle, A. B. (1984). *Ecological communities: Conceptual issues and the evidence.* Princeton, N.J.: Princeton University Press.

Szidat, L. (1969). Structure, development and behaviour of new Strigeatoid metacercariae from subtropical fishes of South America. *J. Fish. Res. Board Can.* **26:**753–786.

Thornhill, J. A., Jones, T. and Kusel, K. R. (1986). Increased oviposition and growth in immature *Biomphalaria glabrata* after exposure to *Schistosoma mansoni. Parasitology* **93:**443–450.

Index